Rooted in field experience and scientific study, Princeton's guides to animals and plants are the authority for professional scientists and amateur naturalists alike. **Princeton Field Guides** present this information in a compact format carefully designed for easy use in the field. The guides illustrate every species in color and provide detailed information on identification, distribution, and biology.

Dragonflies and Damselflies of the West

Dennis Paulson

PRINCETON UNIVERSITY PRESS   Princeton and Oxford

Published by Princeton University Press, 41 William Street, Princeton, New Jersey 08540

In the United Kingdom: Princeton University Press, 6 Oxford Street, Woodstock, Oxfordshire OX20 1TW

Library of Congress Cataloging-in-Publication Data

Paulson, Dennis R.
    Dragonflies and damselflies of the West / Dennis Paulson.
        p. cm. — (Princeton field guides)
    Includes bibliographical references and index.
    ISBN 978-0-691-12280-9 (hardback : alk. paper)—ISBN 978-0-691-12281-6 (pbk. : alk. paper)
1. Dragonflies—West (U.S.)—Identification.   2. Dragonflies—Canada, Western—Identification.
3. Damselflies—West (U.S.)—Identification.   4. Damselflies—Canada, Western—Identification.
I. Title.
    QL520.2.P38 2009
    595.7'330978—dc22

                                                        2008030893

British Library Cataloguing-in-Publication Data are available

This book has been composed in Myriad Pro
Printed on acid-free paper. ∞
press.princeton.edu
Printed in China

10 9 8 7 6 5 4 3 2 1

Contents

Preface

In the past decade, books on dragonflies have proliferated at a rate to match the recent upswing of interest in them. There are good technical manuals dealing with the entire fauna, and regional guides to damselflies, dragonflies, or both are appearing as fast as enthusiasts can snap them up. But there are still no comprehensive field guides to all the Odonata of North America, and this book and a companion volume to come represent an attempt to fill that gap.

The present book has two primary goals. The first is to make it possible to identify any of the 348 species of dragonflies and damselflies now known to occur in the western United States and Canada. The second is to present material about their natural history that will prompt greater interest in their lifestyles. These insects are just as special as another group of well-loved insects, the butterflies, and there is no reason they cannot become as well known. We constantly alter the natural world, with both obvious and not-so-obvious effects, and we can hope that an increased knowledge of dragonflies will help us understand this world better, in particular the ecology and condition of our wetlands.

Global warming is bringing tropical species across our borders, and they are mixing with the species already present. Natural wetlands are being filled and artificial ones created, and the forests that shelter many dragonflies are being cut down in some areas and regrown in others. Dragonflies are indicators of these changes if we understand them well enough. This is an exciting time in the study of the order Odonata, with so much amateur interest in the group and a concordant increase in our knowledge, and I hope these books will not only contribute to the enjoyment of all who choose to learn something about these fascinating animals but also add even more to what we already know about them.

I much appreciate being taken or accompanied in the field in various parts of the continent by John Abbott, Bud Anderson, Richard Bailowitz, Jim Bangma, Giff Beaton, Roy Beckemeyer, Bob Behrstock, Kathy and Dave Biggs, Sheryl Chacon, Jerrell Daigle, Doug Danforth, Marion Dobbs, Bob DuBois, Sid Dunkle, Molly Hukari, Bill Hull, Jim Johnson, Tom Langschied, Greg Lasley, Tim Manolis, Kurt Mead, David Nunnallee, Sumita Prasad, Martin Reid, Martha Reinhardt, Josh Rose, Jennifer Ryan, Jim Sinclair, Gayle and Jeanell Strickland, Mike Thomas, Idie Ulsh, Sandy Upson, Michael Veit, Tom Young, and Bill Zimmerman, all of whom shared their knowledge of odonates. Messrs. Beaton, Behrstock, Danforth, Lasley, and Upson were especially appreciated repeated guides and companions. In addition, many of the other active field workers in North America have given me valuable information about distribution, flight seasons, field identification, variation, habitat, behavior, and phylogeny of Odonata. They include John Acorn, Maria Aliberti, Allen Barlow, Sharon Brown, Paul Brunelle, Rob Cannings, Syd Cannings, Paul Catling, Glenn Corbiere, Dave Czaplak, Nick Donnelly, Cameron Eckert, Tony Gallucci, Rosser Garrison, Bruce Grimes, David Halstead, George Harp, Matt Heindel, John Hudson, Dustin Huntington, Ann Johnson, Paul Johnson, Greg Kareofelas, Ed Lam, Robert Larsen, Mike May, Mike and Barbara Moore, Alan Myrup, Wayne Nordstrom, Janis Paseka, Herschel Raney, Mike Reese, John Rintoul, Don Roberson, Larry Rosche, Tom Schultz, Judy Semroc, Fred Sibley, Bill Smith, Wayne Steffens, Ken Tennessen, June Tveekrem, Steve Valley, and Jessica Ware, and I hope I have not forgotten any of the others who belong on this list. Some of this information was taken from Odonata listserves and web sites.

I greatly appreciate Paul Brunelle's sending me years of field notes that provided much natural-history information. Dustin Huntington sent me his beautiful videos of southwestern odonates that furnished much information about behavior. Jim Sinclair kindly sent me the only specimens of Cream-tipped Swampdamsel I have seen, and Jerrell Daigle furnished other important specimens. Nick Donnelly's dot maps and John Abbott's Odonata Central website furnished the basic information for constructing the range maps presented here, and I thank them for making that information available to all of us. Christy Barton expertly rendered all of my hand-drawn range maps on the computer. I owe much to Linda Feltner for sharing her knowledge of Photoshop and graphic design with me. Natalia von Ellenrieder has my greatest appreciation for her precise anatomical drawings of species after species; without them, many more odonates would be unidentifiable! I also appreciated her encouragement all along the way. In addition, Rosser Garrison very kindly let me use his drawings of dancers (*Argia*) that have graced the pages of other publications. Both of these friends have contributed greatly to the value of this book. Thanks to Elissa Schiff for her careful editing and resulting readability. Working with the staff at Princeton University Press is always a pleasure, and I thank especially Robert Kirk, Dimitri Karetnikov, and Mark Bellis for their advice and cooperation. As always, Netta Smith deserves a world of thanks for accompanying me in the field day after day, for sharing the fun and frustration of photographing dragonflies at all points of the compass, and for emotional and logistic support at all times.

Introduction

This book is a field guide to all the species of Odonata (dragonflies and damselflies) in the western United States and Canada, west of the eastern boundaries of Texas, Oklahoma, Kansas, Nebraska, South Dakota, North Dakota, Manitoba, and Nunavut. State and provincial boundaries were chosen rather than the exact middle of the continent because naturalists' interests and odonate recordkeeping are typically at this level. Thus, many species that occur in the eastern portion of the West as defined here are very much allied with the moist and forested East, and any species that occurs no farther west than midcontinent should be considered an eastern species. However, because odonate diversity is higher in the East, the continent has been divided in this way to allocate similar numbers of species to this western guide and the eastern guide to follow it. The western book treats 348 species, and a book covering the species in the area to the east of it would contain 334 species as of this date. There is much overlap in species covered, as the total North American fauna at present is 453 species. Thus, about one-fourth of the species in each half of the continent do not occur in the other region and are covered in only one book. To provide sufficient text and illustrations for all North American species was considered too much for a single volume.

Numerous species included in this book barely enter its geographic coverage from the east or south, but the great majority of the species covered are resident in the region. A few species recorded from southern Arizona and southern Texas may not have resident populations, or local resident populations may originate and then disappear. Because so few species are likely in these categories, all species have been treated equally.

I should make a few definitions clear at the outset. To geographers and biogeographers, North America includes Mexico, Central America, and the West Indies, but for convenience in this book, I am restricting "North America" to Canada and the United States. A volume on all of North America would include hundreds more species, and we still do not know enough, nor have sufficient photographs, for a book on Mexican and Central American odonates. Numerous additional species that occur in northern Mexico might wander north of the border in especially wet years or as a consequence of increasing global temperatures, and such additions to our fauna have been occurring at the rate of one or two species each year.

Although all Odonata are called dragonflies in other English-speaking countries, in North America many restrict the term "dragonfly" to the suborder Anisoptera and use "damselfly" for the suborder Zygoptera. I follow that practice in this book and use the term "odonate" when referring to both suborders. In the introductory sections, however, I use "dragonfly" to refer to the entire order and "damselfly" when speaking only of that group. The word "ode" is used as a synonym by many odonate enthusiasts, with etymologically mangled modifications of it such as "odophile" (odonate lover).

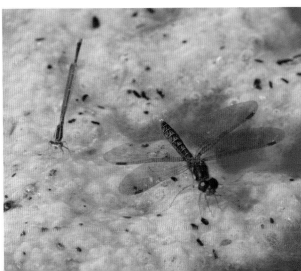

Orange Bluet (left) and Eastern Amberwing males—Hamilton Co., OH, July 2006, William Hull

Odonate species are grouped together by their fundamental similarities into ever larger groups called genera (singular genus), families, suborders, and finally the order Odonata. Wing venation is an important character used to categorize families and genera, but it is de-emphasized here because it is not easily seen in the field. Nevertheless, there are good technical keys to odonates, and an understanding of the venation, as well as the details of the rest of their anatomy, is of value to the odonate enthusiast. Technical anatomical terms are kept at a minimum, but they are necessary from time to time, and they can be learned from the illustrations here.

Because we do not know the exact phylogeny (relationships and order of appearance over evolutionary time) of odonates, much less of most other groups of animals and plants, many authors have chosen to list species in alphabetical order by their scientific or common names rather than attempting to associate them to show their relationships. This is commonplace in odonate books, but alphabetical order always places some closely related genera or species some distance apart in a list. Other books place species by their similarity in appearance (as so many flower books do), which is helpful in a field guide but does not show relationships and often separates closely related species, so the reader is unaware of the relationship. Thus, in this book, I have decided to attempt to place genera and species in a semblance of phylogenetic (taxonomic) order to emphasize their relationships. This is also appropriate because close relatives are often the species to be distinguished in a field guide. There is sufficient literature that in only relatively few cases have I had to use my own judgment. I hope the reader will become familiar with the order of the species, just as birders do in bird books (although it will probably be different in other books!).

See the Appendix for general references to odonates. An extensive list of references for the individual species was compiled but proved too voluminous to include in the published book. A link to these can be found at http://press.princeton.edu/titles/8871.html.

Natural History of Odonates

Bird field guides normally do not discuss the general natural history of birds because that is so well known to naturalists in general and even to lay persons, and it may even not be necessary to know in order to identify the species. But dragonflies are less well known, and they are such interesting animals that all who observe them in the field should know something about their lives. Of course, because they have a larval stage, much of what is important about them goes on out of sight of the usual observer.

Perching Dragonflies perch in many ways, as can by seen by looking at the photos in this book. Most of them, as typified by clubtails and skimmers, perch more or less horizontally, supported by their legs; usually, when flat, the abdomen is held above the substrate. These are *perchers*. Another whole group of dragonflies, typified by darners, hang from a perch, also supported by their legs clutched to the sides of the substrate. These are *fliers*. See below for how this correlates with feeding behavior. Percher dragonflies probably change their postures mostly for thermoregulation (temperature regulation). They can elevate or lower their abdomen with ease, and they can make it perpendicular to the sun's rays in the morning and evening to warm it maximally or point it directly at the sun at midday (*obelisking*) to minimize solar radiation falling on them. This is commonplace, especially in hot climates. We do not know why dragonflies sometimes assume appropriate postures when in an inappropriate situation (e.g., obelisking when not in sunlight).

Wing positions are harder to understand, but among dragonflies they are varied only in the skimmer family. Some skimmers lift their wings when perched on a plant in an open, windy area, lifting them even more when the wind increases, and this is probably for aerodynamic reasons, to keep them stable on their perch. Certain genera of skimmers that routinely perch with wings up (pennants, for example) are those that most commonly perch in open, windy areas on thin, tenuous perches. Many skimmers often seem to droop their wings when they are relaxed, so perhaps that is the most "comfortable" position or takes the least energy to maintain. Dragonflies usually land on a perch with wings horizontal and then may, sometimes in stages, droop them. If you approach such a dragonfly, it may snap its wings back up to horizontal, presumably the best position for a quick takeoff.

Damselflies hold their wings in one of two positions, the open wings of spreadwings (and several other tropical families) and the closed wings of all other damselflies. Open wings seem associated with large damselflies that forage by flycatching, and smaller flycatching damselflies such as dancers may hold their wings closed to avoid being conspicuous. A few pond damsels routinely hold their wings partially open.

Sleeping Odonates usually retire to dense vegetation for the night. Damselflies will fly into a clump of herbaceous vegetation and then crawl within it. They perch with abdomens parallel to plant stems and will move around the stem but not fly if disturbed at night (or at low temperatures). Spreadwings close their wings while roosting. Dragonflies are more likely to move up into shrubs and trees, even well up in the forest canopy, and hang on twigs and leaves on the edges of the vegetation, but some sleep in the weeds with the damsels. Little is known about this aspect of odonate biology.

Flight One of the many special qualities of dragonflies is their superb flight ability. All four wings can be moved independently, as can often be seen in flight photos. The wings can not only be "beat" in the classical sense, moved up and down, but they can be rotated on their own axes somewhat like the feathering of an airplane propeller. This allows great flexibility in just about every way a flying machine needs it. Dragonflies can fly forward at more than 100 body lengths/sec, backward at more than 3 body lengths/sec, and hover, all while keeping their bodies horizontal. They can fly rapidly straight up and straight down, turn on a dime, or move forward or upward slowly, at almost stalling speed, to search for food items. Although at times they seem to move like a rocket, they probably do not exceed 30 miles/hr, the top speed of small birds.

Dragonfly and damselfly in flight

Common Green Darner male—Harris Co., TX, October 2005, Trevor Feltner

Amelia's Threadtail male—Zapata Co., TX, June 2005

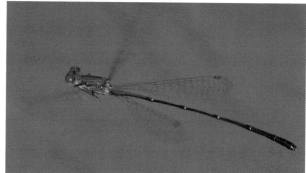

Dragonflies with prey

Common Green Darner male eating Queen Butterfly—Hidalgo Co., TX, November 2005

Eastern Pondhawk female eating Eastern Amberwing—Jasper Co., TX, May 2005

Vision Dragonflies have the finest vision in the insect world. The compound eyes in the largest species have as many as 30,000 simple eyes (*ommatidia*) perceiving the world around them. Because the simple eyes are individual receptors, insect vision is somewhat of a mosaic, and dragonflies are very good at detecting movement. The tiniest movement in the distance stimulates one ommatidium after another. The eyes are so large, especially in darners, that they wrap around the head and afford almost 360° perception. Vision is relatively poor only directly behind and below a dragonfly, and dragonfly collectors learn that to their advantage. Damselflies, with their smaller eyes well separated, perhaps have enhanced depth perception for close-range distinction of aphids from the leaves on which they rest. Dragonflies have a wider range of color detection than mammals and, as other insects, can see into the ultraviolet (UV) range. Many species reflect UV, especially those that look bright blue or bright white to us, and they may look even brighter to other dragonflies.

Dragonfly eye
Ringed Emerald male—
King Co., WA, September
1975, Truman Sherk

Feeding Odonates are all predators, in both adult and larval stages. Adults exhibit three different modes of foraging behavior that have equivalents in bird feeding behavior. Fliers fly around, either back and forth in a confined space or more extensively, and capture other flying insects by *hawking*. All other odonates are considered perchers. Among the perchers, *salliers* watch for flying prey from a perch and fly up to capture it, whereas *gleaners* alternate perching with slow searching flights through vegetation, where they dart toward stationary prey and pick it from the substrate. Gleaners may also flush an insect and chase it through the air. Typical hawkers include darners (which are called hawkers in the UK) and emeralds and certain groups of skimmers, for example gliders and saddlebags. Typical salliers include broad-winged damsels, spreadwings, dancers, clubtails, and most skimmers. Typical gleaners include pond damsels other than dancers.

Most dragonflies take small prey, much smaller than themselves. Tiny flies, leafhoppers, and beetles are common prey. Some species vary these with larger prey, and others seem to be specialists in larger prey. Dragonhunters are well known to live up to their name, and some other clubtails take large prey. Pondhawks commonly and darners more rarely prey on dragonflies up to their own size, and even pond damsels will take another damselfly of the same size, especially when the latter has just emerged and is quite vulnerable. Note that dragonflies that routinely take large prey are probably the fiercest biters when captured!

Damselflies with prey

Rambur's Forktail male eating
Citrine Forktail—Marion Co.,
FL, June 2004

Alkali Bluet male eating
fly—Harney Co., OR,
July 2006

Odonate predators

Great Kiskadee eating Common Green Darner—Hidalgo Co., TX, May 2005 (top left)

Yellow-headed Blackbird eating damselflies—Grant Co., WA, June 1995 (top right)

Orb-weaving spider with spreadwings and meadowhawk—San Juan Co., WA, August 2007, Netta Smith (middle)

Robber fly eating Aurora Damsel—White Co., GA, May 2006 (bottom)

Predators and predator defense Many larger animals prey on odonates. Members of their own order are among the most important predators, with many dragonflies routinely preying on damselflies and smaller dragonflies. The other very important insect predators on odonates are robber flies on adults and ants on tenerals that have just emerged. Spiders are similarly important, both active predators and web builders. Among vertebrates, birds are most important, taking a huge toll on tenerals in wetlands but also capturing many adults. Small falcons and large flycatchers are important predators, and some tropical bird groups (jacamars, bee-eaters) are even dragonfly specialists, their long, slender bills perfect for catching a dragonfly by the wings. Frogs and fish also take their toll at the water surface, and lizards in the uplands.

Dragonflies have their good vision and swift and agile flight to protect them from predators. Many are well camouflaged, especially when perching away from water. The mottled brown body coloration of some species, especially the females, enhances camouflage, as perhaps do the dark wingtips of many forest-dwelling skimmers, again especially the females. When not out in the open, damselflies usually perch along grass stems, grasping them tightly, and sidle around the stem to be on the far side from a potential predator. The well-separated eyes can look around a narrow stem when the body is hidden. A few small dragonflies may effectively mimic wasps, gaining protection from predators that shy away from such well-protected insects.

Sexual Patrol Males of all odonates spend a lot of their time searching for females; natural selection demands it. Each species has a characteristic pattern of sexual patrol. Some species are highly *territorial*, each male staking out a territory and defending it. In the vast majority of species, this territory coincides with an optimal habitat for female egg laying, which of course coincides with optimal habitat for the larvae. Other species do not defend territories, for example, male darners that fly all around the shore of a lake or male damselflies that change perches frequently and do not remain in any particular place. Families with species that defend fixed territories include broad-winged damsels, clubtails, petaltails, and skimmers. Those in which males fly over long beats, not defending fixed territories but aggressive to any other males they encounter, include darners, spiketails, cruisers, emeralds, and some species among the clubtails and skimmers. Most damselfly males defend no more than the perch on which they rest at the moment.

Courtship and Mating Courtship behavior is fairly common in tropical damselflies, but few North American species exhibit it. It is best known in jewelwings of the genus *Calopteryx*. Among western dragonflies, male amberwings come closest to practicing courtship when they attempt to lead females to oviposition sites. In most North American odonates, the male just grabs the female, but she still chooses whether to mate with him or not. Although males will often attempt to mate with females of other species, most females are apparently able to detect by touch whether the male that grabs them is their own species.

Where odonates meet to mate has been called the *rendezvous*. In general, when a mature male odonate encounters a mature (or even immature) female of the same species at the rendezvous, he attempts to mate. In most species, the rendezvous is at the water, but many species, especially those that oviposit in tandem, also mate away from the water and then move to it to lay eggs. The male approaches the female from behind, grabs her with his legs, in some cases even biting her briefly, and immediately attempts to clasp her with his terminal appendages. A male dragonfly clasps the head of the female with two cerci on either side of her "neck" and the epiproct pressed tightly against the top of the head. In some dragonflies, the prothorax may also be involved in this *tandem linkage*. In jewelwings and spreadwings, the male's two cerci are applied to the back of the prothorax, the two paraprocts to the top, holding it firmly. In other damselflies, the cerci contact the mesostigmal plates at the front of the synthorax, the paraprocts on the top and/or side of the prothorax.

Double-striped Bluets in tandem—Edwards Co., TX, July 2004

After the female is firmly clasped, the male will then take a few seconds to transfer sperm from the genital opening under his ninth segment to an organ of sperm storage, the *seminal vesicle*, under his second segment. He will then attempt to swing her abdomen forward to contact those genitalia with the tip of her abdomen. In dragonflies, males of many species may transfer sperm to their second segment before hooking up with a female. When contact is made, the appropriate structures of the two sexes lock in place in dragonflies (but not

Powdered Dancer male transferring sperm—Montgomery Co., AR, May 2006

Copulation

Northern Bluets in copula—King Co., WA, June 2005 (middle left)

White-faced Meadowhawks in copula—Bremer Co., IA, July 2004 (middle right)

Roseate Skimmers in copula—Harris Co., TX, June 2006, Trevor Feltner (bottom left)

damselflies), and the penis transfers sperm through the female's genital pore into her vagina, in which fertilization may take place immediately, or the sperm may pass into one of two types of sperm-storage organs, the *bursa copulatrix* and the *spermatheca*. If sperm from another mating is already present, the male removes or flushes out much of it, thus making it highly likely that his sperm will fertilize her eggs.

At this time the male supports them both on a perch in the majority of species (some copulate only in flight), and in dragonflies, the female's legs grasp the male's abdomen. The copulatory, or *wheel*, position is unique to the Odonata, as is the distant separation of the genital opening and the copulatory organs. That the position looks as much like a heart as a wheel has been repeatedly noted. Copulation lasts from a few seconds, for example in skimmers that mate in flight, to several hours in different odonate species.

Females of many species can retain live sperm throughout their life, essentially fertilizing their own eggs as the eggs travel down the oviduct past the sperm-storage organs.

In a few studies in which most male dragonflies at a pond were removed, some remaining ones mated over 100 times! Males that have mated often have marks on their abdomen where the female legs have scratched them. This is especially obvious in species in which males develop *pruinosity*, as the pruinosity on the midabdomen is scratched off, and the signs are visible at some distance. Female dragonflies that have mated often have marks on their eyes where the male epiproct has scratched or even punctured the eyes.

Egg Laying and Hatching Eggs are laid into the water or bottom material (*exophytic* oviposition) or inserted into plants (*endophytic* oviposition). Females that oviposit exophytically extrude eggs from the genital pore in their eighth segment, usually in flight. Quite a cluster of eggs may be formed before the female taps the water and releases them. The eggs typically

Odonates ovipositing

Cardinal Meadowhawks in tandem—Monterey Co., CA, August 2006, Don Roberson (top left)

Paddle-tailed Darner ovipositing—King Co., WA, September 1997, Netta Smith (bottom left)

Spotted Spreadwings in tandem—San Juan Co., WA, August 2007 (right)

Eggs in situ

Amazon Darner eggs
in cattail—Guanacaste,
Costa Rica, April 1967

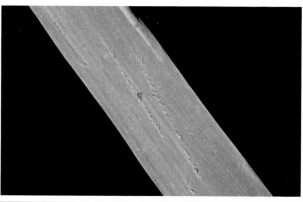

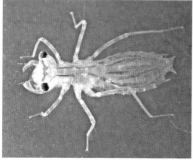

Damselfly and dragonfly larvae

Early instar skimmer larva, just molted—King Co.,
WA, 1973 (top left)

Common Whitetail larva, next-to-last instar—
Summit Co., OH, May 2007, Judy Semroc (top right)

Canada Darner larva, last instar—Thurston Co.,
WA, November 2000, William Leonard
(middle right)

Pond damselfly larva, last instar—Ashtabula Co.,
OH, June 2004, Judy Semroc (bottom right)

sink into a bed of aquatic vegetation or onto the bottom. This method of egg laying is very variable, some females tapping the water in the same spot for an oviposition bout and others flying rapidly over the water and dropping egg clusters at intervals. In some species, eggs are dropped from above the water or even onto dry ground in the case of some meadowhawks. Endophytic species have well-developed ovipositors and include all damselflies and petaltails and darners among the dragonflies. Large numbers of eggs can be laid by a single female, up to the low thousands.

Eggs may hatch after a few days, or embryonic development may take a month or more. In some species, the eggs overwinter and hatch the following spring. Each egg hatches into a very tiny *prolarva* that looks like a primitive insect form, quite different from the larva that will succeed it. When the eggs are laid above the water, the prolarva leaps and flips about until it gets to the water, at which time it will quickly molt into the second larval stadium (each stage is called a *stadium* or an *instar*). When the eggs are laid directly into the water, the prolarval stage may last only a few minutes.

Larval Life History Although they have the standard head, thorax, and abdomen and six legs of all insects, odonate larvae (also called nymphs or naiads) look very different from the adults and have a very different life. One characteristic they have in common is being predators, but the larvae capture their prey by shooting out their *labium*, a sort of lower lip (it has been called a "killer lip"), by hydrostatic pressure and grabbing with a pair of labial palps that open and shut on the prey. The labium then retracts and draws the prey into the mandibles. The larvae of damselflies, petaltails, clubtails, and darners have a flat labium with pointed palps that skewer the prey. The larvae of spiketails, cruisers, emeralds, and skimmers have a spoon-shaped labium with large palps that enclose the prey. Dragonfly larvae respire through gills in their rectum, and the whole rear end of the abdomen is a respiratory chamber that draws water in and out and takes oxygen from it. Because of this ability, the rectum can also be used for jet propulsion under water. And finally, of course, it is the posterior end of the digestive tract, so in dragonflies, the rectum is a unique and multipurpose organ, one of their many unique attributes. Damselflies have three prominent caudal gills that function to extract oxygen from the water, and they can also use these to swim by waving them back and forth like a fish's tail. They are still able to respire after losing their gills, but not as well.

The larvae are very variable in what they do, more so than the adults. Some burrow just below the surface of sand and mud bottoms, grabbing midge larvae they encounter in their semifluid milieu; others squat in the bottom detritus with only their eyes and face exposed, striking out at fellow bottom dwellers; still others climb in the vegetation, stalking their prey as a cat stalks a mouse. Larvae of certain types live right out in the open in temporary ponds that lack larger predators, and some spreadwing larvae swim in stream pools like little fish. As you might guess, their shapes and colors vary in tune with their habits.

Collecting odonate larvae and keeping them in an aquarium is a wonderful way to learn about them. They are quite predatory and even cannibalistic, and to keep them from eating each other, put relatively few in each aquarium and give them lots of vegetation in which to hide. In a warm aquarium, they will grow quickly, even to metamorphosis. If you want to watch emergence, collect larvae in their last stadium and put them in an aquarium or large jar with a stick on which to emerge. Put a screen over it or they will be flying around your house.

Metamorphosis and Emergence The larvae undergo numerous molts, averaging around a dozen, as they grow and feed. In warm tropical pools, a larva may go through its entire development in as little as a month, but most temperate-zone species take at least several months. In colder waters, as in streams, mountain lakes, and at higher latitudes, the larva may take several years to reach metamorphosis, growing through a few molts each summer. During the last few stadia, the adult wings begin forming inside wing pads extending back from the thorax that become more prominent until they are bulging with the wings inside them. During this time, the larva is undergoing a metamorphosis to its adult state even as it remains active—one of the wonders of nature! It finally stops feeding and soon thereafter switches over to aerial respiration; then it leaves the water. Damselflies and clubtails and some small skimmers often emerge just above the

Four-spotted Skimmer emergence—King Co., WA, June 1997

water, but other odonates typically move farther away from it, even climbing up into trees (long-distance travel is typical of river cruisers).

The larva fixes itself to the substrate and then expands its thorax until a split appears in it. It then emerges through the split and hangs backward from the larval skin. After its cuticle hardens for a while and its muscles become stronger, it reaches up and pulls itself out of the exuvia. The wings, folded like accordions, then begin to fill from the base with fluid transferred from the body and fairly soon reach full length. The fluid is then pumped back into the abdomen, and it

Variable Darner exuvia—Apache Co., AZ, July 2007

expands. Finally, the wings open up, and very soon the *teneral* adult flies away. Clubtails manage the same process on horizontal substrates. The cast skin left behind is called an *exuvia*, and looking for exuviae is a good way to find out what species are breeding in an area, as the exuvia is just as good as the larva for identification. In many groups, it is easy to determine the species from the larva or exuvia, but in some, the species are similar enough that identification to genus is more practical.

To find adults emerging, which is very exciting to watch although slow in tempo, you should know something about their emergence times. At temperate latitudes, most damselflies and clubtails emerge during the daytime, usually during the warmer periods at midday, whereas members of other families do so at night, the larger ones leaving the water a few hours after sunset so they are ready to fly as it gets light the next morning. As latitude increases, nighttime temperatures may be too cool for emergence, so darners, emeralds, and skimmers often emerge during the day in northern regions. A cold spell that ensues during emergence can delay or even stop the process. Conversely, with lower latitude, higher temperatures, and even more avian predators, even the clubtails emerge at night, although daytime emergence remains typical of damselflies. On a hot day, a damselfly can go from crawling out of the water to flying away in a half-hour or less, and damselflies probably emerge in the daytime just because high temperature facilitates quick emergence.

Sexual Maturation After leaving the water as a teneral, an odonate slowly continues to harden and color up. The color is often different from the color at maturity, and it changes over a course of days or weeks or even months as the individual becomes sexually mature and returns to the water, completing the cycle. These prereproductive adults are often called immatures (a name that should not be used for the larvae), and they may wander far and wide, even miles from the water. This is especially true of darners, which may fly up mountains much as butterflies do. Both immature and mature individuals of flier dragonflies may form feeding swarms, sometimes of mixed species. After the immature phase, most temperate-zone odonates live a surprisingly short time. Small damselflies live no more than a few weeks, larger dragonflies a month or two. Dying of old age is rarely observed in odonates, but at some lakes with an abundance of bluets, large numbers of dead ones have been observed floating on the surface toward the end of their flight season. Old individuals are often discolored, with tattered wings.

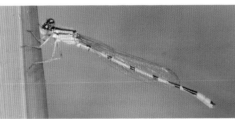

Immature and mature damselfly (above)

Familiar Bluet immature male—Hidalgo Co., TX, May 2005 (top)

Familiar Bluet mature male—Hidalgo Co., TX, June 2005 (bottom)

Immature and mature dragonfly (right)

Western Pondhawk immature male—Harney Co., OR, June 2005 (top)

Western Pondhawk mature male—Grant Co., WA, August 2007 (bottom)

Odonate Anatomy

Understanding color-pattern descriptions of odonates is made easier by understanding their anatomy. The head is made up of the huge eyes (smaller in damselflies) and what might be called the "face" (technically from top to bottom the *frons*, *clypeus*, and *labrum*). There is little else, although sometimes field marks are located on the *vertex* (behind the frons) and the *occiput* (behind the eyes) or even the *labium* (the jointed lower lip). The coloration on the back of the head also varies and can be used as an identifying mark in some cases. Other anatomical features on the top of the head include the *ocelli* (singular *ocellus*), three tiny simple eyes arranged in a triangle that may serve to measure light intensity; and two small *antennae* that probably measure air speed but do not function as olfactory organs as in so many other insect groups. What looks like the "neck" of a dragonfly is actually its *prothorax*, on which are attached the first pair of legs. This tiny segment really does function somewhat as a neck, joining the head quite flexibly to the rest of the animal. The connection between head and prothorax is surprisingly narrow and seemingly flimsy, but it allows the head and body to be moved somewhat independently during flight. Look also at how odonates, at least the perching species, move their heads around while at rest. They are often looking for prey, but they

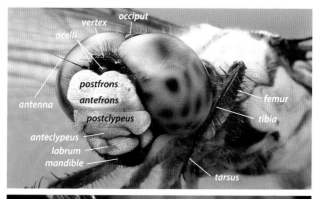

Great Blue Skimmer male—Ascension Par., LA, June 2004, Ronald P. Gaubert

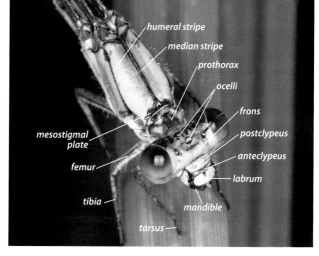

Blue-fronted Dancer female—Ashtabula Co., OH, July 2005, Judy Semroc

can also spot predators, territorial rivals, and mates. Note also how the front legs are often tucked behind the head, perhaps bracing it.

The thorax, also called *pterothorax* or *synthorax*, houses the big important flight muscles. It is actually the fused *mesothorax* and *metathorax*, each with a pair of wings and a pair of legs. The part of the thorax in front of the forewing bases is anatomically the front. Thus, the thorax has a front, sides, and a bottom; the top would be the area containing and between the wings. This is harder to envision in damselflies, as their thorax is skewed so far backward to allow their wings to fold over their abdomen that the front of the thorax looks like the top. I am going to remain consistent with anatomy and call the area in front of the wings the front of the thorax when describing color patterns of damselflies. The thorax is patterned in most species, and the patterns are consistent within and sometimes between families. In damselflies, the dark stripe on either side that extends from the base of the forewing to the second pair of legs is called the humeral stripe; a pale stripe at its anterior (inner, upper) edge is called the antehumeral stripe. These have been called "shoulder stripes" in other books. Some damselflies have additional narrow dark stripes posterior to (outside, below) the humeral. In dragonflies, the stripes take different forms and can be much more complex.

The forewings and hindwings of damselflies are about the same (Zygoptera means *yoked wings*), whereas those of dragonflies are quite different, the hindwing being considerably

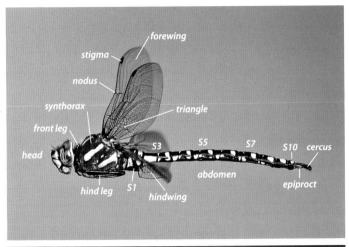

Paddle-tailed Darner male—King Co., WA, August 2007, Larry Engles

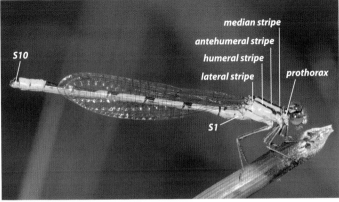

Boreal Bluet male—Grant Co., WA, August 2007

broader and with different patterns of venation at the base (Anisoptera means *unlike wings*). All of our species have prominent markers on the wings, including the *triangle* (dragonflies) or *quadrangle* (damselflies), the *nodus*, and (in almost all species) the *pterostigma*. As do other authors, I use the shortened "stigma" for the pterostigma. Knowledge of venation will be of importance for species identification in some cases.

The 10-segmented abdomen carries the digestive tract and reproductive organs. It is probably longer than is necessary for that purpose just because it acts aerodynamically to

Scans of dragonfly and damselfly wings

Gray Petaltail female— San Jacinto Co., TX, March 2000, Robert A. Behrstock

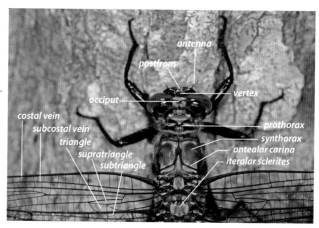

Wandering Glider wings (contrast increased digitally)

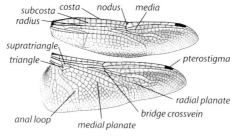

Damselfly wings of three families

Coenagrionidae
Vivid Dancer

Lestidae
Amber-winged Spreadwing

Calopterygidae
Smoky Rubyspot

Dragonfly wings of four families

Petaluridae
Gray Petaltail

Aeshnidae
Paddle-tailed Darner

Corduliidae
Moutain Emerald

Libellulidae
Wandering Glider

put as much weight behind the wings as in front of them. Long abdomens tend to be slender, short abdomens wide. The secondary genitalia of males are carried in S2–3, usually very prominently. **Looking for that basal bulge in side view is the best way to quickly sex an individual**. Females typically have wider abdomens, and these taper less to the rear than those of males, presumably because they carry a load of eggs. Abdomen shape is another important method of sex determination. The tenth segment (S10) carries the *cerci* of both sexes (upper appendages in males) and the *epiproct* (single lower appendage) in male dragonflies

Male dragonfly and damselfly appendages

Comet Darner male appendages—Summit Co., OH, July 2007, Judy Semroc

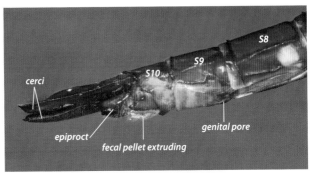

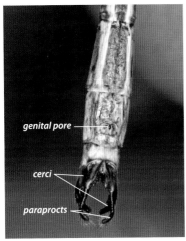

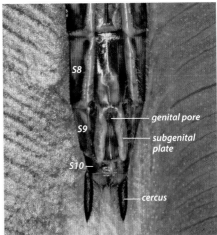

Elegant Spreadwing male appendages— Ashtabula Co., OH, June 2007, Judy Semroc (middle left)

Common Baskettail female abdomen tip—Erie Co., PA, June 2006, Judy Semroc (middle right)

Amazon Darner female ovipositor— Guanacaste, Costa Rica, April 1967 (right)

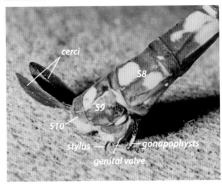

Sexual differences

Mexican Amberwing male and female—Santa Cruz Co., AZ, August 1977

and *paraprocts* (paired lower appendages) in male damselflies. The female reproductive tract opens between S8 and S9 and may be protected by a *subgenital plate* (also called vulvar lamina, present in most dragonflies) or prolonged into a complex *ovipositor* (damselflies, petaltails, darners). Abdomens can be simply striped or spotted, with the spots arranged centrally or laterally, or more complexly patterned with linear and transverse markings.

Odonate Colors

Dragonfly colors are produced in the same way as those of other organisms. Most of the blacks, browns, reds, and yellows are pigments of various types. Most of the blues are structural colors, produced by the microscopic structure of the surface of the cuticle reflecting blue light and letting the other wave lengths pass through. Greens are usually produced by adding yellow pigment to a blue-reflecting cuticle. Iridescent (metallic) colors are also structural. The translucent surface of the eyes makes their coloration particularly glowing. As the adult dragonfly develops within the larva, its pigment patterns are laid down, and they become more evident as it ages after emergence. Sexual maturation is often indicated by a dramatic change in color brought about by the deposition of pruinosity. Note that within a group (a genus or even a family) the great variety of colors and patterns are often part of a common theme, with the variation caused by the presence and relative size of dark markings and the pale colors between them.

Odonate Names

The scientific name consists of a *genus* (plural *genera*) and a *species* name. The name is unique to the species and will allow reference to it in any language. Both words usually have Greek or Latin roots; Greek may actually be more common, so "Latin name" is a misnomer. Anyone with some knowledge of the Classical languages or even of common words in biology will have an insight into the meanings of the names. For example, the Ringed Emerald is *Somatochlora albicincta*. *Somato* means "body" and *chlora* means "green" in Greek; *albi* means "white" and *cincta* means "ringed" in Latin. Thus, *Somatochlora albicincta* is a "white-ringed greenbody." Not all scientific names are so easily translated.

The scientific names and common names used herein are those in most recent use. With species being split (divided into more than one species) or lumped (combined with another species), changes in scientific names may still occur in the North American fauna. The Checklist Committee of the Dragonfly Society of the Americas has taken on the responsibility of keeping the North American checklist up to date by making taxonomic decisions and determining common names for North American species. These decisions are maintained and updated on a website at the Slater Museum of Natural History (http://www.ups.edu/x7015.xml), and the checklist includes both the name of the person who described the species and the year of description.

Until relatively recently, most Odonata of the world lacked common names; odonatologists communicated entirely by scientific names. In 1931, E. B. Williamson (one of the greats of American odonatology) stated "But if there must be common names by edict, would it not be well to go slowly and after some discussion of each proposal so that the feeble-minded and tongue-tied student of the future, reared entirely on a diet of common names, may have bequeathed to him a nomenclator vulgaris as exact, appropriate, and euphonious as possible?" This is exactly what was attempted six decades later, with the caveat that no dragonfly aficionado known to me is either feeble-minded or tongue-tied. An attitude against using common names lasted through much of the twentieth century, but late in that century, Sidney Dunkle and I, discussing the need to encourage wider studies of dragonflies by amateur naturalists, decided to see if we could design a set of "exact, appropriate, and euphonious" names for the North American fauna. That list was then published in *Argia*, the newsletter of the Dragonfly Society of the Americas, and was subject to criticism and suggestions by the membership. Changes were incorporated, and the list was officially endorsed by the society. That list is used herein, with a few recent changes to incorporate new species, newly recorded species, taxonomic changes, and correction of a few inappropriate names. Although stability is the ultimate goal, the naming of living organisms is always an ongoing project. The common names of species are considered "proper names" by many people using them, and thus, their first letters are capitalized, that is, Filigree Skimmer rather than filigree skimmer. Group names are left uncapitalized, for example bluets or mosaic darners.

I strongly encourage all with a serious interest in dragonflies to consider learning both the common and scientific names. Almost all of the scientific literature on Odonata uses their scientific names, and most professional odonatologists communicate by the scientific names. This becomes essential as soon as one moves south of the U.S.–Mexican border and begins to encounter species not known from North America. Furthermore, knowing the genus to which a species belongs is immediately indicative of its relationships, whereas common names do not always provide that information. Genera of odonates are usually well defined, with a suite of characteristics (often morphological, ecological, and behavioral) that allow distinction from other genera. The genera of North American odonates have also been given common names, and those are written in lower case, as they are not to be considered proper names. At times in the text below, I use the real generic name rather than the generic common name, consciously promoting my philosophy. I also very often discuss the genus as a group to be compared with other such groups, as dragonfly genera are usually (but by no means always) discrete and recognizable. Also, the generic name may be as easy to remember and quicker to say than the common name of the genus, e.g., *Orthemis* instead of tropical king skimmer or *Enallagma* instead of American bluet.

Finding Odonates

Because odonates are aquatic animals, the best place to find them, of course, is at the water. Wetlands of all types support populations, but some types are better than others. A warm, productive pond or lake with much aquatic vegetation should have a good list of species, and the species are often common and widespread ones that can be the first to be learned by a beginner. Many odonates are habitat specialists, and to find them you must locate their habitat. The best way to see a diversity of species is to explore a diversity of habitats, not just ponds and streams but as many different kinds of ponds and streams as there may be in the area. Large lakes and rivers have specialists not to be found on smaller wetlands. Tiny trickles and seeps may have still other specialists. Bogs and fens support different species, and some species even find habitats created by humans, such as reservoirs and farm ponds, to be optimal for their ecological needs.

Because odonate presence varies in both space and time, you will also want to check each location at least several times over the flight season. Some species have surprisingly short periods of abundance, with most of the population emerging over a few days' time, present for less than a month, and then disappearing rapidly. Others have long flight seasons. In addition, there is year-to-year variation, with a species appearing at a wetland where it had not previously been present, perhaps remaining in abundance for a few years, then petering out over the next several years.

Most species of odonates leave the waterside when immature, and they can then be found almost anywhere, in some cases even miles (or kilometers, in the metric system) from water. Look for them in sunny clearings or at the edges of roads and trails through forest and field. Walking along an open trail that parallels a stream, for example, may give you access to many species perching in the sun that are only at the stream at certain times of day. Some of the best looks at darners, river cruisers, and emeralds may come here, as you find them hanging up instead of incessantly cruising back and forth. If you see one of these long-bodied flier dragonflies cruising a beat, presumably feeding, watch it for a while, and it may hang up, often when it captures something. You can then get a better look or approach for a photo opportunity. Some species aggregate to feed where their prey is concentrated, often in the lee of trees on windy days. You may be much more likely to find females, as well as immatures, away from water, whereas males are much more easily found at the water. Note also that females often outnumber males at the end of the flight season.

Wherever it gets really hot, it can become too hot for odonates to remain in the sun for any length of time, and they will move into the shade during midday and early afternoon. At such times, it may be profitable to look for them in sheltered shaded areas, either at or away from the water. Species that normally perch out in the open may be hanging under branches to avoid the sun's direct rays. Similarly, if it is cooler, different opportunities present themselves. By visiting a dragonfly locality early in the morning, you may find some individuals sunning themselves, cool enough that they are easily approachable even though perched in the open. As might be expected, the farther north you go, the lower the temperatures at which some dragonfly activity occurs, probably because northern species are specially adapted to exist at lower temperatures.

Binoculars are of great value for getting closer looks at odonates, for example one perched over the water or up in a tree. Close-focus binoculars can be used to examine damselflies right at your feet, and in fact, my criterion for "close focus" is being able to focus on your own feet or closer. With such an optical aid, it might be possible in some cases to identify male damselflies from the structure of their appendages without capturing them.

When searching for odonates, it is best to alternate *near-looking* (even right at your feet if you are walking through vegetation) for damselflies and *far-looking* (up to 20–30 feet away) for the warier dragonflies. Some small damselflies are unlikely to flush from dense vegetation, and gently moving a net or even a stick through grasses, sedges, and other herbaceous vegetation can flush them out and bring them to your attention. Often, I have caught small damselflies by sweeping an insect net through vegetation near water where I had no idea any were present. By this action, you can also examine more closely the damsels' predators (mostly spiders) and prey!

Identifying Odonates

In times past, when faced with an unidentified dragonfly or damselfly, an odonate enthusiast had a simple, if not always easy, task: catch it and key it out. This normally necessitated preservation of the specimen, useful because it allowed the identification to be reconfirmed or possibly changed at any time. Nowadays, there is not only a much stronger desire but also a much greater possibility to identify the same individuals in the field, either with or without the aid of a net. If identification involves capture, the best tool to have, after the net, is a hand lens or other magnifier. A variety of good 10×hand lenses are sold in biological supply houses.

Short of capturing the specimen, the next best way to assure identification is to take what could be called "ID photographs," photos from sufficient angles to capture all important aspects of color pattern, details of wing venation, and even close-ups of appendages, hamules, and ovipositors. Photography has greatly assisted in the identification process, as photos can be taken and examined at leisure with identification guides in hand and, if sufficiently detailed, can serve as a voucher for the identification and for the record.

Short of that option, the observer's best chance for an identification is to observe all these features and either commit them to memory or take notes on them. Obviously if an identification guide is handy, this can be consulted on the spot. Look especially at overall coloration, including differences between thorax and abdomen; color of the eyes and face; and obvious patterning on the thorax and abdomen, which may be complex. Even the coloration of the stigmas, wing veins, legs, and/or appendages can be important to distinguish some species. Behavior can be an important adjunct, one of the first steps to decide whether you are looking at a darner or a clubtail, a pond damsel or a spreadwing.

When trying to identify odonates, remember that males of related species often differ from one another more than females do. So females can be difficult, even very difficult, to distinguish. Often a clue to identification of females comes from the presence of males of only one species of that group. If you see a mating pair, collect or photograph or watch it closely to learn more about what females look like. Bear in mind that males sometimes make mistakes, and mistakes are more likely to be seen in tandem pairs than in copulating pairs. As you become more familiar with the species, you will find these *heterospecific* pairs yourself. *Conspecific* pairs are the rule.

Beyond the simple instructions in what to look for, a few principles should be understood first. One of them is *variation*.

Just as in birds and butterflies, in fact all products of Mother Nature, dragonflies are variable. *Sexual* variation is the most obvious, as the majority of species are dimorphic in color, many of them strikingly so. All odonate species are sexually dimorphic in shape, so an important step in dragonfly identification is to be clear about the general ways in which the sexes differ and then to apply these distinctions to the example under observation. You may immediately see that many of the individuals you initially thought were females were, in fact, immature males. **Being able to distinguish the sexes is a very important, often essential, step in identification.**

Adult odonates also vary in appearance with *age* just as much as birds and more than butterflies (which emerge from their pupae in fully adult colors). Most dragonflies change color as they move toward sexual maturation. The biggest problems in identification will be caused by the youngest individuals, which have not even reached their definitive "immature" coloration.

Immatures are very common, so identification of mature, fully colored adults will have to be accompanied by the recognition that they look distinctly different when younger. Often it is merely a case of a brightening and intensification of the color pattern, for example, immature male bluets going from pale blue-gray to vivid blue or the yellow of an immature male meadowhawk maturing to bright red. Males often go through an immature stage in which they are patterned much like females but then change dramatically at maturity by adding a layer of *pruinosity* (a powdery bloom much like the one we see on plums) to part or all of their thorax and abdomen. Most pruinosity is whitish to pale blue. The aging process may lead to more changes, as when wings become darker and body colors duller. In some skimmers, old females may develop pruinosity and then look more like males. One thing to remember: **eye color will almost always change during maturation**. Interestingly, color change continues to death; individuals may become discolored with advanced age, especially the face and thorax. For example, most male Tule Bluets examined at the end of their flight season at one site had grayish-brown markings on the otherwise blue thorax.

Unlike many birds and butterflies, odonates do not vary *seasonally*, at least not in color. Blue Dashers and Hagen's Bluets in the East become smaller (because the larvae that emerge are smaller) over the course of the flight season, and such variation may be more common than we know. Changes with maturation and aging in whole populations will make for different-appearing individuals at the beginning and end of the flight season.

Only a minority of odonates are known to vary *geographically*, but some of that variation is sufficient that the species involved look rather different in different parts of North America. When obvious, this variation is discussed in the species account. There is also geographic variation that is not so obvious, and much of it has probably not been documented. Perhaps much of the geographic variation evident in odonates is correlated with climate; for example, populations in cooler areas may have more extensive black markings. This can be seen in northern populations of Boreal and Northern Bluets and in the southwestern British Columbia population of Western River Cruiser. Populations of the Pacific Spiketail with the most extensive yellow are in the very arid Great Basin. Populations in hotter areas may have more extensive pruinosity, for example Widow Skimmers and Blue Dashers. There is also some geographic variation in size, mentioned under certain species.

Finally, there is *individual* variation. This is evident in both size and coloration. All species vary in size, as will be evident in each species account. The variation is surprisingly great, much more so than in birds, and is caused by variation in size of the larvae at metamorphosis. Variation in coloration is a bit trickier. Some species seem invariable, presumably truest for species with a minimum of patterning. Any species that is patterned with stripes and spots and dashes and squiggles will vary in the extent of those markings. This is the case for wing as well as body markings. Usually the variation is finite, and differences among species can still be determined. Coloration itself also varies, usually just from a bit paler to a bit darker but sometimes more than that.

Polymorphism is a special kind of individual variation in which individuals appear in two or more discrete color patterns with nothing in between. In North America, this is typical of females of many pond damsels and darners and a few spreadwings, with a brighter morph that looks something like the male of the species (usually with blue markings) and a duller morph quite differently colored from the male (often brownish or greenish). The male-like morph is usually called the *andromorph* (*andro* for male), the one that looks less like a male the *heteromorph* (*hetero* for different), and I use these terms throughout the book. No North American species has polymorphic males.

Odonates are variable. Color patterns, although often indicative, are not always definitive for identification. Most species are separable by their structure—male appendages or hamules, female mesostigmal plates or subgenital plates—and these structures will usually be definitive even when color patterns are not. Anyone who needs positive identification should strongly consider capturing the individuals in question.

Odonate Photography

Dragonflies and damselflies are wonderful photo subjects. So many of them are brightly colored and interestingly shaped that, just sitting still, they are photogenic. Many of them perch in the sunshine in conspicuous places, and walking around a wetland will provide photo op after photo op. Dragonfly photographers usually use lenses that are a combination of macro (for relatively close focusing) and telephoto (for magnification, especially of wary dragonflies). I use a 70–300 mm zoom lens that has macro capabilities at 300 mm, so I don't have to approach too closely and disturb my subject. Damselflies are usually easily approached, but dragonflies can be quite wary. For whatever reason, however, some individuals will be much tamer than others, so just keep trying.

The best photos are taken with cameras on tripods, as you can make sure the dragonfly is in sharp focus and can shoot at a slow enough shutter speed to get a good depth of field on the subject and still gather in background light. Dragonflies may perch on flimsy stems, so they blow in the wind as flowers do, but if your subject is on a solid perch, you can often use a slow shutter speed. The alternative is to use a flash with a higher shutter speed. This works equally well in brightly or evenly lit situations, but the powerful flash on the subject means any distant background will be underexposed, even black. You can get around this when using a flash by photographing dragonflies with backgrounds close enough to be well lit, but it is much better when they provide a smooth background (a dense bed of sedges all the same color, for example) than a cluttered one (a mass of twigs and leaves). However you do it, you will get better photos with the depth of field provided by a diaphragm opening of f/16 or f/18; f/22 or higher is even better. Otherwise, you are restricted to photos perpendicular to the subject and its wings—directly from the side for damselflies, directly from the top for dragonflies. Some photographers with an artistic bent may prefer subjects in partial focus.

The choice between digital and film photography is still available, although digital is clearly becoming the medium for most of us. One advantage of digital photography is that lenses are a bit more powerful than they would be on the equivalent film camera, as the digital CCD is smaller than a 35-mm slide. Thus, on the Nikon digitals that I use, my 300-mm lens is effectively a 450-mm lens, so I have an even better chance of approaching a wary dragonfly within photographic range. Another advantage of digital, of course, is that you can see your results immediately and know whether you have accomplished your photo goals then and there. One of the nicest advantages is being able to download your photos onto your computer and share them with friends and colleagues. Nowadays, a puzzling dragonfly photographed in the field might be identified by an expert at the other end of an e-mail message on the same day.

Always, the hardest odonates to photograph are the "fliers," the ones that perch hanging up and often at some distance from the water. With persistence, by watching many of these dragonflies, you will eventually see one hang up at a place accessible to your stalking. Many of us photograph fliers by catching them and chilling them, but this is necessary only if you have a burning desire for a photo of that species, as posed individuals rarely perch in an entirely natural way.

You may be surprised to learn that we need many more photos of dragonflies and damselflies. While searching for photos for this book, I discovered that some species have not been photographed at all, and in others, males but not females are represented in photo collections. For variable species, there cannot be too many photos. Time after time, I thought I knew the color pattern of a species, only to look at another photo and see unsuspected variation. Descriptions of the species below became longer as I looked at more and more bunches of photos on the Internet or reexamined my own. A good collection of specimens can tell the story of variation, but only in pattern, not in color.

Much of the behavior of dragonflies, of course, involves flying, and they are much harder to photograph in flight. I hope that the present generation of camcorder-wielding birders will pay some attention to odonate behavior. Because they are small and quick, they are more difficult to follow with a lens, and thus, their behavior is more difficult to document than that of birds. But it is worth trying!

Odonate Collecting and Collections

Odonate collections form the raw material that has informed us about them over the last few centuries. This field guide could not have been written without a sizable collection for reference. We are still learning about the occurrence of most species, and voucher specimens are still essential to document distribution, even in this era of superb photographs and knowledgeable field observers. In addition, geographic variation cannot be understood without series of specimens from throughout the range of a species. Also, there are still some questions about species limits that can be settled only by researchers examining series of specimens. Independent of the scientific value of specimens is the heuristic value of handling them. Learning is most memorable by a hands-on approach, so I encourage everyone interested in dragonflies to acquire an insect net.

Although not everyone will wish to do so, the best way to learn to identify dragonflies is by catching them. Like butterflies, many species are easily identified in the field; but also like butterflies, some species may be identified only in the hand. The simplest way to learn their identity is to capture one, identify it in the hand, then either release it or "collect" it for a reference specimen. Take care not to catch tenerals (just emerged) unless you plan to keep them; they are very easily damaged if handled.

If you collect a dragonfly and plan to keep it, there are procedures that you should follow so the specimen is of greatest value to you and possibly to a larger collection to which you may someday donate it. There is really no justification for killing a dragonfly except for education or research, but your personal education or research can fall in those categories. No North American odonate is rare enough that collecting a few specimens for these reasons in any way endangers its populations, but you should be aware that **collecting of any insects without a permit is prohibited in most parks and reserves**. In addition, certain species are officially protected because of their rarity in a particular state, and the Hine's Emerald (*Somatochlora hineana*), an eastern species, is federally protected.

If you wish to make a collection, here is a summary of the steps to follow to collect a dragonfly for a specimen.

1. Catch it.
2. Envelope it.
3. Acetone it.
4. Remove it.
5. Label it.
6. Store it.

Catch it Odonates are subject to predation, particularly by birds, and they are accordingly quite wary. They have quick responses and quick flight, so you must be even quicker. Even damselflies may move away rapidly, although they are sometimes easily approached. Dragonflies are considerably warier.

If your intended "prey" is perched in the open, just sweep the net sideways, trying to center it. If it is in dense vegetation, the best strategy is to come down from above, as a sideways sweep may just flip the vegetation out of the net and the insect with it. Lift the end of the net, and your captive will often fly up into it. Grasp the net below it and turn the net over so you can reach in and grab it. They tend to fly up, and many a dragonfly has escaped instantly when the collector opens the bag to look inside!

For a dragonfly in flight, sweep sideways and quickly flip the net bag over, trapping it inside. It is always best to swing from behind and below, the area of an odonate's poorest vision. Both feeding individuals and those in sexual patrol flights may fly a regular beat, so you can anticipate their flight path. Don't be surprised if they fly out and around you on every pass. I have found that standing next to a shrub or tree apparently makes me less conspicuous.

You can usually grab any dragonfly with impunity, but the larger ones have big mandibles and can pinch. Only a few of the largest species can actually draw blood when they bite, so If you are concerned, just watch what you are doing and grab them by the wings. Their wings are quite strong, without the shedding scales of butterflies. Again, tenerals are an exception to that statement.

Envelope it Place it, always with wings folded back, in a glassine (stamp) envelope labeled with locality and date. To save time and for recordkeeping, I use a field number for each collection based on the year: 08-1, 08-2, etc. You can put more than one individual per envelope, but they may chew on each other, so put their heads at opposite ends. I write "T" on the envelope for pairs in tandem and "C" for pairs in copula, as it is important to keep track of pairs and keep them together in a collection. Make sure you have a pen with indelible ink.

Acetone it After leaving your specimens in their envelopes for a while (for example, until you return home at night) so they can void their intestinal contents, kill them by immersing them briefly in acetone or injecting a drop of acetone into the thorax. Straighten the abdomen of each specimen, arranging the legs so they do not obscure the genitalia on the second abdominal segment of males. Line up each pair of wings so one lies under the other, and separate the forewings slightly from the hindwings to allow easier study of the wing venation. Put them back in their labeled envelopes and leave them submerged in acetone in a tightly closed plastic (e.g., Rubbermaid®) container for 12–24 hr. Cut off a bit of both lower corners of the envelope so the acetone drains when you lift the envelope out.

Acetone extracts fat and water from specimens, and they dry much better and with better color preservation than when merely air-dried. However, you should **avoid breathing acetone fumes**. Work in a well-ventilated setting.

Specimens with extensive pruinosity may become discolored in acetone. In particular, male spreadwing damselflies and pruinose male skimmers such as Common Whitetail and Western Pondhawk change color rather drastically, so I now prefer not to place them in acetone unless the environment is so humid that they may not air-dry adequately.

Remove it Remove your specimens from the acetone and leave them in the open for a few days so the acetone will evaporate. Preferably, have a well-ventilated spot away from people as it evaporates. Try to separate the envelopes for quickest evaporation. All dried dragonfly specimens are stiff and brittle, although those that have been acetoned are stronger and more resistant to breakage.

If you are unwilling or unable to use acetone, an alternative is to let the dragonfly die in the envelope; if it excretes feces from its abdomen, the abdomen dries better. Alternatively, you can kill odonates by freezing them. Make sure the abdomen is straightened rather than curved. Put the specimen in a dry and ant-proof place to dry thoroughly. Then continue as described below.

Label it All specimens should have the following information associated with them: locality, date, and collector (species name can be added later and changed if reidentified). Many

Dragonfly storage envelope

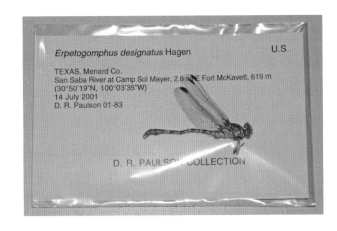

Dragonfly collection

collectors include habitat notes, at the very least something like "pond" or "slow, sandy stream" or "flying over clearing." This information associated with the time collected is of additional value. Pairs in tandem or in copula should always be so indicated.

Store it Store your specimens in a box protected from both humidity and possible pests such as carpet beetles, ants, or mice. For more useful and elegant storage, after they are dry, switch them to "Odonata specimen envelopes" (available from the International Odonata Research Institute) with the collecting information printed or written on a 3×5 card; the specimen can then be examined without removal from the envelope.

You can devise your own storage system for your specimens. Many collectors keep them in boxes of appropriate sizes with a few mothballs in each box, stored on a shelf or in a cabinet or drawer. If you think your home is pest-free, you can take your chances without mothballs, but you should check for bug damage frequently to confirm this.

Logically, the wings of pinned dragonfly specimens are usually spread, but for the most part they are not spread in enveloped specimens. You may wish to store a few specimens with wings spread for photography, drawing, and dorsal or ventral views of the specimens without having to remove them from the envelope. Also, at least a few specimens of odonates with patterned wings should be stored with wings outspread.

To examine the wings of just one side, you can remove the specimen from the envelope and carefully (from front to back) slide a piece of white paper or card between the left and right wing pairs.

The larvae, of course, are collected in entirely different ways. The best way is with a strong dipnet or a metal net like a large tea strainer. Sweep it through aquatic vegetation or drag it over the bottom substrate, swish it around in the water to wash out the sediment, and then poke through it to see what is moving. Many larvae will remain immobile, so you'll have to search for them. A good way to find stream larvae is to stir up the bottom sediment (rocks, sand, mud) while holding your net just downstream. Some larvae can be found by lifting rocks out of the water and scrutinizing them. This is most productive in tropical streams but would also be effective in the Southwest, where rock-dwelling dancers and shadowdamsels live.

Larvae to be retained for study should be preserved in 80% ethanol in vials that seal very well; evaporation is the bane of a preserved collection. Labels should be written with indelible ink on good archival paper; I have used parchment paper for years.

Odonate Threats and Conservation

Everywhere in the world, *habitat destruction* is the greatest threat to dragonflies and damselflies. Wetlands, essential habitats for the larvae and destinations for wandering adults, are being drained and filled at a dizzying rate. Thanks to concern about water birds, these activities have slowed in some regions, and mitigation efforts have provided new wetlands, some of them quite good for odonates. When riparian (waterside) vegetation is removed, odonate faunas shift away from habitat specialists toward the common and widespread species. Rivers themselves are not destroyed, but many factors are involved in their degradation as larval odonate habitats.

Chemical pollution is a bigger problem in eastern than in western North America, but, for example, rivers in the mining country of the northern Rockies are as polluted as those anywhere on the continent. Some are lifeless, so odonates and all other creatures that might have lived there have disappeared. *Siltation* is another problem in rivers, silt brought in from altered and eroded landscapes burying larvae that need clean water from which to extract oxygen. Dams, of course, entirely change the nature of rivers and streams, altering their odonate faunas as well. Furthermore, boat traffic on any water body produces waves that are likely to swamp and kill emerging dragonflies.

On rare occasions, elimination of a particular plant species may have a dramatic effect on odonates. Hyacinth Gliders, for example, seem closely tied to water hyacinths, and the concerted effort to rid southeastern waters of that pest plant will surely reduce populations of the dragonfly.

Drought may turn out to be the most significant factor in reducing odonate populations in the West as well as many other parts of the world. Because of global climate change, many regions are experiencing unprecedented droughts, and that is the case all across North America, especially in the interior West. Shallow ponds all over the region have dried for years at a time, and their odonate fauna disappeared. It remains to be seen how quickly such ponds are recolonized. In the Southwest, rivers are literally being sucked dry by human use, and human populations and water need are growing constantly. With changing climate, not only has drought become more frequent but so have heavy rains, and western streams are always at risk of being reamed out by *floods*. It may take years for a scoured stream to regain its full diversity of aquatic creatures.

Wetlands are often altered by *artificial eutrophication*, the process during which the productivity of a wetland is enhanced beyond the normal range in nature. This condition, usually caused by runoff of fertilizers or other chemicals from the surrounding uplands, eventually results in a system that supports large numbers of just a few competitively successful species. Livestock can pollute water bodies when common, but they also contribute to habitat degradation by *grazing*. Cattle in particular are attracted to streams and ponds and quickly reduce shore vegetation to a stubble. Even if larval habitats are relatively undisturbed, adults have no places to perch. This is especially serious in streams with a narrow band of riparian growth.

One factor we are learning about is the effect of *fish* on dragonfly larvae. There is evidence that quite a few species of odonates survive best (or only) in the absence of fish. These species are adapted to live in fishless waters just as others are adapted to live with fish. But especially in the West, most water bodies, including many that may have lacked fish, have been stocked with hungry spiny-rayed fish (perch, sunfish, and bass among others) from eastern North America, and this presumably has had profound effects on the odonate faunas involved. We will probably never know what these effects have been, as fish stocking has been going on for over a century and continues today. Furthermore, carp in particular muddy up water bodies because of their bottom foraging, and that in turn makes these water bodies poorer habitats for odonate larvae. Now still another series of introductions is taking place, as grass carp are introduced to clean up vegetation in some wetlands—vegetation that is surely important to some odonates. The effects of still other types of aliens are entirely unknown.

Odonates, although large and showy insects, are not noticed as much as birds and butterflies, not as charismatic as mammals, and not as edible as fish, so they have not received wide public attention. But their fate is coupled with that of wetlands, which are as much affected by human activities as any habitat on Earth. There is no reason why the Odonata could not become another of the "poster children" of wetland conservation, along with the waterfowl that have been the subjects of conservation concern for a long time. Dragonflies Unlimited has a nice ring to it!

One final point must be made: **dragonfly populations are not threatened by collecting.** Knowledge of insect population biology makes that clear. Our continued attempt to learn about them, however, is threatened by misguided attitudes that they should not be killed, not even for study. Under "collecting" I include any collecting the objective of which is education and/or research, including that done by school children and amateur naturalists just because of their desire to learn something about nature. Commercial collecting for profit, as has harmed some butterfly populations, has fortunately not been directed toward odonates, which are usually much less attractive than butterflies or beetles as dried specimens. But to support dragonfly conservation, we must support dragonfly research in all its forms, including the continued preservation of specimens.

Odonate Research

What do we still need to learn about western dragonflies? An answer of "everything" would not be too far off the mark. Adult behavior has been studied in detail for only a few species. Even those species have been studied only once or twice, and it is unlikely that a single study of a species describes it throughout its range. We see odonates at the water and assume we have a fairly good knowledge of their habitat preferences, but we do not always know if the abundance of adults corresponds to the preferred habitat of the larvae. Although the larvae of almost all species have been discovered and described, we know very little about what they are doing beneath the water surface. An interesting question to be answered is why so many similar species can be found together, for example, species of bluets and mosaic darners in some lakes and dancers in some streams?

Another fruitful avenue of research would be to document geographic variation in common species to try to gain a better understanding of gene flow and dispersal in different groups. There have been surprisingly few studies of geographic variation in common and widely distributed odonates.

Temperature regulation is an interesting phenomenon in odonates. Are there any damselflies that can regulate their body temperature in ways that have been shown for some dragonflies? Migration is another fascinating phenomenon, so far studied only in Common Green Darners. What would a study of the highly migratory Variegated Meadowhawk reveal?

Another area of odonate evolution that is poorly understood is their coloration. Why do so many species of skimmers have pale spots on segment 7? Why do clubtails often have yellow spots on the sides of their clubs? Why is blue such a common color in males? What is the significance of the striping on the thorax in so many groups? Why aren't they just solid-colored? Why do females of some species exist in two color morphs? This list of questions can be greatly extended.

There is no real line between amateur and professional odonatologists. For better or worse, very few people make a living studying Odonata. Thus, the great majority of those studying and writing about odonates can be classified as nonprofessionals, and they are the ones who have contributed so much to our knowledge of dragonflies. I cannot emphasize enough how much these people (including the reader of these words) can continue to contribute, both by making observations and by sharing them with others. The proliferation of listserves on the Internet has provided a medium by which much of this information is shared, but I will make a plea that you write up your observations for publication, whether in a scientific journal or a local newsletter, so they are preserved with more permanence than may be provided by bits and bytes.

Odonates in the West

Because the geographic scope of this book is extensive, the regions covered by it vary substantially in numerous ways of significance to dragonflies: temperature, rainfall, seasonality, physiography, terrestrial vegetation cover, amount of human settlement, and regional knowledge of Odonata. Dragonflies are sun lovers, and temperature varies by latitude as well as by altitude, so not surprisingly, odonate diversity decreases with latitude and altitude in the West. This latitudinal effect, interestingly, is not as strong in dragonflies as it is in many other groups of plants and animals. Apparently, dragonflies do relatively well in cooler climates as long as they have some warm summer days to sustain activity. Dragonflies need wetlands, so precipitation should be of importance to them. The West is generally dry and perhaps because of that supports fewer species of odonates than the East; for example, Rhode Island has a much larger species list than any western state. The wettest part of the West (the Pacific coast between Alaska and northern California), in fact, is relatively poor for dragonflies, perhaps because it combines high precipitation with cool, cloudy summers. Away from the Pacific coast, rainfall is highest in the West on the eastern edge of the region, especially in Texas, and there the diversity of dragonflies is high. A complex physiography and fairly wet summers may play a part in making southeastern Arizona another region of high odonate diversity.

Explanation of Species Accounts

Size Measurements indicate the range in total length and hindwing length of each species in millimeters (25.4 mm = 1 inch), combining the sexes. They are taken mostly from the handbooks to North American Odonata, with modifications where records were lacking or inadequate. Ranges may not be so extreme at any time within a given population, as there is often seasonal and/or geographic variation, and it must be admitted that greater ranges may be a consequence of more specimens measured. Taking the midpoint between the extremes allows some comparison between species, but size can be used as a field mark only among species that differ considerably, say with no overlap between measurements. Bear in mind that females tend to be a bit bulkier than males and may have slightly longer hindwings, but males usually have a longer abdomen; thus, sexual dimorphism contributes to the range in measurements in each species.

Description A brief characterization of the species may be listed first, often referring only to the male, the sex that is usually more conspicuous to observers and the one that is more easily identified (my apologies for unintended sexism). This is followed by a description that applies to both sexes equally. Then males and females are described separately, often with additional information about sexually immature individuals. Descriptions are oriented toward views from above or from slightly above a side view; the lower sides and underside of most species are pale, without diagnostic color patterns except where mentioned. The first abdominal segment (S1) is not usually described, as it is scarcely visible, but other abdominal segments are referred to by number. This seems more complicated than merely saying "tip of abdomen blue," but exactly which segments have the markings is very important in many cases of field identification. Complex patterns are shown in photos but are often not described except in comparisons for field identification. The photos are intended to convey the most information about appearance, so only brief descriptions of color patterns are included, with comments about variation and sexual differences. The photos augment the descriptions, but a good understanding of odonate structure helps to form a mental picture of each species. It must be emphasized that these descriptions refer to populations in the West; in a few species, those in the East or south of the U.S. border may look different.

Identification Comparison is the theme in this section. It is a good idea to read the generic accounts when trying to work out differences among species in a family. Species are compared only with those that overlap geographically in the West, so use of range maps is essential at all stages of identification. Unlike migratory birds, odonates are not expected outside their normal range, much less far from it. The tropical darners and skimmers that are likely exceptions to this are indicated. If you are concerned with a particular species, find out which other similar species occur in your area. The most similar species will often be distinguished first, followed by species that are less similar but could possibly cause identification problems. Species being compared are shown in boldface: **Canyon Rubyspot**. When comparison is with an entire group, it is shown in capitals: RUBYSPOTS.

Having emphasized distribution, I must add that our level of knowledge is such that the range maps may not be the last word, so if you are outside the range of a species but not too far from it, do not dismiss it yet; comparisons are often made between two species not yet known to overlap. Two categories of distinguishing characteristics are used, although they are not always indicated as such. *Definitive* characters are those that are absolutely distinctive in distinguishing two species. *Indicative* characters are those that are often helpful in distinguishing two species but by themselves are not definitive. With the great amount of attention paid to odonates in recent years, especially by birders, who are likely to attempt to make an identification of every individual seen, we are realizing that there is more individual variation in these insects than we had thought. In some cases, characteristics that were once considered definitive field marks for a species (for example, the size of the club in some clubtails or the amount of black on middle abdominal segments of bluets) have been found to occur rarely in other species. Color characters especially may not always be trustworthy, and anyone who wishes to identify every individual may have to carry a net! Rare variants that have been documented in a species but are scarcely ever encountered are ignored in this book, and I am hoping this will cause a minimum of anxiety to most observers.

Natural History The accounts combine and summarize information from the published literature and my own observations. Statements such as "oviposits in floating sedge stems" or "roosts at chest height" means that I have observed this or someone else has mentioned it for the species but does not necessarily mean it oviposits nowhere else or never roosts higher or lower. Also, such a statement for a species does not necessarily mean it is unique to that species within a genus or family, just that it has been noted. In many cases, statements are based on only a few observations, sometimes coming from scrutiny of my own and other photos; the World Wide Web is now a superb source of information of this sort. In some cases, a broad statement is made in the generic account that applies to all species in the genus, and this is not repeated in the species accounts. For this reason, it is also important to read the generic accounts. When there is detailed quantification, it usually comes from the literature and may be from just a single point in time and space. Because animals vary in behavior just as they do in color pattern, there is without doubt much variation beyond the parameters given (e.g., number of eggs laid, length of copulation). There are so many ways to describe the behavior of animals that I have tried to adopt some standards while trying to avoid having every account sound the same. Perching and flying heights are usually given in comparison with a person: heights of about 18 inches, 36 inches, 54 inches, and 72 inches are described as knee height, waist height, chest height, and head height (if over water, pretend to be walking on it). I encourage the reader to report all natural-history observations that vary substantially from what is written here or that are not described here, either at the generic or specific level.

Habitat A broad description of breeding habitat is given here. For most species, habitat choice was determined from my own observations, but I have used the literature for species less familiar to me. Bear in mind that odonates sometimes breed in atypical habitats and especially that nonbreeding immatures (and mature females) can be found well away from water and often in a great range of terrestrial habitats. The most common determinant of nonbreeding distribution is that, when immature, some species prefer open and others wooded habitats. Much is yet to be learned about the breeding habitats of some species, especially about the larval distribution that determines where we see breeding adults.

Flight Season This information is taken from a variety of well-documented sources, either published, on the World Wide Web, or unpublished from those who keep such records. This information is available for Yukon Territory (YT), British Columbia (BC), Washington (WA), Oregon (OR), California (CA), Montana (MT), Arizona (AZ), New Mexico (NM), Nebraska (NE), and Texas (TX). When no information is available for a species in those regions, information is taken from the closest region adjacent to the area of this book. Rather than list the extreme dates, I list the month in which they fall, so these are rough approximations of the usual flight season. A species listed from June to October, for example, might actually be common from some time in June to some time in October, or it may not appear normally until July, or it may normally disappear in September. Some species seem to occur throughout the year and are listed as such. Others are known to emerge as early as late January and may fly into early December; those are listed as January to December, and they may eventually be found to fly all year. For species that are common, we often have extreme records that are well beyond the normal times of occurrence. These provincial and state dates are guides to when you might expect to see the species, and it is unlikely you will see it in earlier or later months in that region. Note that flight seasons, like distribution, in the western interior are generally poorly documented. Those listed from Nebraska, in fact, are estimates. Nevertheless, if you concluded that flight seasons were both earlier and longer at lower latitudes but relatively short in the northern interior, you would be quite correct.

Distribution Comments about distribution that add to the information presented in the range maps are given here, including a rough idea of the distribution in eastern North America and the limits of the mainland distribution south of this region. In species confined to the West, there are often no further comments.

Comments Included here is information about taxonomy, subspecies, hybridization, name changes, possible additional species, and other matters not covered in the other sections.

Range maps The range maps were constructed from our present knowledge (as of November 2007) of Odonata distribution in the West, primarily from the dot maps published by T. W. Donnelly but also by many additional records established since those maps were published, with many of them published on Odonata Central, others on regional websites, and still others on various regional listserves (see Appendix for information). Some regions are now better known than indicated in any of these sources, however, and many unpublished records that were submitted to me by regional experts have also been incorporated in the maps. Continuations of ranges are indicated in the immediate area to the east of the region covered by this book, but ranges are not drawn in northern Mexico, much of which remains poorly known.

Range maps of this sort should come with two assumptions. (1) The species is likely but not guaranteed to be found everywhere inside the shown range. In particular, it will be mostly limited to its optimal habitats. (2) The species is unlikely to be found anywhere outside the shown range except in poorly known regions, but recall that many parts of the West are in fact poorly known for Odonata. Ranges can be defined fairly well in some states and provinces but not all of them. When records are well separated from one another, the gaps between them are shown, so the maps illustrate what we know of distribution at present, not what we assume is probably the case. The gaps are purposely emphasized to show regions where more fieldwork is indicated, and many of them will doubtless be filled. Note, for example, how many species are known from southeastern Arizona but not southwestern New Mexico. The irregularity of the maps of some species is also a response to the mountainous terrain of the West: a lowland species may be present only in the lowlands, and a montane species only in the mountains where its occurrence is shown. Isolated records at a single locality that seem to be outside the periphery of the range of a species are indicated by dots, as are records of species that barely make it into the West. Some of the dots may represent vagrant occurrences, but this is not always known, and some definitely represent established populations.

I have tried to make the maps detailed enough so that people in each state and province will have a good idea of the known or expected distribution within that political entity. But bear in mind that there may be few records from an area included in the range, range boundaries will certainly have to be modified, and field observers still have much to contribute to what we know about North American odonate distribution. In particular, the far north of Alaska and Canada and parts of the Great Plains remain especially poorly known. Look at maps such as those of the Brimstone Clubtail and Chalk-fronted Corporal. Their complexity probably results from both habitat distribution and degree of knowledge, but there is nothing in nature that dictates a species distribution must be simple!

The photos The photos included with each species account are intended to illustrate both what the species looks like and the specific field marks that are of value for identification. I made considerable effort to find photos of both sexes of every species, but there seem to be no photos available for females of a small number of species. There is quite an admirable goal for this generation of odonate photographers! I have chosen to illustrate variation in species in which variation can cause confusion, for example, the different morphs and ages of female forktails. The great majority of photos are natural, but a small number of individuals are posed; these are indicated. It is my hope that in the near future unposed photos of both sexes of all species will be available. Males are shown before females because they are the commonly seen sex, not because of sexist bias. Finally, to save space I have cut off some dragonfly wings and oriented pond damsels horizontally, no matter their original orientation. My apologies to the photographers and to their subjects for these insults to esthetics. **All photos lacking a photographer's name are by the author.**

The drawings The line drawings, by Natalia von Ellenrieder and Rosser Garrison, were made from single specimens and thus do not necessarily look like the same structures in every individual of the species, but they are representative. They will be important in distinguishing among species of some genera when a specimen or live individual is in hand. They can also be used for confirmation of an identification.

Abbreviations and conventions

TL	total length in millimeters
HW	hindwing length in millimeters
S1–10	abdominal segments 1–10
T1–5	thoracic stripes 1–5 (only in clubtails)

Damselflies *Zygoptera*

Broad-winged Damsel Family *Calopterygidae*

Large, showy damselflies of this family often display metallic bodies and/or colored wings. They are distinguished from other North American damselflies by broad wings with dense venation and no hint of the narrow petiole or "stalk" at the base that characterizes the other families. The nodus lies well out on the wing with numerous crossveins basal to it. Colored wings in this family are heavily involved in displays between males and of males to females. This is the only damselfly family in which individuals point abdomen toward the sun (obelisking) at high temperatures. Closed wings are held either on one side of the abdomen or above it, which may relate to temperature regulation. Leg spines are very long, appropriate to fly-catching habits. Worldwide it is tropical, with a few species in temperate North America and Eurasia. World 176, NA 8, West 6.

Jewelwings *Calopteryx*

These are the most spectacular damselflies of temperate North America and Eurasia, all large with metallic green to blue-green bodies. Different species have wings that are clear, with black tips of different extent, or entirely black. Wing pattern is important for identification. Females are similar to males but usually duller and easily distinguished by white stigmas. All live on clear streams and rivers. This is the group to watch if you wish to see odonate courtship behavior. Watch for wing clapping, wings suddenly opened and shut, which may be communication between individuals or for cooling. World 29, NA 5, West 3.

1 Sparkling Jewelwing *Calopteryx dimidiata* TL 37–50, HW 23–31

Description Slender metallic green damselfly with black wingtips. Noteworthy that this showy species is one of the smallest in its family. *Male*: Eyes dark brown. Entire body metallic green, looks blue when backlighted. Wings with terminal one-fifth black. *Female*: Colored as male or somewhat more bronzy-green, usually with white pseudostigma and with black wingtips less crisply de- fined. Some have only hindwing tips black; others have clear or entirely dusky wings. Immature with reddish eyes, duller body color.

Identification No similar species in its range in the West. Often occurs with **Ebony Jewelwing**, which is larger, with broad black wings. Some female **Sparkling** have darkish wings, never as dark as **Ebony**, and distinctly narrower. Shaped more like RUBYSPOT, but all-green body and black wingtips furnish easy distinction. In flight, looks like glowing green toothpick with flashing black wingtips dancing around, whereas black fluttering wings dominate appearance of **Ebony**.

Natural History Both sexes often together at the breeding habitat, at least at some times of day. Males defend small territories with potential oviposition sites, flutter around and around each other in spiral flight over water, and chase one another along up to 40 feet of stream; spectacular to see, with brilliant green abdomen and flashing black wingtips. Also display to females that approach oviposition site in "floating cross display" by dropping to water and floating for a short distance with wings partially spread and abdomen curled up. This is repeated until female flips wings, showing receptiveness, or leaves the stream. Receptive female then courted with rapidly whirring wings, followed in some cases by copulation, which lasts about 2 min. Female oviposits by walking down leaf until submerged, often quite far, then laying several hundred eggs for about 15–20 min underwater. She then rises to surface, usually unreceptive to further male attention but remaining near water.

1.1
Sparkling
Jewelwing
male—Leon Co., FL,
April 2005

1.2
Sparkling
Jewelwing
female—Marion
Co., FL, April 2005

Habitat Small, sandy forest streams with abundant vegetation and usually swift current; less tied to woodland than Ebony and more often in quite open areas.
Flight Season TX Apr–May.
Distribution Widespread in the Southeast, south to central Florida, and north along the Atlantic coast to New Hampshire.

2 River Jewelwing   *Calopteryx aequabilis*   TL 43–54, HW 27–37

Description Large green damselfly with black wingtips. *Male*: Eyes dark brown. Body glossy metallic green, looks blue in some lights. Wings with more than terminal third black, that on hindwing more extensive than on forewing, younger individuals with paler wingtips. *Female*: Slightly duller or more bronzy; white pseudostigma allows easy distinction from male. Dark of wingtips can be more obscured, inner wing darker, so contrast less evident.

Identification This species unlike any other in its range in the West. Some female **River Jewelwings** have wings dusky enough that wingtips do not contrast much, and a closer look is necessary to distinguish them from female **Ebony Jewelwing** where they occur together on Great Plains. Also, some female **Ebony** have darker wingtips. With experience, **River** are seen to be somewhat narrower-winged. No other *Calopteryx* jewelwings west of Great Plains.
Natural History Rarely seen very far from water. Both sexes at times common on streamside vegetation in optimal habitat. Males often seen in lengthy flights along shores of large rivers

but typically perch over water and defend small territories, interacting constantly with somewhat irregular horizontal circling flights, some of them lengthy. Male stationary display to female spectacular, forewings fluttering and hindwings briefly halted so black tips prominent. Male flutters in front of female with abdomen tip raised, then lands and raises it further, exposing white under tip. Male also dives to water surface briefly with wings outspread. Male and female wing-clap to one another. Copulations lasts a few minutes, oviposition up to 24 min but usually much briefer (average 9 min). Females oviposit on floating, sometimes emergent, vegetation or back down stem and submerge for up to 50 min, usually with male in attendance at surface. Longevity of reproductive adults up to 28 days.

Habitat Clear streams of all sizes and rivers with moderate current, usually with beds of submergent aquatic vegetation. Typically more open and larger streams than those used by Ebony Jewelwing. However, the two often occur together, and River Jewelwings can be common on rather tiny wooded streams as long as there is some sun penetration. Also seen at rocky shores of large lakes in some areas.

Flight Season BC Jun–Jul, AB Jun–Aug, WA Jun–Sep, OR May–Sep, CA May–Jul, MT Jun–Aug, NE May–Sep.

Distribution Widespread in the Northeast, east to Newfoundland and south to Indiana and West Virginia.

Comments Several subspecies described from West not presently considered distinct but indicative of variation in wing pattern.

2.1
River Jewelwing male—King Co., WA, June 2004

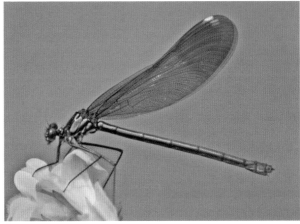

2.2
River Jewelwing female—Malheur Co., OR, June 2000, Netta Smith

Broad-winged Damsel Family   **43**

Description No other North American damselfly is metallic green with black wings. Large size alone is distinctive. *Male*: Eyes dark brown. Brilliant metallic green (blue in some lights) all over, with black wings. *Female*: Slightly duller, with conspicuous white pseudostigma and wings somewhat paler at base than in male.

Identification Females with somewhat paler wing bases could be mistaken for female **River Jewelwing**, but contrast between base and tip usually much greater in the latter. Broader wings should also distinguish **Ebony**. Only other black-winged damselflies in range are some **Smoky Rubyspots**, with much narrower wings and black body.

Natural History Both sexes at and near water much of day, tend to be more at rapids when that habitat is present. Much wing clapping, opening wings slowly and closing them suddenly. Males may have "flights of attrition," bouncing around one another while moving laterally, sometimes surprisingly long distances away from start. Flights persist for many minutes, presumably until one cannot maintain interaction and flies away or lands. Males defend territories for up to 8 days around patches of submergent and floating vegetation in stream, bigger patches being more attractive. Females arrive at water well after males. Courtship display in front of female includes much wing fluttering and often showing of the white under the abdomen tip, then landing on prominent white stigma of female and walking down wings to achieve tandem position. Females return to male's territory, oviposit in rootlets and submergent vegetation of many types, even wet logs, at water surface; also may submerge entirely. Eggs laid at 7–10/min, may total 1800 in a lifetime. Males guard females with which they have mated and often guard additional females that oviposit in their territory, especially when females are at high density (likely because these are attracted to one another). Males that have lost territories may resort to "sneaking," attempting to mate with females on other males' territories. Night roosts may be communal, deep in tall grass. Often seen flying between night roosts in woodland and waterside, even across roads, where fluttery flight is very distinctive. Average longevity 2–3 weeks (including 11 days while immature), maximum 47 days.

Habitat Slow-flowing woodland streams, usually associated with herbaceous vegetation. Occurs on open banks when trees are nearby (trees are essential for roosting at night). May be abundant at small streams in woods where very few other species are present.

Flight Season NE May–Sep, TX Mar–Oct.

Distribution Also throughout eastern United States and southeastern Canada.

3.1
Ebony Jewelwing male—
Price Co., WI, June 2007

3.2
Ebony Jewelwing female—
Fayette Co., IA, July 2004

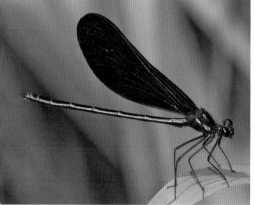

Wings are narrower than in *Calopteryx* jewelwings, usually with small stigma in both sexes. Males are unmistakable with red wing bases; wings vary from clear to black otherwise. Females appear much duller, may lack any hint of red in wings, but are still large, usually metallic damselflies with heavily veined wings. Species of this genus are among the most obvious stream damselflies of the New World tropics. World 37, NA 3, West 3.

4 American Rubyspot · *Hetaerina americana* · TL 38–46, HW 26–30

Description Large metallic damselfly with conspicuous red wing bases. *Male*: Eyes dark reddish-brown, paler below and behind. Mostly metallic red head and thorax, shiny black abdomen. Bright red patches at wing base marked by white veins. Red varies in extent, at greatest almost to nodus; most in Texas, least in southwestern mountains. *Female*: Quite variable. Eyes brown over tan, paler than in male. Duller than male, dark colors of body vary from matte black to metallic green to metallic red (head and thorax only). Wings vary from almost uncolored to diffuse orange wash at base to dark
orange filling same area as red in male. Pale stigma obvious in East, including Texas, smaller but still present (rarely lacking) from west Texas to southeastern California, absent in Pacific coast populations. Stigma typically paler and more contrasty in female than in male.

Identification Larger than other closedwing damselflies. Only rubyspot in most of West. Male unique in much of range. Female distinguished from pond damsels by large size, densely veined wings, and metallic greenish to orange body with conspicuously striped thorax, usually (but not always) orange suffusion in wings. See **Canyon** and **Smoky Rubyspots**.

Natural History Both sexes rest on stems and leaves over water, sexes mixed more than in most damselflies. Commonly perches on small plants in midcurrent, also on rocks. May go into obelisk position in hot sun. Females probably territorial at water as are males. Has been seen to concentrate in large numbers at dusk emergence of mayflies. Flight low and fast over water. Resident males aggressive to intruders, performing horizontal circling flights

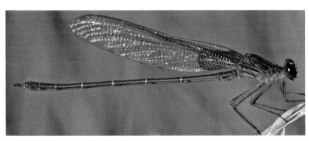

4.1
American Rubyspot
male—Catron Co.,
NM, July 2007;
female—Starr Co.,
TX, November 2005

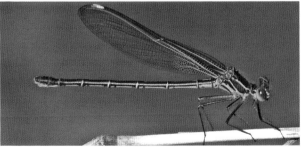

4.2
American Rubyspot
pair—Llano Co., TX,
July 2004

until one leaves; display flights may last for minutes. Red wing spots in males increase in size to about 14 days. Larger spots in more successful territory holders but may reduce hunting success. No courtship, males merely seizing approaching females. Perched females reject inappropriate males by opening wings and curving abdomen upward. Copulation brief, averaging 3 min in one study. Males mate infrequently, averaging less than once per day. Females oviposit on surface vegetation or by submerging entirely down to 3–5 inches, remain in a fairly small area, then emerge after up to an hour. Males remain on alert above their mate, apparently guarding against other males, but relatively seldom re-mate with her, although she may mate with a second male. Typically roost communally at night because of attraction to other roosting rubyspots; males often near their daytime territories. Some may remain on rocks over water for night roost. Maturation period about 6–10 days, average life expectancy about 10–15 days.

Habitat   Clear, swift-running, sometimes rocky streams and rivers of all sizes with shore vegetation for perching and submerged vegetation for oviposition. Common on open streams, also on wooded streams with plenty of sun. Also on flowing irrigation canals in desert areas.

Flight Season   OR May–Oct, CA Apr–Dec, MT Jun–Aug, AZ Feb–Dec, NM all year, NE May–Oct, TX Mar–Jan.

Distribution   Across eastern United States and southern edge of Canada south to northwest Florida; also south in uplands to Nicaragua.

---

5 Canyon Rubyspot   *Hetaerina vulnerata*   TL 36–46, HW 28–32

Description   Large dark damselfly with red or orange wing bases.
Male: Eyes dark brown over light brown. Thorax dark metallic red in front, in duller individuals looking black; narrow antehumeral stripe; and sides dull reddish. Abdomen mostly black, reddish-brown low on sides. Base of wings red, with darker streak in hindwings and contrasting white veins. Extreme wingtips usually brownish, lacks stigma. *Female*: Eyes dark brown over tan. Thorax metallic green in front, striped with brown; sides brown. Abdomen black to metallic green above, light brown below. Wing base not red but often suffused with orange.

Identification A bit larger than **American Rubyspot**, with which it often occurs. Extreme wing tips usually brown in male **Canyon**, not marked in **American** (but sometimes not evident in **Canyon**). **Canyon** always lacks stigma, **American** has it in most populations, including where it overlaps with **Canyon**. However, it may be very small and inconspicuous or even lacking in some of those populations. Amount of red in wings comparable in overlap area, although **American** has more in most parts of range. Top of abdomen in male **American** sometimes with green iridescence, not in **Canyon**, which usually has abdomen base paler than in **American**. Appendages may have to be checked for definite identification. Inner side of cerci in **American** bumpy, in **Canyon** smoothly curved; paraprocts diverge more in **American** than **Canyon**. Females differ a bit more: front of thorax metallic in both but greenish in **Canyon**, often reddish in **American**. Some **American** appear green, however, and distinction might have to be based on presence or absence of stigma. Perhaps best identification feature in side view is that female **Canyon** typically has wider antehumeral stripe and no complete stripe below (posterior to) hindwings, whereas **American** has one or more distinct stripes there.

Natural History Males come to water within a few hours of sunrise to rest on rocks or twigs at riffles in sun or shade; tend to return to same spot or at least immediate area every day for up to 2 weeks. Males engage in brief (average 21 sec) horizontal circling flights over water in defense of territories; sometimes these flights prolonged. Females arrive at water about 2 hr after males. Males mate only every few days, then guard females as they oviposit underwater for a few minutes or up to an hour. Also may take female in tandem again after she emerges from one egg-laying bout, then search for another oviposition site, unusual behavior for damselfly. Then returns to territory and reclaims it. Both sexes sally after aerial insect prey from branches and leaves well above stream in early morning sun and again late in afternoon.

Habitat Wooded canyon streams with rocky riffles; more attracted to shaded streams than American Rubyspot and averaging higher elevation. Disappears from smaller streams that dry up during drought years.

Flight Season AZ Mar–Nov, NM Jul–Oct.

Distribution Ranges south in uplands to Honduras.

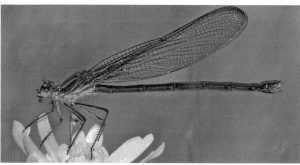

5
Canyon Rubyspot male—Sonora, Mexico, September 2005; female—Sonora, Mexico, July 2006, Doug Danforth

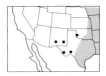

Description Large black damselfly with variably colored wings. *Male*: Eyes very dark brown, almost black. Body entirely black with slight greenish gloss, fine tan stripes on thorax. Usually with red patch in forewing obscured by dark patch in hindwing, most visible in flight. *Female*: Eyes brown over tan, many with conspicuous striped or spotted pattern. Thorax with metallic green markings on pale brown; abdomen mostly black. Wings vary from dusky to black, with no red and contrasty white stigmas. Only rubyspot with greatly varied wing coloration in both sexes, ranging from entirely black to mostly clear, with all in-between types. Extreme wingtips dark, more extensive in individuals with more black at base and tip and base coming together to produce entirely black wings. At least in Texas, individuals later in flight season tend to have more extensively dark wings, but all extremes can be seen throughout season.

Identification Individuals with mostly dark wings easily distinguished from other rubyspots and all other damselflies. Males with most lightly marked wings distinguishable from other rubyspots by black body, from other damselflies by large size, dense venation. Good mark for males that are silhouetted is that line of demarcation between dark base and clear tip of wing usually strongly slanted, almost perpendicular to wing in other male rubyspots. Females differ from female **American Rubyspot** in having less conspicuously striped thorax, markings green and brown, and almost always darker wings.

Natural History Tend to perch higher than American Rubyspot, usually on shaded pools rather than low on open riffles. Females not at water unless mating. Males engage in display flights, circling one another for at least brief periods and moving up- and downstream. Tandem pairs often seen flying about, presumably looking for good oviposition site. Females oviposit underwater for long periods (up to 2 hr) with males perched above driving other males away from spot.

Habitat Slower streams in woodland, tends to be in more heavily shaded areas than American Rubyspot, also less likely to be at rocky riffles. Aquatic vegetation or rootlets from stream-bank trees essential for oviposition.

Flight Season TX Mar–Dec.

Distribution Widespread in East from Wisconsin and Pennsylvania south to southern Florida; also ranges south in lowlands to Costa Rica.

Rubyspots - male appendages

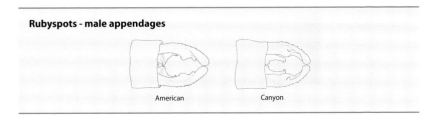

American   Canyon

6.1
Smoky Rubyspot
male—Hidalgo Co.,
TX, June 2005

6.2
Smoky Rubyspot
male—Bexar Co.,
TX, July 2004

6.3
Smoky Rubyspot
female—Starr Co.,
TX, November 2005

These are medium to large damselflies of worldwide distribution that usually hold their wings open, but several genera in the Old World keep them closed. All spreadwings close their wings at night, in bad weather, and when threatened by other odonates (as predators or males harassing females). Those in North America perch with long abdomen inclined downward, even vertically. Most are dark, with top of abdomen metallic and thorax metallic or with metallic stripes or spots, often also with pale stripes. Overall, they are not brightly colored, but males of our genera have blue eyes and face, and some Australian species are colored like bluets. Females of only a few species show blue colors. All have clear wings with stigma longer than in pond damsels, a definitive mark, and long legs with very long leg spines, as befits a predator of flying insects. Most are distinctly larger than pond damsels. World 151, NA 19, West 16.

Stream Spreadwings *Archilestes*

These are large damselflies with outspread wings, both species being larger than any pond spreadwing and much larger than any North American pond damsel. Unlike pond spreadwings, they show conspicuous pale stripes on sides of thorax. Males have blue eyes, females brown or blue (perhaps age variation). Mature males develop pruinosity on abdomen tip. They are found typically on streams but stray to ponds regularly, especially those associated with streams, and sometimes breed in them; they prefer fishless waters, where larvae swim in the open like little minnows. Natural history is much like that of pond spreadwings. Males perch on branches and leaves over water, and pairs oviposit in woody stems, sometimes well above water. Other species occur from Mexico to Argentina. World 8, NA 2, West 2.

7 Great Spreadwing *Archilestes grandis* TL 50–62, HW 31–40

Description Very large spreadwing (largest North American damselfly) with yellow stripe on either side of thorax. Underside of thorax pale, becoming lightly pruinose. Abdomen dark brown to dark metallic green above. *Male*: Eyes and labrum blue. Thorax brown in front with full-length metallic green stripe on either side of midline, half-length stripe at rear edge of brown; sides yellow with another lighter brown stripe along lower sides, whole area developing pruinosity at maturity. Abdomen brown to black above with black apical rings on S3–7; S9–10 pruinose. *Female*: Eyes blue to brown, colored as male but distinguished by bulbous abdomen tip lacking pruinosity.

Identification Great Spreadwing looks twice the size of POND SPREADWINGS, in steady flight over open water easily mistaken for dragonfly, but wings and body much more slender. See **California Spreadwing**, rather similar but smaller and paler, with white side stripe and pale stigmas. Paraprocts diverge under cerci, so barely visible from above (in **California Spreadwing**, paraprocts short and parallel, their rounded tips visible from above).

Natural History Males perch over water, defend small territories. Females seized when they arrive. Pairs oviposit in tandem (or female released during oviposition) in leaf petioles or stems of herbaceous or woody plants, sometimes well above water (perhaps highest known odonate oviposition at 44 feet above water). Oviposition lasts 15–180 min, with up to 230 eggs laid.

Habitat Slow streams, usually with wooded banks; larvae may be seen swimming in open in pools. Less likely to be at ponds than California Spreadwing.

Flight Season CA Mar–Jan, AZ Apr–Dec, NM May–Oct, NE Aug–Oct, TX Mar–Dec.

Distribution Also east to southern New England and north Georgia; ranges south in uplands to Venezuela.

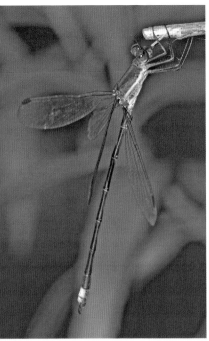

7.1
Great Spreadwing
male—Sonora, Mexico,
August 2006,
Netta Smith

7.2
Great Spreadwing
female—Chihuahua,
Mexico,
September 2005

8 California Spreadwing *Archilestes californicus* TL 42–60, HW 26–35

Description Very large brown spreadwing with white-striped thorax. *Male*: Eyes and labrum blue, thorax brown in front, white on sides, with metallic brown stripe on either side of midline and similar short stripe occupying middle third at rear edge of brown. Abdomen brown above, darker areas slightly metallic dark brown with faint indication of green; S9–10 pruinose. *Female*: Eyes dull blue to brown; colored as male but no pruinosity, abdomen with bulbous tip.

Identification Overall impression one of dullness, with dark areas brown rather than bright metallic. Only **Great Spreadwing** exceeds this species in size, and it is quite similar, but darker overall and more metallic green. Pale side stripe on **Great** usually yellow rather than white as in **California**. Dark metallic stripe before pale stripe extends up almost to wing base (but not always) in **Great**, long oval in midsegment in **California**. Stigmas darker in **Great**, blackish versus tan. Immature **Great** may have white side stripe, must be distinguished by other characters. See **Great Spreadwing** for difference in male appendages.

Natural History Males perch over water, often conspicuously in open on dead twigs, and defend small territories. Wings occasionally twitched closed about 30°, then reopened in two steps; stigmas prominent, perhaps a display. Copulation lasts many minutes. Pairs oviposit in woody tissue, often in willow or alder branches, up to 10 feet above water. With 6 eggs neatly laid, the pair moves slightly downward and repeats process, laying 70–180 eggs at a session. More common at water after midday. Sometimes encountered far from water, even out in open sagebrush.

Habitat Slow streams, sometimes ponds or lakes associated with them.

8.1
California Spreadwing
male—Benton Co., WA,
August 2005

8.2
California Spreadwing
female—Kittitas Co., WA,
August 2002, Dan Bárta

Flight Season WA Jul–Nov, OR Jun–Nov, CA Jun–Dec, AZ Jul–Nov.
Distribution Ranges south in Mexico to Baja California Sur and Sonora.

Pond Spreadwings *Lestes*

Mature males have entirely bright blue eyes (more purplish with age, at least in some species), paler blue to whitish below, and pale blue labrum; male eye color is not included in description unless different. Females usually have brown eyes and light brown to yellow labrum, but in at least some species females occur with blue eyes and labrum; it is not yet known whether this is a function of age or perhaps genetic polymorphism, with brown and blue females as in some pond damsels. Tenerals are brown, then develop darker rings on middle abdominal segments, then become dark metallic above (green in most species), then develop definitive coloration. Males become increasingly pruinose gray (or blue-gray) with age with pruinosity between wing bases and along sides of thorax, and in some species, the thorax becoming entirely pruinose gray. In mature males part or all of S1 and S2 become pruinose; also S9 and usually S10 are pruinose, sometimes extending onto S8. The pattern of pruinosity may be distinctive of the species but also increases with age. Northern species and populations have more pruinosity on the abdomen, typically S8–10 pruinose, whereas southern species and populations often have S9 or S9–10 pruinose. The pattern on S2 is quite distinctive for some species. Females develop less pruinosity, typically with age, on the thorax and sometimes abdomen tip, at its most extreme about as pruinose as males of their species. Mature color pattern is sufficiently variable that many will have to be captured to be sure of identification. Species distinction often must be based on appendage structure, less often on color pattern of the thorax; ovipositor size and shape are important in females. Sex is readily distinguished by shape, with females exhibiting shorter, thicker abdomen with expanded tip.

Pond spreadwings are found on every continent (although barely into Australia); they are diverse in both temperate and tropical latitudes and often are among the most common damselflies at marshy ponds and lakes. Male arrival at water averages later in the day than pond damsels. Individuals are found away from water, often in woodland, where they forage in sunny spots and may take as much as several months to mature. Because of this, immature in-

Table 1   Pond Spreadwing (*Lestes*) Identification

	A	B	C	D
Chalky	3	1	2	2
Plateau	2	1	2	2
Rainpool	5	1	2	2
Spotted	2	2	2	1
Northern	23	1	2	1
Southern	2	1	2	1
Sweetflag	32	1	1	1
Lyre-tipped	2	1	2	1
Slender	2	1	2	1
Emerald	1	1	1	1
Black	2	1	2	3
Swamp	4	1	2	1
Elegant	1	3	2	1
Amber-winged	1	2	2	1

A, Mature male front of thorax: 1, solid green; 2, black or brown with pale stripes; 3, pruinose; 4, green with brown stripes; 5, narrow green stripes.

B, Male appendages: 1, paraprocts half length of cerci or more; 2, half or less; 3, longer than cerci.

C, Female ovipositor: 1, projects beyond abdomen tip; 2, does not.

D, Range: 1, widespread; 2, southern borderlands; 3, California only.

dividuals will often present an identification challenge. Pond spreadwings tend to perch higher than pond damsels and forage by flycatching. They also tend to stay in vegetation, but at times and places males may move out over open water in some numbers, flying low over the surface like pond damsels but never hovering. Oviposition usually is in tandem, and females less often oviposit alone. Eggs are laid in vertical sedge or rush stems above water with the female either staying at one level on the stem or moving up or down. World 84, NA 17, West 14.

9 Chalky Spreadwing *Lestes sigma* TL 39–43, HW 20–23

Description Most highly pruinose of trio of tropical spreadwings of southern border areas. Stigmas somewhat bicolored, conspicuously paler at outer end. *Male*: Mature male with entire thorax pruinose. Abdomen metallic brown above, with S1, two-thirds of S2, and S8–9, often S10, pruinose at maturity. *Female*: Also becomes

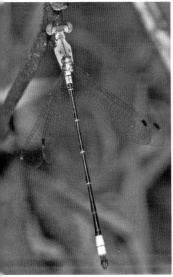

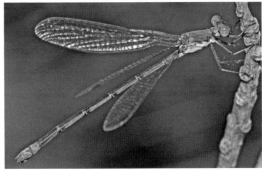

9.1
Chalky Spreadwing
male—Hidalgo Co., TX,
June 2005

9.2
Chalky Spreadwing female—
Hidalgo Co., TX, May 2002,
Robert A. Behrstock

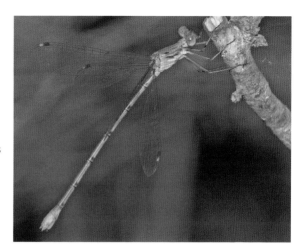

9.3
Chalky Spreadwing
immature female—
Hidalgo Co., TX, June 2005

heavily pruinose at maturity, with blue eyes. Immatures of both sexes with pale thorax, only dark markings a pair of spots in front near wing bases, a wavy and incomplete stripe corresponding to rear edge of dark area of other species, and small spots on lower sides barely visible from side. S8–10 distinctly pale. Dark spots on front and dark stripe on side of thorax often visible through pruinosity on both sexes, especially when front still darker than sides.

Identification Mature males with entirely pruinose thorax unmistakable, not coexisting with any other species showing such heavy pruinosity. Immatures easily distinguished because of largely pale thorax with sparse patterning. Immatures of both sexes with distinctive sharply bicolored stigmas, outer third yellow to orange, but become dark when mature. Occurs with **Plateau, Rainpool**, and **Southern Spreadwings**.

Natural History Males perch in low sedges and grasses at water's edge or in dense vegetation beds; pairs oviposit in same places. Present in nearby woodland when not at water.

Habitat Shallow ponds and marshes with much emergent vegetation, typically at wetlands that fill only during rainy season.

Flight Season TX May–Nov.

Distribution Ranges south in lowlands to Costa Rica.

---

10 Plateau Spreadwing *Lestes alacer*   **TL 34–45, HW 19–25**

Description Distinctively marked species of southern Great Plains and Southwest. *Male*: Thorax with wide median black stripe, relatively wide tan or blue (with maturity) antehumeral stripe with straight edges, and narrow black humeral stripe. Sides pale brown, becoming pruinose in older individuals; pruinosity creeping onto humeral stripe and, in some, mostly obscuring it. From above, S9–10 pruinose, S8 becoming so in older individuals. *Female*: Thoracic pattern similar to male. As in males, oldest females develop blue eyes and blue antehumeral stripes, wide humeral stripe more prominent than in male because no pruinosity on sides. Immature with wide black stripe on front of thorax contrasting with entirely pale sides in both sexes.

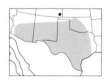

Identification Thoracic pattern of rather wide and straight-edged median stripe, conspicuously pale antehumeral stripe, narrow black humeral stripe, and lighter sides best combination of marks to distinguish **Plateau** from other spreadwings in its range. **Southern** also has pale antehumeral stripes, but they are more irregular, inner and/or outer margin somewhat jagged.

Natural History Males and tandem pairs in marsh vegetation. Males may be common at breeding habitats, immatures common in open woodland. Pairs oviposit in herbaceous vegetation, typically upright sedge and spikerush stems. Mature adults roost, at times communally, in woody vegetation as much as a half mile from water. May spend winter/dry season in woodland away from water.

Habitat Permanent or temporary ponds and seep springs with emergent vegetation, from lowlands well up into mountains. Tolerant of saline conditions.

Flight Season AZ Mar–Nov, NM Jan–Oct, TX all year.

Distribution Ranges south in uplands to Costa Rica.

10.1
Plateau Spreadwing
male—Sonora,
Mexico, September
2005

10.2
Plateau Spreadwing
female—Sonora,
Mexico, September
2005

10.3
Plateau Spreadwing
immature female—
Hidalgo Co., TX,
June 2005

Description Distinctively marked spreadwing, only in Texas. *Male:* Thorax with median stripe consisting of narrow metallic green stripe on either side of midline, then wide bright blue antehumeral stripe (darkens with age), another narrow metallic green humeral stripe, and whitish below. Abdomen mostly dark, variably pruinose (always S9 but may extend to S8 and/or S10). *Female:* Eyes brown, becoming dull bluish or greenish on top. Looks paler than male, thorax pale olive with blue tinge, both dark stripes very thin; lower sides and underside white, becoming faintly pruinose. From side, S8–10 prominently pale and becoming pruinose.

Identification Only spreadwing in range with narrow metallic stripes on either side of midline of thorax, visible at any age. Thus, no prominent dark median stripe, as in overlapping **Plateau** and **Southern Spreadwings**. Median ridge of thorax pale, unlike other species in range, and humeral stripe very narrow, not evident at a distance. Overall effect blue- and green-striped thorax in mature males. Contrasting pale sides of abdominal tip characteristic of mature females.

Natural History Tropical-based species that spends dry season away from water, then returns to rain pools to breed in wet season. All-year flight season in Texas may indicate same life cycle with dormancy in winter. Males and ovipositing pairs can be common in vegetation at shallow ponds, especially ephemeral ones. Pairs oviposit in upright plant stems at water level and up to a foot above it, sometimes female submerging her abdomen.

Habitat Shallow ponds and marshes with much emergent vegetation. Often common at seasonal pools but may occur in permanent waters, both swamps and marshes.

Flight Season TX all year.

Distribution Ranges south in lowlands to Argentina.

11.1
Rainpool Spreadwing
male—Hidalgo Co., TX,
November 2005

11.2
Rainpool Spreadwing
female—Hidalgo Co.,
TX, November 2005

Description Rather dull brown spreadwing with bicolored thorax showing prominent spots on underside. *Male*: Thorax brown, somewhat metallic in front, with very narrow tan antehumeral stripe and pruinose lower sides. Black humeral stripe wider above, usually widening abruptly in two steps to reach rear of hindwing base. Pair of black spots on each side of underside of thorax, just visible from side. Abdomen metallic dark brown above, S1–2 and S8–10 pruinose with maturity. *Female*: Eyes brown, paler below. Patterned as male but with pruinosity only on sides of thorax.

Identification In hand, and perhaps in side view when close, two dark spots on either side of thorax distinctive; other species may show one spot on each side, regularly in superficially similar **Northern**, **Southern**, and **Sweetflag** in same range. Very narrow antehumeral stripes, so front of thorax looks dark brown or even black and contrasts strongly with whitish sides in mature individuals of both sexes. Characteristic stepped line of demarcation between black front and pale rear of thorax distinctive but shared by **Black Spreadwing** in its limited range and quite different-looking **Swamp Spreadwing**. **Black** very similar to but somewhat stockier than **Spotted** and less likely to be spotted under thorax; male paraprocts longer and female ovipositor larger. **Black** also has much earlier flight season, teneral **Spotted** appearing in July

12.1
Spotted Spreadwing
male—Grant Co.,
WA, August 2007

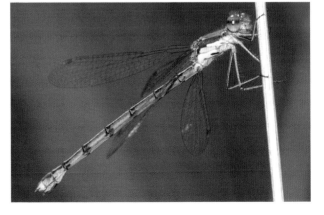

12.2
Spotted Spreadwing
female—Grant Co.,
WA, August 2007

when **Black** season finishing. **Spotted** appears to be only species in range with lower surface of eyes brown in mature males. Pruinosity at abdomen tip not quite as pale and conspicuous in **Spotted** as in **Emerald**, **Lyre-tipped**, and **Northern** at same times and places.

Natural History Roosts mostly in woodland to 10 feet or more above ground, even well up in canopy of tall forests; in open country, in tall grass and shrubs. Most likely of northwestern pond spreadwings to be in woods. Males can be abundant in tall emergent vegetation, pairs also reach high density. Many pairs mate away from water, then fly to oviposition site in tandem. Pairs oviposit from water surface to several feet above it, usually in slender stems and commonly over land at dried-up lake edges. Although some variation in substrates used for egg deposition, dead stems of bulrushes chosen in many areas. Pairs usually move downward on stem while ovipositing, single eggs deposited every few millimeters. Females often continue laying eggs after being released by male, especially in late afternoon.

Habitat Ponds and lakes of all sizes with at least some emergent vegetation. In West, usually only spreadwing at large lakes with cattails and bulrushes. Can live in quite saline lakes as long as vegetation present.

Flight Season YT Jul, BC Jun–Oct, AB Jul–Oct, WA Jun–Nov, OR May–Nov, CA May–Dec, MT Jul–Oct, AZ Jun–Sep, NM Apr–Oct, NE Jun–Oct.

Distribution Ranges across northeastern North America south to northern Alabama and Virginia.

13 Northern Spreadwing *Lestes disjunctus* TL 33–42, HW 18–23

Description Common pond spreadwing all across northern North America. *Male*: Thorax in mature individuals varies from dark in front, pruinose on sides, with narrow blue antehumeral stripe and sometimes very narrow blue median stripe, to entirely pruinose (much less common). These differences seem characteristic of populations or at least at one time and place. Abdomen metallic dark green above, S1–2 and S8–10 becoming completely pruinose with maturity. Pruinosity may vary with locality, age, or individual; not known. *Female*: Polymorphic, eyes brown or blue. Thorax with wide dark median stripe, wide dark humeral stripe widest at upper end, greenish or tan (blue in andromorph) antehumeral stripe narrowing at upper end, white lower sides and underside. Abdomen entirely blackish above.  Some females become almost as pruinose as males, in same areas of thorax and abdomen, perhaps limited to andromorphs.

Identification Looks superficially exactly like several other pond spreadwings, especially **Southern** and **Sweetflag** (see those species). Also much like **Lyre-tipped,** but both sexes differ in color of rear of head and stigmas (see that species) as well as differently shaped paraprocts that can be seen in hand and, with good view, in field. Remaining species in range all have differently colored thorax.

Natural History Mostly in herbaceous vegetation, may be some distance from water. Males frequent beds of dense emergent vegetation, perching from just above water to waist height, and can be very common there. Also present in more scattered vegetation and flying back and forth over open water. Sometimes breeds in shallow vegetated ponds that dry up each summer, more typical habitat of Emerald and Lyre-tipped Spreadwings. Females and tandem pairs arrive at water at midday, ovipositing pairs common through afternoon. Copulation lasts about 15 min. Pairs oviposit on live stems of bulrushes and sedges or dead stems of rushes and up to several feet above water, placing up to 6 eggs in one incision. Pairs also seen ovipositing entirely under water, coming up for air at intervals and then submerging again; unusual behavior in spreadwing. Sexual maturation in 16–18 days.

Habitat Well-vegetated ponds and lakes of all kinds; common in boggy situations.

13
Northern
Spreadwing
pruinose male—150
Mile House, BC, July
2006, Netta Smith;
striped male—
Kittitas Co., WA,
August 2005;
female—Kittitas Co.,
WA, September 2007

Flight Season YT Jun–Sep, BC Jun–Oct, AB Jul–Sep, WA Jun–Oct, OR May–Sep, CA Apr–Oct, MT Jun–Nov, AZ Jul–Sep, NM Jul–Sep, NE May–Aug.

Distribution Mountains of Southwest. Ranges of this and **Southern Spreadwing** incompletely known, may overlap on Great Plains. Also from northern Ontario and Labrador south to Indiana and West Virginia.

14 Southern Spreadwing *Lestes australis* TL 36–46, HW 18–25

Description Common, brightly patterned eastern species. *Male*: Front of thorax metallic brown-black with light brown antehumeral stripe becoming blue with maturity. Light yellowish below, becoming whitish pruinose. From above, S9 heavily pruinose, S10 becoming lightly so in older individuals; S8 pruinose only low on sides. *Female*: Eyes usually brown, may be blue-tinged with maturity. Colored much like male but not pruinose, antehumeral stripe usually pale tan.

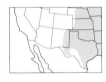

Identification Overlaps with very similar **Sweetflag Spreadwing** and perhaps with **Northern Spreadwing** at eastern edge of region. Distribution of spreadwings poorly known on plains, but **Southern** might coexist with **Northern** and **Sweetflag** in Nebraska. Probably not distinguishable except in hand from **Northern** by slightly larger size, males by blunt distal spine on cerci (sharper in **Northern**), slightly curved paraprocts (straight in **Northern**). **Southern** even more similar to male **Sweetflag** but differs in slightly narrower apical notch on S10 and slightly curved rather than straight paraprocts. In addition, male **Southern** never seem to develop completely pruinose thorax, as happens in some **Northern** and **Sweetflag**, and typically lack pruinosity on top of S9 (**Sweetflag** usually has it). **Southern**

14.1
Southern
Spreadwing
male—Dade Co., GA,
May 2006

14.2
Southern
Spreadwing
female—Hidalgo Co.,
TX, June 2005

female easily distinguished from **Sweetflag** by much smaller ovipositor, but extremely similar to female **Northern** and may not be distinguishable.

Natural History Adults spend about 2 weeks away from water in sexual maturation, then another 10 days at water (maximum 50). Males occupy perches in grass and shrubs at water's edge for long periods, but little aggression is shown among individuals. May spend all day at water, visit only in morning, or show both morning and afternoon visits. Females come to water in afternoon, and mating peaks in later afternoon. Copulation takes 6–19 min, and pair spends an hour in tandem. Oviposition in standing reed stems above water. Both sexes average two matings during lifetime.

Habitat A wide variety of ponds and lakes with aquatic vegetation.

Flight Season TX Mar–Dec.

Distribution Ranges of this and Northern Spreadwing incompletely known. May overlap on Great Plains. Also throughout much of eastern United States.

Comments This species was long considered a subspecies of *Lestes disjunctus*, under the name Common Spreadwing. With some structural differences and a somewhat different flight season, it probably deserves its rank as a full species, but genetic differences between the two are less than those between most species of spreadwings.

Description Northern spreadwing frustratingly similar to several others, best identified in field by looking for large ovipositor on females in pairs. *Male*: Thorax black in front with narrow blue-green antehumeral stripe, pale below, may be entirely gray pruinose in mature individuals. Abdomen metallic green-black above, becoming pruinose on S1, basal two-thirds of S2, and S8–10. Pruinosity heaviest of any northwestern North American spreadwing, less pruinose on plains and eastward. *Female*: Eyes blue as in male. Thoracic pattern of broad dark median and humeral stripes and fairly broad pale antehumeral stripes as in similar spreadwings but

15.1
Sweetflag Spreadwing
male—Pend Oreille Co.,
WA, August 2004

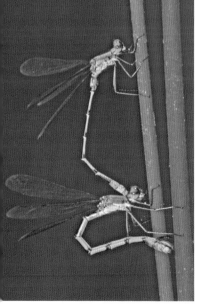

15.2
Sweetflag Spreadwing
pair with pruinose
female—Pend Oreille Co.,
WA, August 2003

15.3
Sweetflag Spreadwing
nonpruinose female—
Walker Co., GA, July
2006, Marion Dobbs

more likely to become covered by pruinosity; also S8–10 become pruinose. Thus, female colored as male, but shaped very differently.

Identification In northwestern part of range, usually separable from **Northern Spreadwing** by slightly larger size and much heavier pruinosity, occupying all of thorax in mature males and many females. Occasional male or female **Northern** is heavily pruinose, so trait may not be definitive but often indicative where they occur together. Farther east, unfortunately, neither sex of mature **Sweetflag** seems to attain this level of pruinosity. At close range, look at S2 from above; entirely pruinose in **Northern**, apical third not so in **Sweetflag**. This character quite indicative but might not be definitive in all populations. In hand, easiest structural difference to see in male is shape of notch at rear of S10, rounded and wider in **Sweetflag** and pointed and narrower in **Northern**. **Sweetflag** also has longer hamules, 1.7 mm long or more in ventral view, with stalk longer than narrow pointed blade. You may have to compare species to see these microscopic differences. Also overlaps with **Southern** locally in Great Plains; see that species. Females easily distinguished from **Northern** and all other North American spreadwings by ovipositor size, tip of ovipositor valves extending to or usually beyond tip of cerci unique in this species. Keep in mind that male pond spreadwings are well known to achieve tandem with wrong species!

Natural History Males and pairs can be common in appropriate habitat; mating and oviposition occur mostly in afternoon. Pairs usually oviposit in live stems of upright bulrushes, rushes, and cattails in fairly open stands up to several feet above shallow water or mud. Also in low sedges and buckbeans in floating mats. Heavily pruinose females may perch in open at waterside like males.

Habitat Ponds and lakes with emergent vegetation, often associated with bogs or fens. Sometimes in temporary ponds that dry during late summer.

Flight Season YT Jul–Aug, BC Jun–Aug, WA Jul–Sep, NE May–Jul.

Distribution Range doubtless much more extensive in West than shown from verified records, as identification long confused with Northern Spreadwing. Also throughout East from Ontario and Nova Scotia south to Arkansas and Virginia.

16 Lyre-tipped Spreadwing   *Lestes unguiculatus*   TL 31–44, HW 17–24

Description Common northern spreadwing of temporary wetlands. *Male*: Thorax dark metallic brown in front, with narrow blue-green or tan median line and humeral stripes; sides pale blue to whitish with pruinosity. Abdomen metallic dark green or brown above; S1, sides of S2, sides and sometimes top of S8, and S9–10 pruinose with maturity. Some males at maturity, however, lack pruinosity; pairs of both types seen together. *Female*: Eyes brown or blue. Thorax metallic brown in front with narrow median line and wider antehumeral stripe pale blue-green or yellow, sides pale blue-green or yellow. Abdomen entirely metallic brown to green above.

Identification Looks about like **Northern** and **Sweetflag Spreadwings** in field, similarly variable. In hand, lyre-shaped paraprocts provide definitive identification. Note that long, straight paraprocts of **Northern** might be crossed at ends. In both sexes (important to identify females), **Lyre-tipped** has rear of head pale, others dark, and stigmas usually with pale ends in **Lyre-tipped**, entirely dark in others (many spreadwings have just the veins at either end of the stigma white). Note that light pruinosity eventually covers rear of head in mature males of all three. Another distinction of males is that **Northern** and **Sweetflag** typically have upper surface of S2 mostly or entirely pruinose, whereas pruinosity appears only on side of this segment in **Lyre-tipped**. Typically, pattern of pruinosity in **Lyre-tipped** forms dark V in top view of S8, whereas in the other two, this segment is entirely pruinose, but some variation makes this an indicative character. Females and im-

16.1
Lyre-tipped
Spreadwing
male—Winneshiek
Co., IA, July 2004

16.2
Lyre-tipped
Spreadwing
female—Winneshiek
Co., IA, July 2004

mature males may have dark areas quite metallic green, need to be distinguished from **Emerald Spreadwing** by head and stigma color, male paraprocts, and smaller ovipositor.

Natural History Males spend much time at water resting on vertical stalks but do not defend perch sites, more commonly move from perch to perch. Approaching females taken in tandem immediately. Copulation lasts about 25 min but is often broken and resumed, accompanied by short flights in tandem. Pair then explores potential oviposition substrates for about a half-hour, then oviposits for over an hour, usually over relatively dry substrates rather than over water and typically backing down stem as eggs are laid. Pairs more and more common through afternoon. Female sometimes continues by herself. Living sedge, bulrush, bur-reed, and pitcher plant stems common substrates, and few hundred eggs laid at about 2/min, 1–2 per incision.

Habitat Shallow marshes and marshy edges of ponds and lakes, often in completely open areas and typically drying up in midsummer. These are exactly the habitats affected by drought, and this species declines wherever drought prevails in the West. However, also quick to colonize newly flooded areas, including farm ponds and other artificial wetlands, and usually most common spreadwing in prairie potholes. Immatures often abundant in grassy meadows.

Flight Season BC Jun–Aug, AB Jul–Sep, WA Jun–Sep, OR May–Sep, CA Jun–Sep, MT Jun–Aug, NE May–Oct.

Distribution Widespread in Northeast from southern Ontario and Nova Scotia south to Arkansas and Maryland.

Description Long-bodied spreadwing, S9 less than half as long as S7; males distinctly longer than females. Vein around extreme wing-tips conspicuously pale. *Male*: Thorax black in front with rather wide blue antehumeral stripes, unmarked yellow on sides and underside. Abdomen typically lacking any pruinosity, but sometimes S9 pruinose, apparently more likely in northwestern part of range. Heavily pruinose between wing bases as all spreadwings. *Female*: Eyes blue or blue over yellow. Pattern on thorax and abdomen as in male, no pruinosity.

Identification Males distinguished from other spreadwings by virtual lack of pruinosity on abdomen as well as looking longer than any other species of similar body bulk. Pale vein at extreme wingtips distinctive of both sexes. Not metallic like larger **Amber-winged**, **Elegant**, and **Swamp Spreadwings**, colored more like **Lyre-tipped**, **Northern**, and **Sweet-flag,** but reduced pruinosity distinctive. Absence of pruinosity always leaves pale blue antehumeral stripes good for identification in both sexes. Abdomen length in female close enough to that of **Lyre-tipped** and **Northern** that separating these three is difficult. **Lyre-tipped** usually has pale-tipped stigmas, but **Northern** and **Slender** are colored about the same. Presence of males might have to be used for identification.

Natural History Males in shrubs and low tree branches in swampy woodland, also dense herbaceous vegetation at lake shores. Can be abundant in woodland during maturation. Maturation in color in about 2 weeks, reproductive activity at 3 weeks. Mating may take place away from water, and females oviposit solo, unusual in spreadwings, and about a foot above water. Eggs commonly laid in cattails, 1 egg per incision. After maturation may live for 6 weeks.

17.1
Slender Spreadwing
male—Wayne Co., OH,
July 2007

17.2
Slender Spreadwing
female—Winneshiek
Co., IA, July 2004

Habitat Lakes and ponds with abundant emergent vegetation, usually associated with forest.
Flight Season NE May–Oct.
Distribution Widespread in the East from Ontario and Nova Scotia south to Louisiana and north Florida.

---

18 Emerald Spreadwing   *Lestes dryas*   TL 32–40, HW 19–25

Description Rather stocky spreadwing with front of thorax emerald-green. *Male*: Thorax metallic green in front with narrow pale antehumeral stripe present or not, pruinose white on sides. Abdomen metallic green above, at maturity becoming pruinose on S1–2 and S9–10, often extending to S8. *Female*: Brown or blue eyes at maturity. Thorax metallic green in front with narrow pale median line and antehumeral stripes. Abdomen entirely metallic green above, without pruinosity.

Identification Emerald green thorax and abdomen diagnostic along with stocky build (other green species such as **Elegant Spreadwing** are larger, longer), but bear in mind that in several

18.1
Emerald Spreadwing male—150 Mile House, BC, July 2006, Netta Smith

18.2
Emerald Spreadwing green female—Skamania Co., WA, August 2005

18.3
Emerald Spreadwing bronze female—Chelan Co., WA, June 2004

species, even those in which adults lack green, dark part of abdomen may be green in immatures, and green highlight may show up on thorax. None is brilliant green like this species. Only species with similar structure is **Black Spreadwing**, so similar that it may be same species. Front of thorax usually dark brown to black in latter species. Thorax and abdomen of **Emerald** sometimes look brownish, especially younger individuals, and oldest males can look quite black above. Note widened tips of paraprocts of male.

Natural History Can be very common in shrublands and forest near breeding ponds. Breeding males and pairs tend to stay over dry rather than flooded parts of habitat. Pairs oviposit in live stems of sedges, grasses, and horsetails and hanging willow leaves, high above ground in the latter. One egg inserted in each incision.

Habitat Shallow ponds, marshes, and fens, often those that dry up in late summer; typically densely vegetated. Also found at edges of permanent wetlands but may not breed successfully where aquatic predators are common. Widely distributed from hot sagebrush steppe to cool boreal forest.

Flight Season YT Jun–Aug, BC May–Aug, AB Jun–Aug, WA Jun–Sep, OR May–Nov, CA Apr–Oct, MT Jun–Aug, AZ May–Jun, NM Jul–Aug, NE May–Jul.

Distribution In mountains in southern part of range. In East from southern Canada south to Iowa, Kentucky, and Maryland, also all across northern Eurasia.

Comments See under Black Spreadwing.

19 Black Spreadwing *Lestes stultus* TL 35–44, HW 21–26

Description Dark spreadwing of California, much like **Emerald** but blackish instead of green. *Male*: Thorax metallic bronzy-black in front with narrow pale greenish antehumeral stripes. Lower sides and underside whitish. Abdomen dark bronzy-brown to greenish above, pruinose on S1, basal two-thirds of S2, and S8–10 at maturity. *Female*: Eyes brown with bluish tinge above. Color pattern as male but no pruinosity.

Identification Much like **Emerald Spreadwing** in both sexes, but slightly larger and front of thorax dark brown to black, not emerald

19.1
Black Spreadwing male—
Glenn Co., CA, June 2004

19.2
Black Spreadwing female—
Butte Co., CA, June 2004

green. Antehumeral stripe evident in male **Black**, usually not in male **Emerald**; female **Black** usually with that stripe complete, incomplete at upper end in **Emerald**. There is variation in both sexes, however. The two may meet and intergrade in southern Oregon, but in addition, green color may show up on **Black** south of range of **Emerald**. In California, **Black** is early species of lowlands, **Emerald** later flier of mountains; not known to occur together. **Black** distinguished from other species occurring with it by short paraprocts in male and large ovipositor in female. Colored much like **Spotted**, but thorax with no spots on underside and dark side stripe not as wide at upper end (does not extend below base of hindwing); female ovipositor larger.

Natural History Males perch on stems and leaves of emergent vegetation over shallow water. Pairs oviposit on live rushes or sedges.

Habitat Ponds, small marshy lakes, and slow streams with abundant emergent vegetation.

Flight Season CA Mar–Sep.

Comments This species is barely separable from the Emerald Spreadwing, and seeming intermediates between the two species have been found in southern Oregon. They probably should be considered of no more than subspecies rank, but no official change of status has been suggested.

---

20 Swamp Spreadwing *Lestes vigilax*   TL 42–55, HW 23–27

Description Large metallic green spreadwing. *Male*: Eyes dark blue-green with pale blue highlight over yellow-green, looking somewhat bicolored. Thorax entirely metallic green to bronze in front or with narrow reddish-brown antehumeral stripe. Sides and underside pale yellow, becoming whitish pruinose, with pruinosity obscuring border between dark and light parts of thorax, originally a wavy line. Abdomen metallic green to bronze above, becoming pruinose on S1 and S8–10 at maturity; sides of S2 also pruinose. *Female*: Eyes brown over yellowish, blue in some (oldest?) individuals.

20.1
Swamp Spreadwing
male—Murray Co., GA,
June 2005, Marion Dobbs

20.2
Swamp Spreadwing
female—Cumberland
Co., ME, July 2006

20.3
Swamp Spreadwing
immature female—
Rabun Co., GA, July
2004, Giff Beaton

Colored as male, but always with narrow pale reddish-brown antehumeral stripe. Abdomen duller than in male, with no pruinosity in most, but pruinose tip in a small percentage.

Identification Longer-bodied than most other species in range, bulkier than **Slender Spreadwing** and metallic green rather than dark brown above. Most like **Elegant Spreadwing**, which see.

Natural History Males usually perch in sheltered areas in shade, often in tangled vegetation, and are difficult to find. Tandem pairs oviposit in pickerelweed stems and other plants at water surface, even well out from shore. Dull immatures in dense herbaceous and shrubby vegetation near water.

Habitat Wooded ponds and lakes with abundant emergent vegetation, often where shrubs grow in shallow water. Slow streams and bog-margined lakes included in this description.

Flight Season TX May–Nov.

Distribution In East from southern Ontario and Nova Scotia south.

21 Elegant Spreadwing *Lestes inaequalis*   TL 45–58, HW 25–31

Description Large metallic green spreadwing. *Male*: Eyes dark blue-green with bright blue highlight over pale blue-green, strongly bicolored. Thorax metallic green to bronze in front, may show fine reddish-brown midline. Sides and underside pale yellow, rarely becoming whitish pruinose. Abdomen metallic green to bronze above, becoming pruinose on S9 and then S10 at maturity; sides of S1–2 also pruinose. *Female*: Eyes dark green over dull yellow to light green, strongly bicolored. Colored as male, but often with narrow pale reddish-brown antehumeral line. Minimal or no pruinosity on

most, but underside of thorax and abdomen tip may become pruinose in oldest individuals.

Identification Impressively large size and mostly metallic upper side at maturity distinctive. Eyes more strikingly bicolored (actually tricolored) than in any other species but **Swamp Spreadwing**. Both sexes distinguished from that quite similar species by pale rear of head and pale tibiae (dark in mature **Swamp** but pale in immatures). Pruinosity in male **Elegant** usually not obscuring sharp border between front and sides of thorax as it does in **Swamp**. At close range, look for distinctive long paraprocts in male, extending beyond tips of cerci.

21.1
Elegant Spreadwing
male—Holmes Co., MS,
May 2005, Giff Beaton

21.2
Elegant Spreadwing
female—Fayette Co.,
IA, July 2004

Females distinguished from **Swamp** by slightly larger ovipositor valves, extending beyond lower edge of S10 and dark below (valves in **Swamp** entirely pale). Basal plate of ovipositor pointed in **Swamp**, squared off in **Elegant**. Typically, pale antehumeral stripe in female narrower in **Elegant** than in **Swamp**. Much longer-bodied than **Emerald Spreadwing** with similar green thorax.

Natural History Males conspicuous, perching low in shrubs and other emergent vegetation and flying over open water. Often in shade and more active later in day. Usually not very common. Reported to oviposit in water lily leaves, unusual for a spreadwing.

Habitat Lakes, ponds, and slow streams with abundant vegetation, in or out of woodland. Most likely pond spreadwing at edge of slow streams.

Flight Season TX May–Aug.

Distribution In East from southern Ontario and Quebec south to north Florida.

22 Amber-winged Spreadwing *Lestes eurinus* TL 42–52, HW 26–30

Description Large green and yellow spreadwing with amber-tinted wings and distinctive markings on sides of thorax. *Male*: Thorax metallic greenish in front without pale stripes; lower sides and underside yellow with irregular dark stripes, becoming whitish pruinose. Abdomen metallic dark green, S1 and S9–10 (more rarely S8) pruinose above. *Female*: Eyes blue above, yellow below. Thorax as in male but does not become pruinose.

Identification No other spreadwing has amber wings, and no other spreadwing has dark stripes across bright yellow sides of thorax. Bulkiest of pond spreadwings, although **Elegant** and **Swamp** are as long. Superficially most like those two species because of large size, green thorax, and sparse pruinosity but differs in wing color and very short paraprocts. Most like **Swamp**, mature male differs in pruinosity usually on S9–10 rather than S8–10, although probably total overlap. Best mark short paraprocts when they can be seen, as well as wing coloration. Differs further from **Elegant** in having entirely dark tibiae. Bicolored blue and yellow eyes of female also distinctive, although shared with female **Swamp Spreadwing**.

Natural History Males more active flyers along lake shores and out over open water than other spreadwings, impressively large and fast, perhaps while searching for females. Appears more likely to take larger prey such as other damselflies, including teneral spreadwings. Females oviposit in tandem or solo from just above water to several feet up on sedge, cattail, rush, and bur-reed stems, also on top of water lily leaves. Eggs laid in clusters, averaging 6–7 per cluster.

Habitat Variety of permanent lakes and ponds with at least some emergent vegetation; has been found in everything from bog lakes to pasture ponds.

Flight Season MO May–Aug.

Distribution In Northeast from southern Ontario and Newfoundland south to Missouri, northern Georgia, and Virginia.

22.1
Amber-winged
Spreadwing
male—Sussex Co.,
NJ, June 2005, Allen
Barlow

22.2
Amber-winged
Spreadwing
pair—Grafton Co.,
NH, June 2006

Pond Spreadwings—male appendages

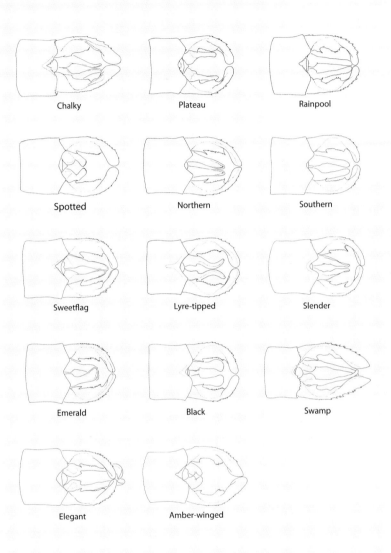

Chalky

Plateau

Rainpool

Spotted

Northern

Southern

Sweetflag

Lyre-tipped

Slender

Emerald

Black

Swamp

Elegant

Amber-winged

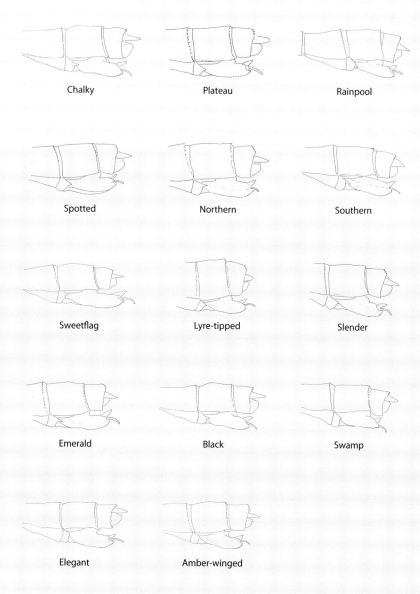

Chalky

Plateau

Rainpool

Spotted

Northern

Southern

Sweetflag

Lyre-tipped

Slender

Emerald

Black

Swamp

Elegant

Amber-winged

Pond Damsel Family *Coenagrionidae*

This largest family of damselflies is the second-largest family of odonates. Almost anywhere in the world, they are usually the most common damselflies in open ponds and marshes. There are also many species on streams, but other families often dominate on tropical streams, and most damselflies have narrow wings; thus, a commonly used name "narrow-winged damsels" is not used here. They are generally small species, a few as large as spreadwings. The wings are held closed in all but a few species in which they are held half-open, less open than in spreadwings. The color pattern has similar elements in almost all genera. The eyes in most are black or dark brown above, brightly colored below, and it is that color that is used in their description. The "face" (labrum, anteclypeus, and clypeus) is often paler than the top of the head, and most species have postocular spots, pale spots contrasting with the dark head just in from the back of each eye; they may be connected by a pale line of the same color or not. Typically, the thorax has a dark median stripe, a pale antehumeral stripe, and a dark humeral stripe, the sides and underside pale. Sometimes the pale stripe is interrupted. In males, the abdomen is all dark or all pale or some combination of dark and pale, often with a contrastingly colored tip. Females usually have slightly thicker abdomens than males (they carry all those eggs), and an ovipositor is present but not usually as prominent as in spreadwings. Females typically but not always share the male's head and thorax pattern but usually have the abdomen darker above, with less pattern. Females in some genera are polymorphic, one morph colored more like the male, the other duller. World 1082, NA 103, West 87.

Eurasian Bluets *Coenagrion*

This diverse group of temperate Eurasian species with a few North American representatives look essentially like typical blue American bluets, *Enallagma*, but may or may not be closely related to them. No field characters differentiate males collectively, so they will have to be distinguished individually from various American bluets as well as each other. From above, black on female abdomen covers the entire upper surface of each segment except for a narrow basal ring, whereas black on middle segments of many American bluets bulges at the posterior end and tapers at the anterior end somewhat like a torpedo. Black markings fall well short of front of segment in *Enallagma*, leaving a more conspicuous ring. Also, female *Enallagma* usually have at least one distinct dark stripe visible around the middle of each eye, but *Coenagrion* lack these stripes. In hand, male *Coenagrion* have forked paraprocts in side view, shared by a few *Enallagma*, and females lack vulvar spine, present in *Enallagma*. In habits they resemble *Enallagma* but tend to be more common in dense herbaceous vegetation. World 40, NA 3, West 3.

23 Prairie Bluet *Coenagrion angulatum* TL 27–33, HW 16–22

Description Mostly black-bodied bluet of plains. *Male*: Eyes black over pale greenish. Thorax with wide black median and moderate humeral stripe, blue antehumeral about as wide as humeral; also fine but conspicuous black line low on side of thorax. S1–2 blue, S2 with black subapical bar; S3–6 black with basal blue rings, widest on S3 and becoming narrower to rear; S7 almost entirely black; S8–9 blue, S10 blue on sides. *Female*: Polymorphic, pale areas greenish or blue. Thorax striped as in male. Abdomen entirely dark  above, conspicuously pale on sides, with narrow pale rings at end of each segment, slightly wider and distinctly blue on S7–9; S8 with pair of dorsal blue spots at base.

Identification Few other male bluets in its range with blue S1–2 and S8–9 but mostly black middle abdominal segments. Barely overlaps with **Skimming Bluet**, which is smaller and has top of S2 and virtually all of S3–8 black. FORKTAILS in its range have green-striped thorax. **Stream Bluet**, in different habitat and more slender, differs in same way in color

23
Prairie Bluet male—
Ward Co., ND, June
2003, Tom D. Schultz;
female—Devon, AB,
July 1995, John Acorn

pattern as **Skimming**. Of closely related species, female much like **Taiga Bluet** but pale spots at base of S8 distinctive. Much less pale color on abdomen than vividly ringed **Subarctic Bluet**. Co-occurring species of AMERICAN BLUETS (*Enallagma*) such as **Alkali, Boreal, Familiar, Northern, River**, and **Tule** also have broader rings of pale color on middle segments and characteristic torpedo shape of black on each segment rather than straight edge of Eurasian bluet. However, two *Enallagma*, **Hagen's** and **Marsh Bluets**, are rather similar to *Coenagrion* because the abdominal markings are less obviously different. Nevertheless, both have bulges that hint at the torpedo pattern; they also lack the basal spots on S8 and have smaller postocular spots than **Prairie Bluet**. Always check for dark horizontal stripes around the eyes to be sure if you have an *Enallagma* (striped) or a *Coenagrion* (not).

Natural History Extremely abundant at some prairie wetlands. Males common in dense grass rather than over open water. Sexual maturation takes about a week. Copulation for 20 min or more. Oviposition typically in tandem on submergent vegetation at surface. In emergent vegetation, pair backs down under water and may remain for up to 30 min, then floats to surface and flies away, still in tandem. Females lay 150–200 eggs in each of several clutches.

Habitat Prairie lakes, ponds, sloughs, and slow streams, usually with much marsh vegetation, some of them sufficiently shallow to go dry at times.

Flight Season BC Jun, SK May–Jul.

Distribution Also east to James Bay and Iowa.

24 Subarctic Bluet *Coenagrion interrogatum* TL 28–32, HW 17–21

Description Far northern bluet with divided antehumeral stripes. *Male*: Eyes black over blue. Blue with extensive black markings. Moderate median and humeral stripes, blue antehumeral slightly wider than humeral and divided near upper end, looking something like exclamation mark (although *interrogatum* means to question!). Conspicuous black stripe on side of thorax expanded at upper and lower ends. Abdomen blue with prominent black U on S2 with arms wider than base, S3 with black apical ring, then each subsequent segment with more and more black, so S4 appears half black, S5 three-quarters black, S6 seven-eighths black, and S7 with very narrow blue basal ring; S7 also has blue tip, S8–9 blue, and S10 black above. *Female*: Polymorphic, either blue like male or green. Eyes brown over pale green. Thoracic stripes as in male. Abdomen mostly black above, but S1 pale, S2 extensively pale on sides, forming black torpedo but with exaggerated base; conspicuously pale basal ring on S3–8, also larger area on S8–9, and S10 pale.

24
Subarctic Bluet male,
female—Oneida Co.,
WI, June 2005, Mike
Reese

Identification Divided antehumeral stripe distinctive, as is conspicuous black stripe low on sides of thorax with bulges in it like string of pearls. Extensive blue on abdomen base and tip not like any other western bluet, and easily distinguished because of thoracic pattern in any case. Female distinguished from other bluets by divided antehumeral stripe and much pale color at abdomen tip. No other bluet in range has most of S9–10 blue as well as tip of S8.

Natural History Males and pairs in tandem in dense vegetation, not usually over open water; copulating pairs often perch in shrubs up to head height. Pairs or lone females have oviposited in floating sedge and grass leaves and stems and upright grass stems.

Habitat Boreal fens and bogs, usually associated with sphagnum and other aquatic mosses but often in shrubs.

Flight Season YT May–Aug, BC May–Aug, AB May–Aug, WA Jul, MT May–Jul.

Distribution Ranges east, mostly in Canada, to Maine and Newfoundland.

25 Taiga Bluet   *Coenagrion resolutum*   TL 27–33, HW 15–20

Description Widespread and common northern bluet of sedge marshes with much black on abdomen and U-shaped mark on S2; lower sides of thorax sometimes greenish. *Male*: Eyes black over blue-green. Thorax with wide median and moderate humeral stripe, relatively narrow antehumeral stripe sometimes broken into exclamation point. Fine black line just behind humeral stripe from wing base halfway down thorax. Black U on S2 with base wider than arms; S3–5 about half blue, half black; S6–7 almost all black; S8–10 blue, with black on top of S10 and often paired apical markings on S9. *Female*: Polymorphic, either light pinkish-brown or bright greenish to blue-green. Entire top of abdomen black from S2 to S10, with  fine pale basal rings on each segment; narrow apical rings on S7–9 blue in both morphs. Immature males can have fully blue markings on abdomen but tan thorax.

Identification Males with more black on abdomen than other small bluets that occur with them, in characteristic pattern of U-shaped mark on S2, then two short and one long black section before blue tip. Black mark on S2 different proportions than similar U-shaped mark on **Subarctic Bluet**, which also always has interrupted antehumeral stripe. Northern populations of **Boreal Bluet**, with much more black than elsewhere, may have U-shaped mark on S2, but larger than **Taiga** and differs in other ways. No other bluet has paired black apical markings on S9 or fine but conspicuous black line just behind humeral stripe. No other male bluet shows green on thorax as this species often does. Females of other Eurasian bluets have extensive pale color on S8 or S8–9, as do those of some AMERICAN BLUETS. AMERICAN BLUETS of ponds and lakes with no pale color on S8–9 have middle abdominal segments extensively marked with pale brown or blue and never have apical blue rings on S7–8. Some

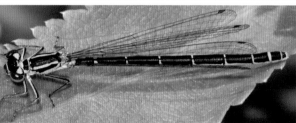

25
Taiga Bluet
male—Kittitas Co.,
WA, July 2005; brown
female—Kittitas
Co., WA, June 2007,
Netta Smith; green
female—Williams
Lake, BC, July 2006

stream bluets, not likely to overlap in habitat, more similar. Female **Rainbow Bluet** most like female **Taiga** in color pattern but has mostly blue S10, orange face and orange stigmas, and much smaller postocular spots. Female **Stream Bluet** has most of S9–10 blue.

Natural History Males abundant in appropriate habitat, females and pairs generally harder to find than those of American bluets. Males and tandem pairs stay within dense grasses and sedges, rarely if ever over open water, and fly less strongly than American bluets. Also on open sphagnum mats. Males cruise through vegetation rapidly in search of females. Pairs oviposit on both floating and emergent plant stems, usually just below surface.

Habitat Sedge marshes and fens and well-vegetated pond and lake edges, at large lakes in sedge beds. Often in stands of water horsetail. Common in habitats of both other North American representatives of group. In shadier, cooler habitats in southern parts of range.

Flight Season YT May–Aug, BC May–Aug, AB May–Sep, SK May–Jul, WA Jun–Sep, OR Jun–Aug, CA May–Aug, MT May–Aug, AZ Jun–Jul, NE Jun–Aug.

Distribution Restricted to higher elevations in southern part of range. Also through eastern Canada and south to Iowa, Pennsylvania, and Massachusetts.

American Bluets *Enallagma*

This large genus of small to medium-sized damselflies is found primarily in temperate North America, but a few are seen in Eurasia. Typical species are mostly blue, the amount of black variable with species. Another group occurs with abdomen mostly black, tip variably blue, also exhibiting much variation. A different group contains yellow to red species; still another has mostly violet species. Postocular spots vary from small and isolated ("spots") to large and connected or almost connected by a line ("dumbbell") to narrow and connected ("line"). Pond species typically show much blue on the abdomen, whereas stream species exhibit a black abdomen and blue tip. The thorax is typically patterned with black stripes, wide median and narrow humeral; the pale antehumeral stripe is usually wider than humeral. Males rarely and females more often show a pale line on the median carina dividing median stripe, but this line is characteristic of some species. Blue bluets show a "torpedo" pattern from above on most abdominal segments: pointed at anterior end, then parallel-sided and often constricted just before bulging at posterior end. Many species have polymorphic females, either brown (sometimes green) or blue. Eurasian bluets (*Coenagrion*) are very similar to mostly blue American bluets, but males have paraprocts that are slightly forked in side view, and females lack a vulvar spine, characters visible only in hand. Otherwise, they are best distinguished by characteristics of included species.

Stream bluets typically perch facing the shore at tip of a branch or leaf projecting from the shore; they also commonly hover for long periods over water, both of these behaviors distinguishing them from dancers that share this habitat. Both copulating and tandem pairs are frequently seen. In tandem oviposition, males grasp substrate or, if no substrate is available, they lean forward whirring wings, not resting still in vertical position as typical of dancers. Blue bluets are distinguished from dancers that are mostly blue usually by having black on top of the eyes (a black "cap" usually absent in dancers), black on top of S2 (on sides in dancers) and S10 (no black in dancers), and almost all lacking black stripes low on S8–9 (most dancers have them). Female bluets are easily distinguished from female dancers if eyes can be seen; those of most bluets have at least one horizontal stripe, not so in dancers (vague stripe in female Sooty Dancer). Many bluets can be distinguished from one another in the field by relative proportions of blue and black on various abdominal segments, but some of mostly blue species have to be captured to be distinguished by appendages. Some females are distinguished only by close examination of mesostigmal plates. World 47, NA 37, West 25.

Table 2   Bluet (*Enallagma, Coenagrion*) Identification

	A	B	C	D	E	F
Prairie	1	3	1	1	1	2
Subarctic	1	3	1	4	2	3
Taiga	1	2	1	1	12	2
Arroyo	1	3	1	1	2	1
River	1	2	1	1	1	1
Claw-tipped	1	1	1	2	2	1
Atlantic	1	1	1	1	23	2
Familiar	1	1	1	1	12	1
Tule	1	3	1	1	2	1
Skimming	1	3	1	2	1	2
Azure	1	3	2	3	4	1
Big	1	1	1	1	23	3
Alkali	1	2	1	2	12	2
Northern	1	1	1	2	12	3
Boreal	1	1	1	2	2	3
Marsh	1	1	1	1	12	2
Hagen's	1	1	1	1	12	2
Neotropical	2	3	1	2	2	1
Baja	1	2	1	2	3	1
Stream	1	3	13	4	3	1
Attenuated	1	3	2	4	5	3
Rainbow	1	3	3	1	3	1
Turquoise	1	3	1	4	23	2
Slender	1	3	1	5	25	1
Double-striped	1	2	1	4	3	1
Burgundy	3	3	5	1	3	1
Orange	4	3	4	14	3	1
Vesper	5	3	3	14	3	1

A, Thorax color: 1, mostly blue; 2, mostly violet; 3, mostly red; 4, mostly orange; 5, mostly yellow.

B, Abdomen color: 1, middle segments mostly pale; 2, middle segments about half pale, half black; 3, middle segments mostly black.

C, Abdomen tip: 1, S8–9 blue on top; 2, part of S7 as well as S8–9 blue; 3, only S9 blue on top; 4, S9 orange; 5, S9 mostly black.

D, Female abdomen: 1, mostly dark to tip (or S10 pale); 2, S8 all or mostly pale; 3, S7 pale; 4, S9–10 mostly pale; 5, S8–10 mostly pale.

E, Postocular spots: 1, spots; 2, dumbbell; 3, line; 4, continuous with back of head; 5, top of head mostly blue.

F, Male cerci: 1, obviously longer than paraprocts; 2, about same length as paraprocts; 3, obviously shorter than paraprocts.

Description Small southwestern bluet with much black on abdomen. Postocular spots large, forming dumbbell; narrower in female. *Male*: Eyes blue with black cap. Median stripe wide, some with incomplete blue on carina; antehumeral stripe considerably wider than narrow humeral. Abdomen with much black, including large apical spot on S2 and torpedo pattern covering three-fourths or more of S3–6; all of S7 above with narrow blue basal ring, and S10 above. *Female*: Polymorphic, brown or blue. Eyes greenish-tan with brown cap. Thorax as male. Abdomen entirely black above,  with torpedo pattern on S3–6 producing basal pale rings visible from above.

Identification Male with more black on middle segments than other blue bluets with which it occurs, at most extreme looking mostly black like **Skimming Bluet**, but always conspicuous blue rings around segments. Long cerci usually apparent but not quite as conspicuous as in rather similar **River Bluet**, which averages larger and with more blue on middle segments. Somewhat like male **Tule Bluet** but a bit smaller, often with more black on middle segments, and with characteristic cerci. Female much like female **Familiar**, **River**, and **Tule Bluets** but smallest of these. Probably indistinguishable except in hand. Male could be mistaken for male FORKTAIL at quick glance, but all FORKTAILS in its range with either green thorax or front of thorax with pale dots instead of stripes.

Natural History Males common in emergent vegetation at waterside but never in numbers like some pond species. Females seldom found except in mating pairs. Pairs oviposit in floating vegetation; male may release female, and she submerges headfirst to continue egg laying.

Habitat Slow-flowing streams or lake margins with emergent vegetation.

Flight Season CA Apr–Nov, MT Jun–Aug, AZ Feb–Nov, NM Feb–Nov, NE Jun–Aug, TX Mar–Nov.

Distribution Ranges south in uplands to Honduras.

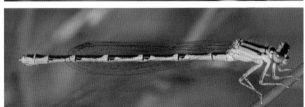

26
Arroyo Bluet
male—Kinney Co.,
TX, July 2004; brown
female—Sonora,
Mexico, September
2005

Description Large bluet of running water with long appendages. Postocular spots usually without connecting line. *Male*: Eyes black over blue. Thorax with wide median and narrow to wide humeral stripes, in some populations narrowed toward upper end and then widened into spot. Black subapical spot on S2, in some connected to lateral markings as U. Black ring extending forward as point on S3 and beyond, black more extensive but point shorter on each succeeding segment until black filling most of S7 and cut straight across; S8–9 blue, S10 black. Upper part of cerci conspicuously long and straight. *Female*: Polymorphic, brown or blue. Eyes brown over tan. Thorax as male. Abdomen black above on all segments, on S3–7 pointed forward into pale basal ring; black on S9–10 not fully covering segment.

Identification Male resembles many other bluets, especially **Alkali** and **Tule Bluets**, because of much black on middle segments; former more likely to occur with it. Looks a bit larger than **Tule** when comparison possible. Postocular spots usually bridged by line in **Tule**, not so in **Alkali** and **River**. Male also colored like smaller and more slender **Arroyo Bluet**, larger size usually adequate distinction; black on S3 usually terminal ring on **River**, usually more extensive on **Arroyo**. Long cerci of male, with ventral process, distinctive from all other bluets. Narrow humeral stripe distinctive where found; female may be distinguishable from similar **Arroyo**, **Familiar**, and **Tule Bluets** by this mark, otherwise only in hand. **Alkali Bluet** has similar humeral stripe but usually entirely pale S8. Usually only bluet common on open rivers.

Natural History Both sexes in open herbaceous vegetation, often perch on ground like dancer. Males perch on stems at outer edge of vegetation beds. Pairs or single females oviposit on emergent and floating vegetation; females may go entirely under water, down to several inches deep, for as long as a half-hour.

Habitat Streams and small rivers, mostly in open country but often with riparian borders; also flowing irrigation canals.

Flight Season AB Jun–Jul, OR Jun–Oct, CA Jun–Sep, MT May–Oct, NM Apr–Aug, NE May–Sep.

Distribution Also in East in narrow band from Iowa to southern Ontario.

Comments Has hybridized with Tule Bluet.

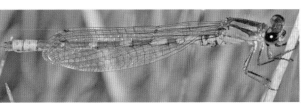

27
River Bluet
male—Teton Co.,
WY, September 2007;
brown female—Inyo
Co., CA, June 2003

---

28 Claw-tipped Bluet *Enallagma semicirculare* TL 29–33, HW 15–18

Description Slender blue to purplish-blue bluet with long clawlike cerci that barely enters region from Mexico. Postocular spots large, forming dumbbell. *Male*: Eyes blue with black cap. Antehumeral stripe wide, humeral stripe narrow. Color varies from typical bluet blue to bluish-purple. Abdomen with usual black marks, including subapical bar on S2, black apical rings on S3–5, entire top of S6–7 except narrow basal ring, and top of S10. *Female*: Eyes brown above, tan below. Head and thorax patterned as male with pale blue colors. Abdomen black above, extensively blue on sides of S1–3, bluish tinged on sides of S4, rest tan except for S8, small area on S9, and tip of S10 blue.

28

Claw-tipped Bluet
male—Sonora,
Mexico, June 2005,
Doug Danforth;
female—San Luis
Potosí, Mexico,
October 2003,
Robert A. Behrstock

Identification Superficially like other largely blue bluet in its range, **Familiar Bluet**, but easily distinguished by long slender cerci if those can be seen. Transverse bar on S2 also distinctive if abdomen tip not visible. Most individuals distinctly more purplish than bright blue **Familiar**, less so than **Neotropical Bluet**, which occurs with it south of border. Female distinguished from most coexisting bluets (**Arroyo**, **Familiar**) by blue S8. Very similar to blue female **Neotropical**, and coexistence in the United States may yet be documented. The latter has black on rear edge of S8 and conspicuous blue markings on sides of S9, whereas **Claw-tipped** has no black on S8 and almost no blue on S9. Blue also brighter and more extensive at abdomen base of **Claw-tipped**, especially on S3, where blue covers much of segment, and torpedo mark much narrower than on S4.

Natural History Males perch on leaves and twigs over water as other bluets. Otherwise unknown.

Habitat Typically at pools of slow streams in woodland or open, also at shallow ponds. Found in abundance at rainwater ponds in northern Mexico.

Flight Season AZ Apr–Nov.

Distribution Ranges south in Mexico to Chiapas.

29 Atlantic Bluet *Enallagma doubledayi*   TL 28–37, HW 16–21

Description Common bright blue bluet of Southeast that strays into east Texas. Postocular spots form transverse line. *Male*: Eyes blue with slight amount of black above. Thorax with usual bluet stripes, humeral slightly narrower than antehumeral. Black markings include large round apical spot on S2, larger narrow rings on S3–6, S6 often prolonged forward into long point; S7 black above with narrow blue basal ring; S8–9 blue, S10 black above. Midabdomen black rings usually narrow but  sometimes more extensive, extending forward as much as midsegment. *Female*: Polymorphic, brown or pale bluish. Eyes brown over pale greenish or tan. Thorax as male. Abdomen with all segments continuously black above, only slight indication of torpedo-shaped markings.

Identification Only other all-blue bluets in range in West are very common **Familiar Bluet** and mostly coastal **Big Bluet**. Normally considerably smaller than **Big**, should be recognizable by size alone, but smaller individuals of **Big** complicate this. **Big** usually has median stripe on thorax divided by fine blue line, generally dependable field mark. Otherwise, capture and scrutiny of appendages are necessary. **Familiar** more easily distinguished from **Atlantic** by good look at large cerci of former; also postocular spots often larger and without interocular bar but overlap. **Atlantic** sufficiently rare in region so that in-hand identification is recommended. Females of these three very similar, **Atlantic** distinguished from **Familiar** by narrower postocular spots, often no wider than interocular bar, and from **Big** by smaller size and lack of fine pale line down front of thorax. Mesostigmal plates always worth checking.

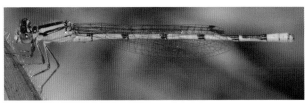

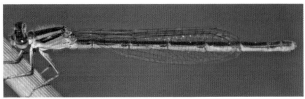

29
Atlantic Bluet male,
brown female—Lake
Co., FL, April 2005

Natural History Males range all over typical well-vegetated ponds, usually perching on stems above water. Pairs oviposit in tandem on horizontal stems or female solo on vertical stems, moving underwater headfirst and briefly while male waits above.

Habitat Shallow grassy ponds, less often lake borders, in open or open woodland.

Flight Season TX May.

Distribution Few isolated records in Texas, populations should be sought. Widespread in East, mostly on Coastal Plain and Piedmont, from New Hampshire to Mississippi.

Description Abundant and widespread bright blue bluet. Postocular spots not connected in most populations, can be very small or even absent in North; form narrow dumbbell in Southwest. *Male*: Eyes blue with small black cap. Thoracic stripes typical of bluet, widest black median, somewhat narrower blue antehumeral, and quite narrow black humeral. Abdomen blue with black apical markings on S2–7, beginning as spot on S2, pointed and somewhat wider spots on S3–5, covering two-thirds of S6 and almost all of S7; S8–9 blue, S10 black above. *Female*: Polymorphic, brown or blue. Eyes tan to greenish tan with brown cap; may be very pale in Southwest, almost whitish. Thorax as in male. Abdomen black above, covering all segments, but typical bluet torpedo pattern evident.

Identification Largely blue middle segments should distinguish male from other bluets with more black. Most, including the same-sized **Boreal** and **Northern**, best distinguished by distinctively shaped cerci clearly longer than paraprocts (reverse in other two) and black marking on S2, a spot touching rear of segment (bar separated from rear in other two, where they overlap with **Familiar**). In areas where most overlap takes place, **Familiar** usually has distinctly smaller postocular spots. **Familiar** larger than **Hagen's** and **Marsh** that overlap with it widely but colored very similarly. **Familiar** often has relatively wider antehumeral and narrower humeral stripes than the two smaller species, thus showing more blue in front view of thorax, but they overlap. In Texas overlaps with **Big Bluet** (usually larger, with more black on middle segments and divided median stripe on thorax as well as barely perceptible cerci) and very rare **Atlantic** (can be very similar; look at cerci). Female much like those of other blue bluets with black on entire abdomen, will have to be identified in hand. Fortunately, tandem pairs are common!

Natural History Extremely abundant at many sites. Males spend much of day at water but do not visit each day while reproductively mature. Males arrive at water in late morning and

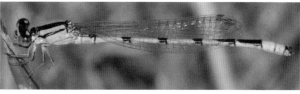

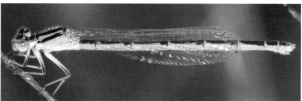

30
Familiar Bluet
male—Hidalgo Co.,
TX, June 2005; blue
female—Burnet
Co., TX, July 2004

remain at peak numbers until midafternoon. Mating peaks by midday. Copulation lasts about 20 min. Tandem pairs oviposit in soft plant tissues of all kinds at and below surface, up to a few inches above water in shrub stems. Pair engages in exploratory flights in which female tests substrate and often lays some eggs, pair then moving again. These flights are surprisingly lengthy, averaging 34 min and moving substantial distances. Then, at some appropriate site, female backs underwater to continue ovipositing, and male releases her before his head gets wet. Female may also back down stem, then lay eggs as she ascends. Underwater oviposition bouts last about 10–30 min, female then popping to surface. Male typically waits for her there and grabs her again, pair moving to new site. Females often reject attempts at second tandems, whether by first mate or another male. No underwater oviposition at some sites, probably because of lack of appropriate plants. Entire course of oviposition may last 2 hr. Both sexes average just over one mating in lifetime.

Habitat Lakes, ponds, open marshes, and slow streams, even margins of rivers, as long as emergent vegetation is present. Broad habitat tolerance, including freshly created wetlands, may explain widespread abundance.

Flight Season OR May–Oct, CA Feb–Dec, MT May–Aug, AZ all year, NM Feb–Nov, NE Apr–Oct, TX all year.

Distribution Ranges south through Central America at increasingly higher elevations to Venezuela. Also widespread throughout eastern United States and far southern Canada, along coast to Newfoundland.

Comments Has hybridized with Tule Bluet.

31 Tule Bluet *Enallagma carunculatum* TL 26–37, HW 14–22

Description Common bluet of marshes with much black on abdomen. Postocular spots usually form narrow dumbbell. *Male*: Eyes blue with black cap. Median stripe broad, humeral stripe narrow, antehumeral wider than humeral. Abdomen with large black apical spot on S2; apical black mark on S3–6 somewhat pointed forward and occupying more than half of segment, longest on S6; most of S7 black; S8–9 blue, S10 black above. Unusual variants have most segments almost entirely black, thus abdomen looking black with blue tip. *Female*: Polymorphic, brown or blue; blue may be absent from middle segments, which may be golden-yellow. Eyes tan with brown cap. Thorax as in male. Abdomen black above, black expanded apically and narrowed basally, extending forward in point on S3–8, with narrow but varying pale ring at base of each. From above, black on each segment shaped like fat candle in small candle holder.

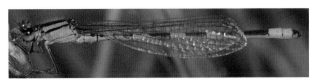

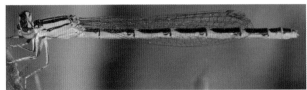

31
Tule Bluet
male—Coconino
Co., AZ, September
2005; brown female,
blue female—
Benton Co., WA,
August 2005

Identification Male easily distinguished from most coexisting bluets (**Boreal, Familiar, Hagen's, Marsh, Northern**) by much black on all middle abdominal segments, exceeding blue on all segments. **Alkali** and **River** quite similar but usually more blue than black on S3–4, also more likely to have postocular spots not bridged. Like **Arroyo** but latter a bit smaller and more slender, a bit more black on middle segments, and longer cerci that can be seen at close range. Female looks like other bluets with black covering most of abdomen (**Arroyo, Familiar**, and **River** and slightly smaller **Hagen's** and **Marsh**), could be distinguished only by looking at mesostigmal plates in hand. In parts of range, occurs only with **Northern** and/or **Boreal Bluets**, then distinguished by S8 almost all black (other two usually with much pale color on S8).

Natural History Typical bluet in habits, with males and tandem pairs perching all over shore and emergent vegetation and flying well out over open water, even of very large lakes. Very common in optimal habitats. On average, copulation lasts 21 min, exploration for oviposition sites 11 min, surface oviposition 58 min, and underwater oviposition 20 min. Pairs and solo females oviposit in standing bulrushes and presumably other plants. Mayflies and small flies common prey.

Habitat Marshy and open lakes, ponds, and slow streams and rivers, occurring commonly at larger and more eutrophic lakes than some of its relatives. In Pacific Northwest, often common at lakes and ponds with tall cattail growth where other bluets are lacking. Also found in some saline lakes, but not as characteristic of extreme environments as Alkali Bluet.

Flight Season BC May–Sep, AB Jun–Sep, WA Apr–Nov, OR Apr–Nov, CA Feb–Oct, MT Jun–Oct, AZ Apr–Nov, NM Apr–Dec, NE May–Sep, TX Jul–Sep.

Distribution Ranges south in Mexico to northern Baja California. Also in northeastern North America from Ontario and Missouri to Nova Scotia and Maryland.

Comments Has hybridized with Familiar and River Bluets.

Description Small bluet of lakes and ponds with brown eyes and mostly black abdomen. Postocular spots small, isolated. Smallest individuals in east Texas. *Male*: Eyes brown over tan to greenish, small streak of blue across front. Thorax with broad median and humeral stripes, antehumeral no wider than humeral. Abdomen mostly black with blue at base forming wavy line down S2 (because black both on top and sides of segment) onto base of

S3; in some, line interrupted by black on top and sides in contact; otherwise S8–9 blue but with black low on sides. Faintly indicated basal rings on S3–7 either blue or whitish. *Female*: Eyes brown over tan. Thorax as male. Abdomen entirely black above, black extending well down sides of segments. S8 above with paired blue squarish marks or entirely blue. Surprisingly, females during copulation and oviposition may replace blue markings with dull gray-brown.

Identification Several aspects of male coloration unusual among blue bluets. Blue streak on eyes otherwise brown above is unique, as other blue bluets have mostly blue eyes. Entirely black S10 is shared with **Turquoise Bluet** but no other blue bluet. Perhaps most like **Turquoise** but smaller, less elongate, and typically on ponds rather than streams. **Turquoise** has blue eyes, joined postocular spots, no black on sides of S2. Male **Slender Bluet** differs in same ways as **Turquoise** but also has narrow humeral stripes and more blue on abdomen base, thorax looking overall bluer. Perhaps most like **Lilypad Forktail** in habitat and habits, as well as black body with blue thorax and abdomen tip, but **Forktail** has mostly blue eyes, a black ring around blue abdomen base, and more blue on abdomen tip (a bit on both S7 and S10). Female **Skimming Bluet** differs from other similar-looking damselflies in its range with black abdomens in having pale markings only on top of S8. Closest might be young andromorph **Rambur's Forktail**, still with blue thorax, but in that species entire S8 is blue, not just top. Although not known to overlap in range, also similar is female **Neotropical Bluet**, rarely as bright blue as **Skimming** and with narrower median and humeral stripes and usually blue spots on S9.

Natural History Males perch on lily pads or emergent grasses and sedges and fly quickly from one perch to another, low over water as name indicates. Often hover over open water, unlike Lilypad Forktail. Pairs and solo females oviposit in floating debris and vegetation.

Habitat Lakes, ponds, and slow streams with clear water, usually with abundant beds of water lilies or other floating vegetation.

Flight Season TX Mar–Sep.

Distribution Ranges east across southern Canada and United States to New Brunswick and north Florida.

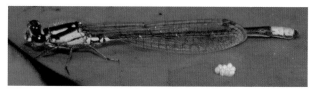

32
Skimming Bluet
male—Taylor Co.,
GA, May 2006;
female—Marquette
Co., WI, July 2005,
Mike Reese

33 Azure Bluet *Enallagma aspersum*   TL 27–34, HW 15–20

Description Slender bluet of ponds with more blue on abdomen tip than in other species. Postocular spots very large, joining blue color on back of head. *Male*: Eyes blue, black cap reduced. Black and blue thorax with moderate median stripe and narrow humeral stripe. Abdomen mostly black above with blue on sides of base and extensive at tip: S1 blue, S2 blue with black apical spot, S3 blue at base and on sides, S4–6 with narrow basal ring, S7–10 blue with black basal spot on S7, S10 with much black above. *Female*: Eyes black over brown.

Body black and blue, no heteromorph. Thorax blue with wide median and relatively narrow humeral stripes, antehumeral stripe slightly wider than humeral. Abdomen mostly black above, conspicuously blue on sides of S1–3, entire base of S7 (with narrow black line down segment), and pair of blue basal spots on S8.

Identification Male distinguished by mostly black abdomen with long blue tip, including distal part of S7, and extensive blue at base as well. Other slender bluets with mostly black abdomen (**Skimming**, **Stream**, **Turquoise**) have less blue on base and only S8–9 at most blue at tip. Female **Azure's** pattern of big blue spot on base of S7 unique in its range.

Natural History Males perch on edge vegetation and fly out at intervals. Copulation lasts 10–20 min. Pairs oviposit on surface vegetation. Some females climb down stem headfirst, released by male immediately. Submerged solo oviposition, sometimes rather deep, lasts 5–25 min. Blue colors in female turn gray during oviposition. Males remain nearby and seize females when they emerge, but females refuse to copulate again.

Habitat Ponds and lakes of all sizes with much emergent vegetation; may be especially common at small boggy ponds. Usually restricted to fishless waters.

Flight Season NE May–Sep, TX May–Aug.

Distribution Also widespread in far southeastern Canada east to Nova Scotia and eastern United States south to Arkansas and South Carolina.

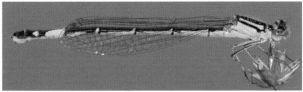

33
Azure Bluet
male—Tuscarawas
Co., OH, July 2005,
Greg Lasley;
female—Dade
Co., GA, May 2003,
Giff Beaton

Description Largest blue bluet in region, most common on large lakes and in coastal region. Postocular spots form very narrow transverse line. *Male*: Eyes blue with small black cap. Thorax typical of blue bluets but dorsal carina usually blue, bisecting wide black median stripe. Abdomen blue with black markings, a large apical spot on S2, apical spots (almost rings) on S3–6 narrowly pointed toward front, top of S7 except narrow basal ring, and top of S10. *Female*: Polymorphic, brown or blue. Eyes tan, darker above. Thorax as male, also with pale line dividing median stripe. Abdomen entirely black above with usual torpedo pattern on S3–7.

Identification Typical individuals distinctly larger than other bluets, with dark markings of middle abdominal segments often more extensively drawn out in long point toward front than in **Atlantic** and **Familiar**, only other blue bluets in range in West. Unfortunately, some individuals smaller, some with less black on abdomen, so capture and scrutiny of appendages may be necessary for certainty. **Familiar** easily distinguished by long cerci equaling S10 in length (barely visible in **Atlantic** and **Big**). Female resembles females of those two

34
Big Bluet male—
Wakulla Co., FL,
April 2005; brown
female—Liberty Co.,
FL, April 2005

species, perhaps separable by size and pale median carina of thorax. **Familiar** females more likely to have isolated postocular spots, **Big** a fine transverse line. Otherwise, mesostigmal laminae must be examined in hand.

Natural History Males perch on grass and sedge in beds of same. Females oviposit under water, head down, while male guards above.

Habitat Large sandy lakes and lower reaches of rivers, even extending into brackish estuaries. Most common bluet at some large water bodies near coast.

Flight Season TX Mar–Sep.

Distribution Ranges south on Gulf coast of Mexico to Tamaulipas and east along Gulf and Atlantic coast, sometimes farther inland, to Maine.

35 Alkali Bluet *Enallagma clausum* TL 28–37, HW 16–23

Description Large bluet characteristic of alkaline lakes but occurring in other habitats. Postocular spots large, forming dumbbell. *Male*: Eyes blue with black cap. Wide median and antehumeral, narrow humeral stripes. Abdomen with much black: subapical spot on S2 almost touching end of segment, apical markings on S3–7 reaching half-length of segment on S3–5, more than half on S6, and four-fifths of S7, typically increasing regularly to rear; S8–9 blue, S10 black on top. *Female*: Polymorphic, brown or blue. Eyes tan or pale greenish with brown cap. Abdomen typical of blue bluets, S2 with anchor-shaped marking usually extending entire length of segment; torpedo markings of equal length on S3–7, with pale basal  rings prominent and S8 entirely pale (rarely with black middorsal line); black on S9–10 slightly reduced in comparison with other species.

Identification Male marked most like **River Bluet** in having S3–5 almost equally blue and black; **Boreal**, **Familiar**, and **Northern** have less black, **Arroyo** and **Tule** more black. At close range in side view, male **Alkali** distinguished from **Familiar** and **River** by short cerci, from **Boreal** and **Northern** by shorter, more curved paraprocts. Female has more pale color on abdomen than all others of this group, S8 entirely pale, and much pale color at base of all middle segments. Female **Northern** and **Boreal** usually with some black on S8, other female bluets with S8 entirely black above. Very narrow humeral stripe, often widened into dot near upper end as in some dancers, distinctive except from **River Bluet**, female of which usually has mostly dark S8. In female **Alkali**, lower end of middorsal carina on thorax constricted and slightly elevated, unique among this group.

Natural History Males perch on lakeside rocks and grasses or fly over water; perching on bare ground more likely than in other bluets. Perhaps in absence of competition, reaches tremendous densities in preferred lakes, sufficiently abundant in some areas to produce

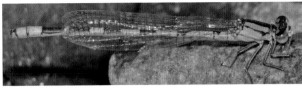

35
Alkali Bluet
male—Kittitas Co.,
WA, September 2006,
Netta Smith; blue
female—Okanogan
Co., WA, August 2003,
Netta Smith

bluish cast to lake surface, even hundreds of tandem pairs visible at once. Females and pairs oviposit directly on floating alga mats and alga-covered rocks in lakes with no emergent vegetation, both at water surface and well beneath it, at least to a foot deep.

Habitat In West, occupies alkaline lakes and ponds, some of them too alkaline for any other odonates. Also may be at edge of large, slow-flowing rivers. Emergent vegetation present or not. Farther east, may occur in large lakes of more normal chemistry.

Flight Season BC May–Aug, WA May–Sep, OR May–Aug, CA May–Sep, MT Jul–Aug, NM May–Jul, NE Jun–Aug.

Distribution Also from Minnesota and Iowa to southern Ontario and Quebec.

36 Northern Bluet   *Enallagma annexum*   TL 29–40, HW 17–24

Description Typical mostly blue bluet that can be very abundant at northern latitudes. Postocular spots large, forming dumbbell. *Male*: Eyes black over blue. Thorax with wide median stripe and narrow humeral stripe. Black subapical, often crescent-shaped, bar on S2, black rings on S3–5, widest on S5; most of S6–7 black; S8–9 blue, S10 black above. Populations in Alaska and Canada from Rockies west look darker because of more extensive black. They may have broader humeral stripes and considerably more black on abdomen (large spot on S2, most of S4–5 black) than populations to south and east. Also, U.S. Pacific coast populations average more black than those of interior. *Female*: Polymorphic, either brown or blue. Eyes dark brown over light brown. Thorax as male. Abdomen mostly black above; pale basal rings on S3–8, brown or blue often more extensive on S8, at most may fill segment; S9–10 black.

Identification Often occurs with **Boreal Bluet**, impossible to distinguish except in hand. Very similar to **Familiar Bluet**, but that species has cerci longer than paraprocts, reverse of that in **Northern**, and black marking on S2 usually larger, touching rear of segment (separated from it in **Boreal** and **Northern** where they overlap in range with **Familiar**). Also very similar to **Alkali Bluet**, but most individuals have less black on S5. **Hagen's** and **Marsh Bluets** colored very similarly but smaller and with smaller postocular spots, most noticeable when comparison possible and especially evident in females. In most areas of overlap (not in far North), bar on S2 in **Northern** distinguishes it from both smaller species, which have larger spot on that segment. Female much like female **Alkali** and **Boreal Bluets**, will have to be distinguished in hand by examination of mesostigmal plates. Differs from **River** and **Tule Bluets** by usual presence of pale color on top of S8, but at least in California, some females with all black S8.

Natural History Typical of bluets, males at water in large numbers, perched in all vegetation types, although more commonly at edges rather than inside dense stands, and flying over open water. Mating usually takes place in sunny clearings near water, tandem pairs

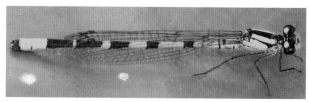

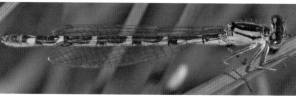

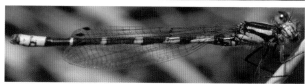

36
Northern Bluet
male—King Co.,
WA, June 2004; blue
female—Okanogan
Co., WA, June
2005; very dark
male—Fairbanks, AK,
July 2002, Wim Arp

then flying to water. Copulation lasts 10–27 min. Pairs oviposit flat on floating vegetation, not up on emergents, and vascular plants are preferred over algae. Much exploratory oviposition by pair, followed by period of actual egg laying in tandem. Pair eventually separates, and female moves below surface headfirst. Male remains guarding for some time but often gives up. Submerged oviposition up to 90 min.

Habitat Marshy and somewhat open lakes, ponds, and slow streams in West, perhaps only streams in California. Not limited to fish-free water bodies as in East.

Flight Season YT Jun–Aug, BC Apr–Sep, AB Jun–Aug, WA May–Oct, OR May–Oct, CA Apr–Nov, MT May–Sep, AZ Jun–Aug, NM May–Oct, NE May–Aug.

Distribution Ranges south in Mexico to Baja California Sur, also widespread in Northeast to Newfoundland and south to Indiana and West Virginia.

Comments Known in all previous North American literature as *Enallagma cyathigerum*. That name is now considered to be restricted to the Eurasian species that looks much like *annexum* but is genetically distinct from it.

Bluets - male appendages

Northern   Boreal

---

37 Boreal Bluet   *Enallagma boreale*   **TL 28–36, HW 17–22**

Description Typical mostly blue bluet, abundant at northern latitudes. Postocular spots large, forming dumbbell. *Male*: Eyes blue with black cap. Thorax with wide median stripe and narrow humeral stripe. Abdomen with black subapical bar on S2 (larger in eastern populations), black rings on S3–5, widest on S5; most of S6–7 black; S8–9 blue, S10 black above. Northern populations, especially uplands of Alaska and Canada, can have much more black, with broad humeral stripe, narrow antehumeral stripe sometimes divided, large spot on S2 touching rear margin, and extensive black on middle abdominal segments (more than half of S4–5). At the most extreme, some populations have prominent irregular black markings on sides of S8. *Female*: Polymorphic, either brown or blue. Eyes dark brown over light brown. Thorax marked as in male. Abdomen mostly black above; pale basal rings on S3–8, brown or blue usually more extensive on S8, at most may fill segment; S9–10 black.

Identification Indistinguishable from **Northern Bluet** except by structural differences apparent with magnification. This is one pair of species that in the field will have to be lumped (they have been called "borthern" or "nobo" bluets). Also very much like **Familiar Bluet**, with which it overlaps over a relatively small region; **Familiar** has cerci longer than paraprocts, reverse of that in **Boreal**; larger postocular spots; and black marking on S2 larger, touching rear of segment. **Hagen's** and **Marsh Bluets**, which often occur with **Boreal**, are slightly smaller, usually have slightly smaller postocular spots, and also differ by larger size of black spot on S2. **Boreal** also much like **Alkali Bluet**, but males have slightly less black on middle abdominal segments. Occurs with **Tule Bluet** in many areas, males easily distinguished by mostly blue middle segments and very slightly paler blue coloration. Fortunately, these other species not present in northern regions where **Boreal** shows much more black. Females differ from most female bluets by having extensive pale color on S8; other species so colored (**Alkali**, **Northern**) will have to be distinguished in hand by examining mesostigmal plates.

Natural History Males fly over open water and perch on emergent vegetation, at some places in prodigious numbers, producing bluish film over water surface. Pairs form at or away from water, sometimes superabundant along open corridors in woodland. On average, copulation lasts 23 min, exploration for oviposition sites 11 min, surface oviposition 67 min, and underwater oviposition 23 min. Pairs oviposit at water surface. Mayflies and small flies common prey. Mean life expectancy of reproductive adults 4 days, maximum 17 days.

Habitat Ponds, open marshes, and lake margins with much emergent vegetation. Ubiquitous in some areas, for example, common from boggy mountain lakes to alkaline prairie

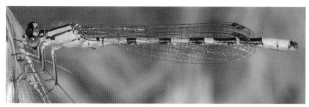

37
Boreal Bluet
male—Grant Co.,
WA, May 2005;
brown female—
Baker Co., OR,
August 2004; blue
female—Cowlitz
Co., WA, September
2006

ponds in Pacific Northwest. Boreal and Northern Bluets occur together very commonly in parts of the West.

Flight Season YT May–Jul, BC Apr–Sep, AB May–Aug, WA Apr–Oct, OR May–Sep, CA Apr–Sep, MT May–Sep, AZ Jun–Oct, NM May–Sep, NE Jun–Jul.

Distribution Ranges south in uplands of Mexico to Durango; at higher elevations in southern part of range. Less common along immediate Pacific coast than Northern Bluet, more common in some arid regions. Also widely in eastern Canada and south in United States to Iowa and West Virginia.

38 Marsh Bluet *Enallagma ebrium* TL 28–34, HW 16–21

Description Small very common bluet of northern regions. Postocular spots large, forming dumbbell; rarely separated spots. *Male*: Eyes blue with black cap. Typical bluet thoracic stripes. Abdomen with black markings including large apical spot on S2, rings on S3–5 becoming slightly wider to rear, top of most of S6 and all of S7, and top of S10. *Female*: Polymorphic, brown or blue. Eyes brown over tan or pale greenish. Thorax as in male. Abdomen entirely black above, with torpedo markings.

Identification Most like **Hagen's Bluet** in both sexes, not distinguishable except in hand, although difference in male cerci may be visible with close-focus binoculars. Male colored also like **Boreal**, **Familiar**, and **Northern Bluets** with which it occurs, distinguishable by smaller size and, from **Boreal** and **Northern** in much of area of overlap, by spot on S2 usually touching rear of segment. Capture and scrutiny of appendages always advised if any doubt remains. Females of both **Hagen's** and **Marsh** distinguishable from all other bluets by wide mesostigmal plates visible in hand, then distinguished from each other by those plates in **Hagen's** raised well above thorax at their posterior end and those of **Marsh** lying flat.

Natural History Males can be abundant perching on shore vegetation or algal mats, tend to stay in vegetation rather than out over open water. In Northwest, females much less often seen than in larger Northern and Boreal Bluets, but pairs appear at midday and later. Females oviposit in tandem or solo on floating or emergent aquatic plants, also lone females descend below surface as far as a foot or more.

Habitat Lakes and ponds of all sizes, typically bordered with abundant emergent vegetation, in open or woodland. Less likely than Hagen's Bluet at bog ponds and other acid waters.

Flight Season YT Jun–Jul, BC May–Aug, AB Jun–Aug, WA Jun–Sep, MT Jun–Aug, NE Jun–Sep.

Distribution Ranges east to Newfoundland, south to Tennessee and West Virginia.

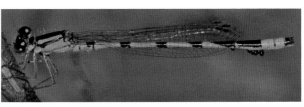

38
Marsh Bluet
male—Merrimack,
NH; June 2006; blue
female—Coos Co.,
NH, June 2006

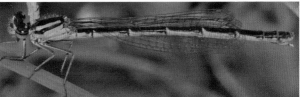

Description Small very common bluet of northern regions. Postocular spots large, forming dumbbell. *Male*: Eyes blue with black cap. Typical bluet thoracic markings. Abdomen blue with black apical spot on S2, black rings on S3–5, S6–7 mostly black above with anterior point penetrating basal ring; S10 also black above. *Female*: Polymorphic, tan, green, or blue. Eyes tan or greenish with brown cap. Thorax as male. Abdomen entirely black above, middle segments with torpedo markings.

Identification Marsh Bluet most similar, males distinguished only by shape of cerci, which can be seen at close range. Differs from other similarly colored bluets in exactly same way as **Marsh Bluet** (see that species). Still other bluets that occur with **Hagen's** differ by having more black on midabdomen. Female identical to **Marsh** except for mesostigmal plates, to be examined in hand. In **Hagen's**, wide plates tilt upward toward rear, well elevated above thorax, whereas in **Marsh**, plates are flat. Also check other coexisting bluets with almost all of dorsal surface of abdomen black: **Familiar**, **River**, **Taiga**, and **Tule**.

Natural History Males in marsh vegetation and over open water, resting on stems or algal mats. Mating takes place near water, and tandem pairs appear at water in afternoon. Copulation averages 22 min, then pairs in tandem 58 min before oviposition. Females oviposit in floating dead and live plant stems, submerging and crawling around underwater stems for about a half-hour. Males usually remain above their submerged mates, but other males often grab them when they surface, leading to remating and another bout of oviposition. Males act as lifeguards, pulling floating females from water (of course to mate with them).

Habitat Open marshes, lakes, and ponds with abundant emergent vegetation, including bog ponds. Often most common species at large northern lakes.

Flight Season BC Jun–Jul, AB Jun–Aug, MT Jun–Aug, NE Jun–Sep.

Distribution Ranges east to Nova Scotia and south to Indiana and Maryland, in mountains to Georgia.

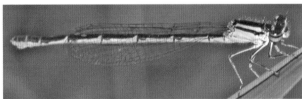

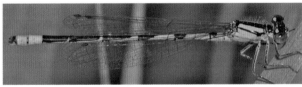

39
Hagen's Bluet male—Horsefly, BC, July 2006, Netta Smith; green female—St. Louis Co., MN, August 2004, Jim Bangma; blue female—Androscoggin Co., ME, July 2006

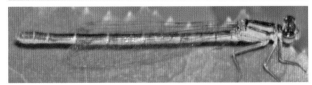

Description Violet bluet of Texas streams. Postocular spots large, forming dumbbell. *Male*: Eyes dark brown or violet over pale greenish or tan. Thorax violet in front, blue low on sides, with narrow median and humeral stripes. S1–3 violet, S2 with prominent U-shaped black mark, and S3 with black apical ring. Middle segments black with very narrow pale blue basal ring; S8–9 mostly blue-violet but 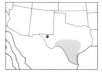 divided by narrow black apical ring of S8, also some black low on sides of each segment; S10 black. *Female*: Polymorphic, either dull greenish (sometimes with blue intermixed) or blue. Eyes tan with brown horizontal stripes above. Black lines on thorax as in male. Abdomen black above except S1 and square spot on top of S8 pale; S9–10 mostly black from above but with some pale markings.

Identification No other bluet like male, with violet thorax and abdomen base, black abdomen with largely blue tip. Violet-colored dancers all have mostly violet abdomen and blue S10 as well as S8–9. Unique among southwestern bluets in interrupted blue on S8–9 and U-shaped black marking on S2. Female one of few damselflies in its range with conspicuously pale S8. Female **Skimming Bluet** similar but smaller, on lakes, and barely if at all overlap in range.

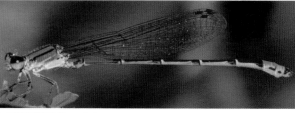

40.1
Neotropical Bluet male—Sonora, Mexico, August 2006, Netta Smith; green female—Kinney Co., TX, July 2004

40.2
Neotropical Bluet pair with blue female—Kinney Co., TX, July 2004

Others are female **Rambur's Forktail**, which has all of S8 and sides of S9–10 blue. Female **Painted Damsel** also with conspicuous pale area on abdomen, but on S7 and not S8. Similar to female **Claw-tipped Bluet** but no known overlap in the United States (see that species).

Natural History Males commonly perch at tips of leaves overhanging water, alternating perching with hovering for long periods over water. Pairs oviposit in emergent vegetation.

Habitat Typically on pools of slow-flowing streams and rivers, often with grass beds along shore, may also be in areas of swift current. Also on well-vegetated seeps.

Flight Season TX Apr–Dec.

Distribution Ranges south mostly in lowlands to Argentina.

41 Baja Bluet *Enallagma eiseni* TL 27–34, HW 13–19

Description Very distinctive bluet of southwestern border, with black on front of middle abdominal segments. Smaller in Arizona than in Baja California. *Male*: Eyes mostly blue, with small black cap. Postocular spots form narrow dumbbell. Typical bluet thoracic stripes, but median stripe prominently divided. Abdomen blue with black markings including dorsal stripe on S2, black apical rings and rearward-pointing "spearpoints" on S3–6; S7 all black above, S10 with fine black middorsal stripe. *Female*: May be polymorphic, either brown or blue on thorax and abdomen; blue females definitely known. Abdomen typical bluet, black above, apical ring on S3–7; S8 mostly pale except median dorsal line, S9 mostly black but pale apical lateral spots.

Identification Males unlike any other species, with large basal black markings on middle abdominal segments. Because of amount of black, closest to **Arroyo Bluet** at first sight. Might be mistaken for mostly blue DANCER, also because S10 with so much blue, but dorsal black markings quite different from lateral black markings of DANCERS. Females distinguished from all co-occurring bluets by mostly pale S8 with dark median line. Most similar to female **Claw-tipped Bluet**, but that species has S8 blue above and black below.

Natural History Males on waterside emergent vegetation, at times in numbers; both sexes also wander well away from water into desert scrub. Breeding behavior unrecorded.

Habitat Shallow rocky and sandy streams in arid country, mostly on pools.

Flight Season CA Jun, AZ Jun–Oct.

Distribution Found so far in United States only in Tia Juana River Valley in far southern California and at Quitobaquito Springs (closed to public), Arizona, also just across border in Sonora. Previously thought to be endemic to Baja California.

41
Baja Bluet male,
female—Pima Co.,
AZ, October 2006,
Doug Danforth

Description Slender bluet of streams with mostly black abdomen and distinctly greenish tinge on head and thorax. Postocular spots form narrow dumbbell. *Male*: Eyes turquoise with black cap. Thorax greenish-blue with usual bluet stripes, humeral distinctly wider than antehumeral. Abdomen black above S2–7, black extending on point almost full length of S8. Sides of S2–3 conspicuously blue or tinged with green; blue basal rings on S4–7, sides of S8 and S10 and all of S9 blue. *Female*: Polymorphic, either green or blue on thorax. Eyes greenish with black cap. Thorax as in male, but humeral stripe often divided lengthwise by brown stripe, losing distinctness of stripe in extreme cases. Abdomen black above, conspicuously greenish or blue on sides, S9–10 blue with pair of black basal spots on S9. Paler individuals with S8 blue at tip, S9–10 entirely blue.

Identification Other male bluets with abdomen mostly black above include **Azure**, **Skimming**, **Slender**, and **Turquoise**, last one most likely in habitat of **Stream**. **Turquoise** most similar but with more blue at abdomen tip and never greenish on head and thorax. **Slender** with narrow humeral stripes, also S8–9 entirely blue as in **Turquoise**. **Azure** with much more blue on thorax and abdomen base and **Skimming** smaller, again with S8–9 entirely blue. Female's greenish head, thorax, and abdomen base together with blue abdomen tip distinguish her from most other damselflies. See **Rainbow** and **Turquoise Bluets**.

Natural History Males hover a foot over water of pools for long periods, then perch on stems of herbaceous plants growing in or extending over water. Pairs often common. Mating takes place at water or in nearby woodland. Copulation lengthy for bluet, lasting 55–119 min, then pair remains in tandem for some time after that before oviposition begins. Egg laying may begin before or after pair separates, and males may accompany females for part or all of oviposition, which can be completely underwater for 15–31 min. Pairs and single females also oviposit at surface on beds of submergent vegetation.

Habitat Medium streams to large rivers with slow to moderate current, often with much water-willow. Also vegetated lake shores in northern part of range.

Flight Season NE May–Sep, TX Apr–Sep.

Distribution Ranges south in lowlands of eastern Mexico to Hidalgo, also across much of East to Nova Scotia, not on southern Coastal Plain.

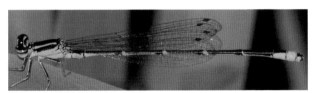

42
Stream Bluet
male, green
female—Bremer
Co., IA, July 2004;
Blue female—
Buchanan Co., IA,
July 2004

Description Very long, slender bluet with scarcely any black on head, thorax, and abdomen tip. Postocular spots large, head almost entirely pale. *Male*: Eyes blue over green. Thorax with median and humeral stripes very narrow, at quick glance appearing all blue. Abdomen black above, blue on sides of S2 and base of S3, distal third of S7 and all of S8–10 blue. *Female*: Eyes tan, darker above. Thorax as male. Abdomen black above, scarcely any basal rings evident; S9–10 blue.

Identification Size and slenderness set it off immediately. No similar species in West, although check **Slender Bluet** as most similar. Female might be mistaken for rather slender female **Furtive Forktail** of same habitat, slightly smaller and with front of thorax and tip of abdomen black.

Natural History Males perch in and fly through shrubby thickets and tall grasses, higher than most other damselflies and moving leisurely along with much hovering. Long abdomen in damselflies probably facilitates hovering. Pairs in tandem move through grass and shrubs at same levels, impressively long as a pair, then drop to lower vegetation to oviposit. Pair backs down herbaceous stem, then male releases female and she may go well below surface.

Habitat Shrubby borders of wooded lakes and swamps. Always associated with woodland.

Flight Season TX May–Jun.

Distribution Also scattered through Southeast, most common on Coastal Plain.

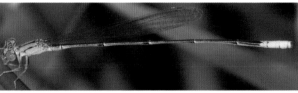

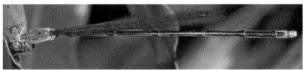

43
Attenuated Bluet
male, female—San
Jacinto Co., TX, May
2005

Description Stream-dwelling bluet unmistakable because of orange face, greenish thorax, and blue abdomen. Postocular spots form narrow interrupted line. *Male*: Eyes orange in front, green behind, yellow below. Face orange, postocular spots blue. Thorax bluish-green, median and humeral stripes wide, antehumeral stripe quite narrow and yellowish or chartreuse; median stripe may be divided by very narrow pale stripe. Legs mostly yellowish. Abdomen black above, blue on sides of S1–3, light green on sides of S4–7, narrow blue basal rings on S3–7; black stripe on top of S8 narrows to rear, shows bright blue sides, S9 all blue, S10 black above. *Female*: Eyes brown above, yellow below. Thorax, legs, and sides of abdomen yellowish to yellow-green. Abdomen black above with terminal blue rings on S7–8, much of S9–10 bluish.

Identification No other western damselfly colored like male. Even with rear view, yellowish legs give it away. Could easily be mistaken for FORKTAIL because of greenish thorax and mostly black abdomen, but orange face and legs peg it as this species. Female much like

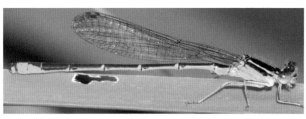

44
Rainbow Bluet male,
female—Bremer Co.,
IA, July 2004

female **Stream Bluet** but yellower, especially on face, and less blue on abdomen tip than in that species (**Rainbow** has top of S9 mostly black, **Stream** mostly blue).

Natural History Males perch on stems at outer edge of vegetation at edges of pools. Pairs oviposit in grass at water surface; female reported to descend below surface.

Habitat Slow streams and rivers lined with beds of emergent vegetation, typically in open country. Also at ponds along stream courses.

Flight Season MT Jun–Jul, NE Jun–Sep.

Distribution Also east to Vermont, south to Kentucky and West Virginia.

45 Turquoise Bluet *Enallagma divagans*   TL 27–36, HW 17–22

Description Slender blue bluet of streams with prominent blue abdomen tip. Postocular spots form narrow dumbbell. *Male*: Eyes blue with small black cap and horizontal dark stripe. Thorax typical bluet but with antehumeral and humeral stripes about same width. Abdomen black with blue on S1, sides of S2, and sides of base of S3; S8–9 entirely blue, S10 entirely black. *Female*: Eyes brown above, tan below. Thorax blue like male or blue mixed with dull greenish, much of humeral stripe may be entirely obscured by brown. Abdomen black above, blue along sides, variable blue at tip (S8 all black or with extensive blue tip; S9 all blue or with basal black spots; S10 all blue).

Identification Male most like **Stream Bluet** and often found in same habitat, but S8–9 all blue, whereas **Stream** has black covering most or all of top of S8 (some eastern populations of **Turquoise** may show some black). Blue usually visible on side of S10 in **Stream** but not in **Turquoise**. **Stream** also shows narrow blue abdominal rings, **Turquoise** abdomen entirely black above. **Turquoise** quite blue, but head and thorax of **Stream** may show greenish tinge. Also much like **Slender Bluet**, which differs in having more extensive blue coloration: larger postocular spots, narrower black humeral stripe, more blue at abdomen base.

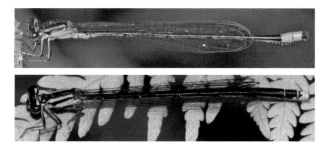

45
Turquoise Bluet
male—Liberty
Co., FL, April 2005;
female—Middlesex
Co., MA, June 2004,
Tom Murray

Female much like female **Stream Bluet** but brown on humeral stripe when present covers most of stripe rather than dividing down its length as in **Stream**. **Turquoise** female usually blue rather than green as typical of female **Stream Bluet**, and eyes usually brown (**Stream** with green to turquoise eyes). Immature females of orange bluet group (**Florida**, **Golden**, **Orange**, **Vesper**), with light blue thorax and blue abdomen tip, look much like female **Turquoise**, but all have less blue at abdomen tip, much black on top of S9.

Natural History Males perch low at stream margins or hover over water a few inches up for long times, reminiscent of threadtail. Pairs oviposit in tandem in submergent vegetation at surface. Female may back down stems by herself for up to 30 min, with male waiting above.

Habitat Small woodland streams with moderate current, also swampy areas with slight current and wooded shores of large lakes and reservoirs.

Flight Season TX Apr–May.

Distribution Ranges east to southern Maine and northern Florida.

46 Slender Bluet *Enallagma traviatum* TL 29–32, HW 15–19

Description Slender bluet (of course!) of pond habitats with rather fine thoracic stripes. Postocular spots large, head largely pale with black lines dividing pale areas. More blue on prothorax than related species. *Male*: Eyes blue over blue-green, no black cap. Thorax with relatively limited black markings; median stripe only moderate width and, in southern populations, may be split lengthwise  by pale stripe and so reduced as to be almost lacking; humeral stripe quite narrow. Abdomen black above, prominently blue on sides of S2–3 with full-length black mark on S2 somewhat buoy-shaped (becoming wider toward rear then suddenly contracted); narrow basal blue rings on S3–7, S8–9 blue, and S10 black above, blue below. *Female*: Eyes greenish or bluish over tan, rather colorfully striped. Thorax blue like male but dark stripes may be even more reduced, greenish-brown rather than black. Abdomen pale bluish on sides, black above; black marking on S2 as in male and blue basal rings and typical torpedoes on S3–7; S8–10 all blue except black basal spot on S8 of variable size, pointed behind. Some females strongly greenish on thorax and all but tip of abdomen (polymorphic).

Identification Limited black on thorax of male distinctive, somewhat like **Attenuated Bluet**, but that species found in swamps and is longer and more slender with more blue on abdomen tip, including part of S7. Superficially like **Azure Bluet**, which has more black on head and much more blue at both base and tip of abdomen. Female also like **Azure**, but both sexes of **Slender** easily distinguished by narrow humeral stripe. **Turquoise Bluet**, of different habitat, has darker head and eyes, wider humeral stripe, and no blue on S10.

Natural History Males perch in shore vegetation or fly and hover well out over water. Tandem pairs also hover for lengthy periods over water. Copulation lengthy and may be broken and resumed. Female oviposits in tandem with male supported in air or solo, at surface or submerged.

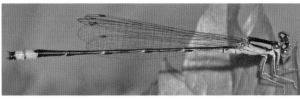

46
Slender Bluet male—
Floyd Co., GA, May
2006; female—Essex
Co., MA, July 2004,
Tom Murray

Habitat Lakes and ponds, open or with abundant vegetation.
Flight Season NE Jun–Aug, TX May.
Distribution Also east to New Hampshire and northwestern Florida.
Comments All western populations are subspecies *E. t. westfalli*.

Description Tiny bluet with doubled thoracic stripes. Postocular spots form narrow interrupted line, dashes at ends may be small, isolated. *Male*: Eyes blue. Both median and humeral stripes divided longitudinally into pair of narrow stripes. Abdomen with S1–2 with black stripe above, expanded subapically on S2; S3 with narrow black line ending in apical ring; S4–6 with apical ring sharply pointed at front; S7 mostly black above but also pointed in front; S8–9 entirely blue, S10 black above. *Female*: Eyes tan. Thorax tan with stripes as in male. Black along top of all abdominal segments except S9–10  (sometimes tip of S8), which are blue. Sides of middle segments also blue, brighter toward rear.

Identification Besides very small size for bluet, doubled median and humeral stripes in both sexes unique among group, thorax looking finely instead of coarsely striped. In males, entirely blue eyes, with no black "cap," unusual among bluets, although shared by slender, pale eastern bluets and many DANCERS. Narrow black line on S3 on largely blue abdomen also unique. Females distinctive among typical bluets in gradually becoming bluer toward rear, with tan head, thorax, and abdomen base but increasing blue on sides and then tip of abdomen. No DANCER colored exactly like this either. Male **Double-striped** a bit darker blue than **Familiar Bluet**, with which it often occurs.

Natural History Males and tandem pairs can be very common, almost swarming among beds of vegetation and flying well out over open water. Both spend much time hovering over open water, characteristic of species. Females oviposit in tandem or solo after being released by male, on floating sedges or submergent vegetation they can reach from surface. Females also descend underwater, breaking tandem before male is submerged and, in at least some cases, male hovering above.

Habitat Ponds, lake margins, and slow streams with much emergent vegetation.
Flight Season CA May–Sep, AZ Mar–Oct, NM Mar–Nov, NE May–Sep, TX Jan–Nov.
Distribution Ranges south in Mexico to Baja California, Chihuahua, and Tamaulipas, and east to Connecticut and northwestern Florida.

47
Double-striped Bluet
male—Kinney
Co., TX, July 2004;
female—Caldwell
Co., TX, May 2005

Description Black and red bluet with purple thoracic stripes. Pos-tocular spots form very narrow transverse line. Smallest individ-uals in east Texas. *Male*: Eyes red, purplish above. Thorax red with purple-black stripes, both median and humeral wide; narrow dark stripe on side of thorax as well. Abdomen black with red-purple gloss at base and tip, orange to red-orange low on sides of S1–3 and S7–10. *Female*: Eyes orange above, yellowish below. Thorax marked as male but pale color orange instead of red. Abdomen entirely black above, dull pale orange on sides. Im-mature with pale colors pale blue.

Identification Nothing else like male, with red head and thorax and black abdomen, in this region. Female rather like female **Orange Bluet** but smaller, with much wider humeral stripe with metallic overtones and no orange at abdomen tip. **Citrine**, **Furtive**, and **Rambur's Forktails** with bright orange thorax show no thoracic stripes.

Natural History Males perch in dense grass or on floating leaves (often water lilies) or fly for extended periods over open water; very difficult to see over dark water. Pairs oviposit on water lilies (often on underside through hole in leaf) and other floating vegetation. One oviposition bout may take up to 30 min.

Habitat Sandy ponds with abundant shore vegetation, usually dense grass, and/or beds of floating vegetation, especially water lilies. In open or surrounded by woodland or forest. Also at slow streams edged by herbaceous vegetation.

Flight Season TX Apr–Sep.

Distribution Ranges east, mostly on Piedmont and Coastal Plain, to Delaware and northern Florida.

48
Burgundy Bluet
male, female—San
Jacinto Co., TX,
May 2005

Description Slender orange bluet with orange abdomen tip. Postocular spots form trans-verse line. *Male*: Eyes orange. Thorax orange with moderately wide black median and hu-meral stripes, pale antehumeral about as wide as humeral. Abdomen mostly black with orange sides of S1–2, very narrow basal rings on S3–7, low on sides of S8, all of S9, and sides of S10. *Female*: Polymorphic, may be orange, duller than male; blue (perhaps only younger individuals); or greenish. Eyes brown over greenish or tan. Thoracic stripes as in male, hu-meral may be a bit narrower. Abdomen black above, most of S9 and all of S10 pale. Imma-ture of both sexes with pale colors pale blue, gradually turns orange.

Pond Damsel Family **99**

Identification Male like nothing else in range. Somewhat like male **Vesper Bluet** but orange instead of yellow and blue. Colored most like male **Orange-striped Threadtail**, which has ruby-red eyes, less orange on thorax with mostly black sides, prominent white rings on middle abdominal segments, and very little orange at abdomen tip, also even more slender than bluet. Female should be distinguished from female **Vesper Bluet**, which has much narrower humeral stripe (see that species for more information), and female **Burgundy Bluet**, which has much wider humeral stripe  and no pale color on tip of abdomen. Blue female also like female **Turquoise Bluet** but less pale color at abdomen tip and no brown in humeral stripe. When faced with puzzling individual, recall that immature **Orange Bluet** of both sexes can be fairly bright blue.

Natural History Away from water, may be up in trees. Males and pairs in tandem usually at outer edge of grass and sedge beds, fly well out over open water when disturbed. Also common where water lilies cover surface. Although at times seen earlier, most common at water later in day, with late afternoon peak and activity at least until dark, averaging not quite as late as Vesper Bluet. Good example of a species for which midday surveys may be inadequate, although mating has been observed as early as midday. Pairs oviposit in water lilies and floating grass and on algal mats; female or even pair may submerge completely for 10–20 min. May take 12 days to mature, reproductive life as long as 3 weeks.

Habitat Lakes, ponds, and slow streams, quite broad habitat choice.

Flight Season NE May–Oct, TX all year.

Distribution Also eastward across southern Ontario to Nova Scotia and south to southern Florida.

49
Orange Bluet pair with orange female—Dade Co., GA, May 2006; male—Monroe Co., GA, June 2004, Giff Beaton; immature female—West Feliciana Par., LA, March 2003, Gayle and Jeanell Strickland

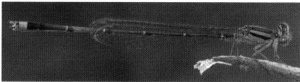

Description Slender bluet with yellow thorax and restricted blue abdomen tip, active in evening. Postocular spots narrow dumbbell or
transverse line. *Male:* Eyes yellow, more orange above. Thorax yellow
with narrow dark brown median stripe, sometimes with pale carina,
and quite narrow brown humeral stripe. Abdomen black above, sides
of S2–3 bright yellow and narrow pale yellowish basal ring on S3–7; S9
all blue, S10 black above, blue below. *Female:* Eyes orange-brown
over yellow. Surprisingly variable and probably polymorphic, thorax
usually greenish to pale turquoise but can be yellow to orange; per-

haps geographic variation also involved. Abdomen black above except sides of S9 and all of
S10 either blue as in male or more rarely yellow-orange; this variation may not be in all populations. Immature of both sexes with pale colors light blue.

Identification Combination of yellow-orange thorax and blue abdomen tip makes male and
blue-tipped female unique in region. Orange-tipped female like **Orange Bluet**, as are immatures of both sexes, but **Vesper** has narrower humeral stripe. At close range, it can be seen that
Vesper's antehumeral stripe pale to lower end, whereas that of **Orange** narrowly but clearly
bordered by black at lower end. Both species have pair of pits on top of prothorax where male
paraprocts contact female; those in **Vesper** right in middle, those in **Orange** a bit closer to
front. Immature females also look much like female **Turquoise Bluet**, although they only
rarely occur together; **Vesper's** narrow humeral stripe, never interrupted by brown, again
furnishes distinction. **Turquoise** also has more blue on tip, involving S8 as well as S9–10.

Natural History Spends most of day in vegetation, even up in trees, not far from water. Activity of males at water starts in midafternoon and can peak after sunset, later than most
other damselflies are flying. Males perch on low vegetation at water, much attracted to lily
pads; also commonly fly out over open water. Pairing and oviposition peak still later, and
tandem pairs and single females may continue to oviposit while completely dark! Females
may mate while still in immature color.

Habitat Typically lakes, usually where woodland available at or near shore.

Flight Season NE Jun–Oct, TX Mar–Oct.

Distribution Ranges east across most of eastern United States and far southeastern Canada
to Nova Scotia.

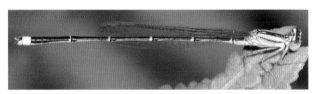

50
Vesper Bluet male,
blue female, orange
female—Polk Co.,
FL, April 2005

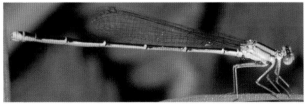

Bluets - male appendages

Prairie

Subarctic

Taiga

Arroyo

River

Claw-tipped

Atlantic

Familiar

Tule

Skimming

Azure

Big

Alkali

Northern

Boreal

Bluets - male appendages (*continued*)

Marsh

Hagen's

Neotropical

Baja

Stream

Attenuated

Rainbow

Turquoise

Slender

Double-striped

Burgundy

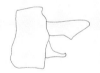

Orange

Vesper

Bluets and Yellowface - female mesostigmal plates

Prairie

Subarctic

Taiga

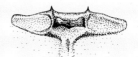

Arroyo

River

Claw-tipped

Atlantic

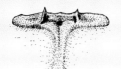

Familiar

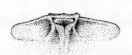

Tule

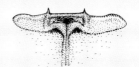

Skimming

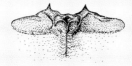

Azure

Big

Alkali

Northern

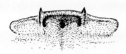

Boreal

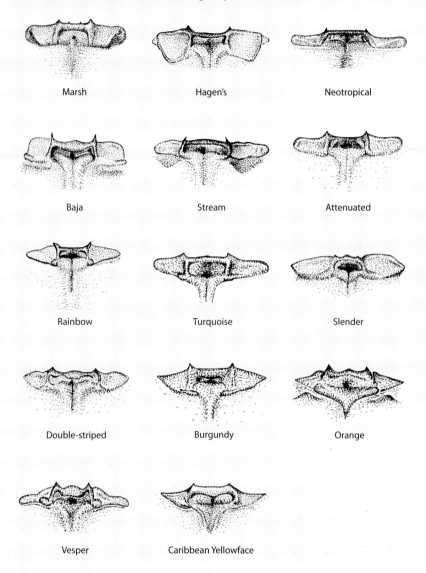

Marsh

Hagen's

Neotropical

Baja

Stream

Attenuated

Rainbow

Turquoise

Slender

Double-striped

Burgundy

Orange

Vesper

Caribbean Yellowface

This genus has two very similar species, the other one in western Mexico. They are much like *Enallagma* bluets and presumably are related to them, but the yellow face of the males easily distinguishes them. The very long cerci are longer than in most bluets and conspicuous at close range. World 2, NA 1, West 1.

51 Caribbean Yellowface *Neoerythromma cultellatum* TL 27–31, HW 13–16

Description Mostly black-bodied bluet-like pond damsel with bright yellow face. Postocular spots large, not quite connected by narrow bar that itself can be divided in middle. *Male*: Eyes yellow in front, black above, greenish below. Black humeral stripe broad, antehumeral stripe about as broad, often yellow-green; sides of thorax blue. Most individuals have prominent narrow black side stripe and spot between it and humeral stripe. Abdomen black, with blue S1–2, basal rings on S3–6, posterior sides of S7, and S8–9. Black V-shaped mark covers much of top of S2, and S10—unlike many bluets—entirely black. *Female*: Face dull greenish. Eyes brown over greenish-tan. Thorax as male, but antehumeral stripe blue. Abdomen black above, extensive blue on sides of S1–2, less extensive on S3–7, and extending over base as narrow ring; S8 blue with large black subapical spot or black with blue basal ring, S9 blue with black basal spot or black with blue only low on sides, S10 entirely blue.

Identification No other damselfly of our area is blue and black with a bright yellow face. Very long cerci of male also distinctive. Females are much like some species of BLUETS but show a more obvious black stripe on side of thorax, together with a small spot there, and have a mixture of blue and black on S8–9 different from any other. Fortunately overlaps in range with very few BLUETS, which are almost absent from tropical lowlands. Female **Familiar Bluet** most likely to be seen with it, but that has black on top of posterior abdominal segments and no stripe or spot behind humeral stripe on side of thorax.

Natural History Often rests flat on lily pads but also perches on emergent grasses and sedges at edge of water. Males usually associated with open water rather than dense vegetation. Commonly fly out over open water, much like many bluets, and stay just above surface. Pairs and single females oviposit on floating or emergent vegetation at water surface, often laying eggs in semicircle under floating leaves.

Habitat Large open ponds or canals, usually with narrow band of grasses or sedges at shore, but also ponds fairly densely vegetated with floating vegetation.

Flight Season TX Apr–Jan.

Distribution Ranges south in lowlands to Venezuela, also in far southern Florida and Greater Antilles.

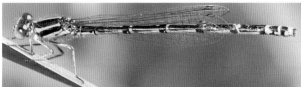

51
Caribbean Yellowface male—Cameron Co., TX, November 1999, Robert A. Behrstock; female—Tamaulipas, Mexico, September 2007, Marion Dobbs

Forktails are very small to small damselflies of worldwide distribution, including many oceanic islands. Often in dense vegetation, a few species also fly over open water. "Forktail" refers to a forked projection at the end of S10 in males of most species. Eyes in mature individuals appear with a dark cap, horizontal stripes are often but not always evident and are usually lacking in fully mature individuals. Males typically have a blue-tipped black abdomen, an easy distinction from the mostly blue bluets and dancers, but some species of both of those groups display a similar pattern. Dark abdominal markings are usually parallel-sided, covering top of the abdomen, but appear rather bluet-like in Desert Forktail. The combination of green thorax and blue-tipped abdomen is mostly typical of forktails. Females of many species are polymorphic, heteromorph orange at first and then becoming dark or pruinose, and andromorph brightly colored like the male; heteromorph is always more common. Forktails tend to stay near water even when immature, unlike most other damselflies. Copulating pairs are frequently seen in some species, infrequently in others; the sightings presumably are correlated with length of copulation, which varies greatly in forktails. Oviposition is usually into plant tissues at and just below water surface and not in tandem in most forktails, unlike most other pond damsels. This is probably because the sexes gather at the water together, and females have effective ways of discouraging male attention. A number of species pairs or trios include species very similar to one another. In all these cases, two of them are found together at some times and places. These include Desert and Rambur's; Pacific and Plains; Black-fronted and San Francisco; and Mexican, Western, and Eastern. In most cases, the species are distinguished only by looking at structural details in hand, but there are diagnostic field marks in a few cases. Males of five other species—Citrine, Fragile, Furtive, Lilypad, and Swift— are very distinctive among forktails and among all coexisting damselflies. However, females will always be a problem in parts of the West, and many will be identifiable only in the hand. World 66, NA 14, West 14.

Table 3 Forktail (*Ischnura*) Identification

	A	B	C	D
Swift	1	3	1	1
Lilypad	1	3	3	1
Furtive	1	5	5	1
Rambur's	1	41	1	1
Desert	1	1	1	1
Pacific	2	2	1	1
Plains	2	2	1	1
Mexican	1	2	2	1
Western	1	2	2	1
Eastern	1	2	2	1
Black-fronted	3	1	1	2
San Francisco	3	1	1	2
Fragile	4	6	4	1
Citrine	5	7	5	1

A, male thorax: 1, green, striped; 2, black in front with four pale dots; 3, solid black in front; 4, exclamation points in front; 5, yellow, striped.

B, male abdomen tip: 1, S8–9 blue; 2, S8–9 blue with black stripe; 3, part of S7 and/or S10 as well as S8–9 blue; 4, only S8 blue; 5, only S9 blue; 6, no blue on abdomen; 7, abdomen yellow.

C, female morphs: 1, hetero orange to dull, andro remains blue; 2, both hetero and andro pruinose at maturity; 3, orange becomes blue or green and pruinose at maturity; 4, andro dark at maturity; 5, hetero orange to dull at maturity.

D, oviposition: 1, solo; 2, tandem.

Description Large blue forktail of open water. *Male*: Eyes green with black cap, face dark, large postocular spots blue. Thorax blue with black median and humeral stripes and narrow black lines along side. Abdomen black with S1, tip of S2, much of S7, and S8–9 blue. Few individuals with almost no blue on S7. *Female*: Polymorphic, andromorph colored just like male (eyes duller). Heteromorph with eyes green, thorax bright orange at first, changing to green with age, and with black median and humeral stripes and entirely black abdomen. Abdomen in older individuals may become pruinose.

Identification Considerably larger than any other forktails with which it occurs and with black-striped blue thorax, male more likely to be mistaken for a BLUET. But no Pacific Coast BLUET has black abdomen with blue base and tip (see also **Exclamation Damsel**). Heteromorph female bluet-sized but differs in having abdomen entirely black above, with no narrowing of black toward front of segment as in BLUETS. Orange thorax of immature heteromorph immediately differentiates from all BLUETS, large size from other forktails with that color. Female **Taiga Bluet** somewhat similar to green and blue females but smaller and scarcely if at all overlaps in distribution and habitat.

Natural History Although not as restricted as similar Lilypad Forktail in East, commonly associated with water lilies. Both sexes perch flat on leaves of water lilies and pondweeds with abdomen tip touching substrate. Also perch on ground, rocks, and logs much more than other forktails. Females often found far from water, more like bluet than forktail. Rapid low flight much like that of Lilypad Forktail characteristic of species; also bend tip of abdomen down in flight, then straighten it when they land. Males display at each other with blue abdomen tip raised. Copulation lengthy, can last an hour or more (84 min in one case). Males seem especially likely to attempt mating with females of other damselfly species and can successfully achieve tandem with some. Females oviposit solo in floating stems, pondweed leaves, and other plant parts; may descend below surface while doing so. One of earliest odonates over much of its range, almost gone by midsummer.

52
Swift Forktail male, andromorph female—Grays Harbor Co., WA, May 2005

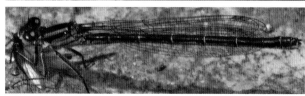

52
Swift Forktail orange female—Grays Harbor Co., WA, May 2005; green female—King Co., WA, May 2005

Habitat Ponds, open marshes, and lake margins with abundant shore vegetation but open water also present. Beaver ponds and bog ponds much used, also slow-flowing ditches. Perhaps more restricted than other forktails to habitats with clear water.

Flight Season BC Apr–Aug, WA Apr–Sep, OR Apr–Aug, CA Feb–Sep.

Distribution A lowland or lower-mountain species throughout its range.

53 Lilypad Forktail *Ischnura kellicotti* TL 24–31, HW 12–18

Description Medium-sized brightly colored forktail inextricably linked to lily pads. *Male*: Eyes dark blue with large black cap. Postocular spots large, blue. Thorax blue with wide median and humeral stripes. Abdomen black, bright blue on S1, much of S2, tip of S7, and much of S8–10. *Female*: Eyes green with black cap; postocular spots blue. Thorax and abdomen base as male, abdomen tip with less blue (top of S8–10). Abdomen increasingly pruinose with age, pattern difficult to make out. Immature female patterned as mature female, but all pale areas bright orange; intermediates rarely seen. Change from orange to blue with age quite unusual among odonates. Also noteworthy is wing-color change in both sexes, from bright amber in just-emerged individuals to clear as in most damselflies, with blue upper surface of forewing stigmas in male.

Identification Very distinctive species, most likely to be mistaken for co-occurring **Skimming Bluet**. **Lilypad** differs from **Skimming** by its bright blue eyes, larger postocular spots, and blue on top of S10. Note that blue on S2 crosses rear part of segment in **Lilypad**, only front part in **Skimming**. Orange or heavily pruinose females might be mistaken for immature or mature female **Eastern Forktail**, respectively, but orange abdomen tip or blue pruinosity rather than white or gray makes for easy distinction. Then there is typical **Lilypad Forktail** behavior; **Eastern** rarely if ever rests on lily pads.

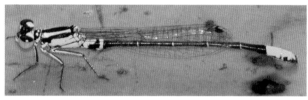

53
Lilypad Forktail
male—Clay Co., FL,
June 2004; orange
female—Lowndes
Co., GA, November
2004, Giff Beaton;
blue female—Leon
Co., FL, April 2005

Natural History Both sexes almost invariably perch flat on lily pads, with abdomen curled down near end to touch leaf (for support? predator detection?). Flight rapid and low between perches, streaking rather than fluttering. Both orange and blue females mate, but mating seldom seen, so perhaps female mates only once. Larvae live under water lily leaves. Few odonates so closely tied to a single type of plant.

Habitat Ponds and lakes with extensive beds of water lilies.

Flight Season TX May–Aug.

Distribution Also widely in the East from Wisconsin to Maine and south to Florida.

54 Furtive Forktail *Ischnura prognata*   TL 30–37, HW 14–20

Description Long, slender forktail of southern swamps. *Male*: Eyes green, tiny postocular spots blue, thorax green with black stripes, abdomen black above, yellowish on sides; S9 blue. *Female*: Imma-ture with eyes dull orange; face, postocular spots, thorax except thin black midline, and S1–3 bright orange; rest of abdomen black above. With maturity, eyes green, head and front of thorax dull metallic brownish, sides of thorax pale, whitish to greenish or bluish. Often a small dark triangle on sides of thorax at wing base just behind dark median stripe. Abdomen black above with varying amounts of gray pruinosity.

Identification Only forktail so long and slender. If size and shape not sufficient, presence of blue only on S9 is a good mark for males (**Rambur's** has blue on S8 also). Mature females distinguished from other dark species with at least some abdominal pruinosity (**Citrine**, **Eastern**, **Fragile**) by often pale bluish or greenish sides (others are usually gray to white), stigmas graded from dark proximally to light distally (others all dark). Dull-colored mature female **Rambur's** also looks like **Furtive** but less elongate, rarely if ever overlap in habitat. **Rambur's** typically has pale postocular spots; **Furtive** lacks them. Immature females col-ored much like immature female **Citrine** but much larger; also **Citrine** shows much more orange on abdomen, S1–5 and S9–10. Immature female **Rambur's** has less orange, usually only on S1–2 (S1–3 in **Furtive**).

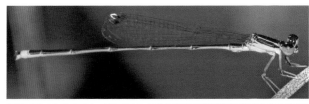

54
Furtive Forktail male,
immature female,
female—Highlands
Co., FL, April 2005

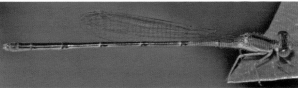

Natural History Both sexes, mature and immature, usually present together in preferred habitat; apparently do not wander away from water and never move into open habitats. Move slowly through dense beds of vegetation in flight, even up into lower tree branches; long abdomen probably adaptation to hovering. Reproductive behavior not recorded.

Habitat Swamps and swampy borders of slow streams, always under canopy. Usually associated with dense growth of herbaceous plants such as tall grasses, water smartweed, and lizard's tail.

Flight Season TX May.

Distribution Occurs in the East from Arkansas and Indiana to Massachusetts and south to Louisiana and Florida.

---

55 Rambur's Forktail   *Ischnura ramburii*   TL 27–36, HW 15–19

Description Large, brightly marked forktail very common in South.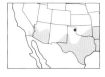
Male: Eyes green with black cap, tiny postocular spots blue. Thorax green with wide median and humeral stripes; antehumeral stripe narrow, rarely broken. Abdomen shiny black above, golden on sides; S8 blue, S9–10 black above, blue on sides. Populations at the far western edge of the U.S. range with all of S9 blue also. *Female*: Polymorphic. Heteromorph with postocular spots, thorax, and S1–2 bright orange with black midline, abdomen otherwise entirely black above, pale greenish on sides of S3–7 and orange on sides of S8–10. Thorax becomes duller and eventually orangey-brown to greenish-brown with maturity, with faint paler antehumeral stripe; sides of abdomen pale tan throughout. Andromorph colored just like male, also varying geographically, with blue thorax when immature.

Identification Males and andromorph females easily distinguished from all other damselflies in range by large size (for a forktail) and blue S8 and sides of S9 (but also covering S9 in Southwest). Other forktails in range with green, striped thorax and blue abdomen tips are

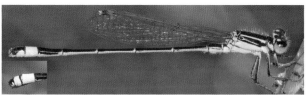

55.1
Rambur's Forktail male—Starr Co., TX, November 2005; inset andromorph female—Clark Co., NV, October 2005, Steve Potter; immature heteromorph female—Monroe Co., FL, April 2005; heteromorph female—Hidalgo Co., TX, May 2005

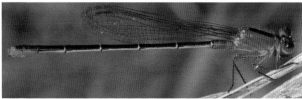

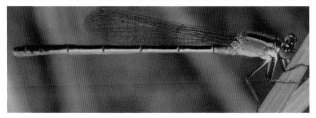

55.2
Rambur's Forktail pair with andromorph female—Bexar Co., TX, July 2004

smaller and have S8–9 blue, S8 with black stripe on side (**Desert**, **Eastern**), or longer and more slender with S9 blue (**Furtive**). **Lilypad Forktail** and several bluets superficially similar, but all with blue thorax. Immature andromorph female **Rambur's** also with blue thorax, but no others with abdomen tip similarly colored. See also female **Furtive Forktail**.

Natural History Both sexes at or near water most of time. Males and sometimes mating pairs can be very common in preferred habitats, often more out in open than other forktails, at edge of vegetation beds. Both female morphs commonly seen in pairs, orange immatures relatively rarely. Copulation very lengthy, averaging 200 min and up to 7 hr, majority occurring in afternoon. Females of copulating pairs typically grab substrate, unlike most pond damsels but apparently characteristic of at least some forktails. Abdomen has to be very flexible to do this. Oviposition solo, mostly in late afternoon, in floating leaves, stems, and debris. Effective predator on other small damselflies.

Habitat Occurs in great variety of habitats in lowland range—lakes, ponds, marshes, ditches, even brackish waters, as long as some shore vegetation such as grasses and sedges present. Also found at high elevations at large lakes in Mexico and Central America and one such locality in Arizona, Point of Pines Lake in ponderosa pine zone at 6200 feet.

Flight Season CA Feb–Dec, AZ Feb–Dec, NM Apr–Oct, TX all year.

Distribution Ranges south, mostly in lowlands, to Chile and Paraguay, and throughout the Southeast north to Indiana and Maine.

56 Desert Forktail *Ischnura barberi* TL 28–35, HW 14–19

Description Large forktail of arid Southwest with much pale coloration. *Male*: Eyes blue-green, tiny postocular spots blue and connected by a narrow line. Thorax green, with black median and humeral stripes; antehumeral stripe often paler green or golden. Abdomen shiny black above, with well-developed torpedo markings, conspicuously golden-orange on sides; S8–9 and sides of S10  blue (blue may also extend forward on side of S7). *Female*: Polymorphic, heteromorph with eyes brown, paler below, large dumbbell-shaped orange postocular spots; surprisingly variable, with thorax either entirely pale orange to tan or with very narrow or relatively narrow black median stripe. Rarely have trace of dark humeral stripes, even more rarely complete stripes (perhaps only in northern California) that make them look just like bluet! Abdomen black, orange to tan on sides, with S1, often S2, and most of S8–10 same pale color (usually black spot on S9). Uncommon variants have S8–10 all dark or all blue! Older individuals become dull brownish all over, even somewhat grayish with pruinosity on sides of thorax. Andromorph colored just like male.

Identification Only species of similar size and color is **Rambur's Forktail**. Usually distinguishable because **Desert** has S8–9 entirely blue, **Rambur's** only S9, but populations of **Rambur's** in southern Nevada, western Arizona, and southeastern California, in midst of range of **Desert**, also have S8 blue (both have sides of next segment blue). In male and andromorph female **Desert**, pale salmon coloration prominent on sides of middle abdominal segments and extends up at rear end of each middle segment, almost connecting across top, so black reduced in view from above; no such extension in **Rambur's**. Antehumeral stripe conspicuously broader than in **Rambur's**; that and abdominal pattern should allow easy identification in field. Antehumeral stripe often different color from side of thorax, but some individuals exactly like **Rambur's** in this. In addition, small isolated postocular spots in **Rambur's**, slightly larger spots connected by line in **Desert**. Heteromorph females much paler than orange immature **Rambur's** females and differently colored than greenish-brown mature females, with much narrower median stripe on thorax or, commonly, lacking that stripe entirely. Female **Rambur's** always show wide black median stripe. Abdomen tip usually with less black than **Rambur's**, often entirely pale. Rare bluet-patterned individuals distinguished by comparing their structure with real BLUETS.

Natural History Typically perches in sedges and grasses at shoreline, sometimes at high densities. Mating pairs common in afternoon. Female oviposits alone at surface, often laying on underside of floating leaves.

Habitat A great variety of lakes, ponds, marshes, canals, and ditches, all having in common somewhat alkaline or even saline water. Larvae survive drying of ephemeral playa lakes in New Mexico in masses of damp stonewort. Common in some tidal marshes in south coastal California.

Flight Season CA Mar–Nov, AZ Mar–Oct, NM Mar–Nov, NE Jun–Sep, TX Apr–Oct.

Distribution Ranges south in Mexico to northern Baja California and Sonora.

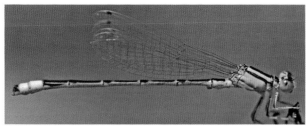

56
Desert Forktail
male—Sonora,
Mexico, April 2005,
Doug Danforth;
andromorph female,
heteromorph
female—Graham Co.,
AZ, August 2007

Description Small, slender forktail of far West with pale dots on thorax. *Male*: Eyes green with black cap. Head black with small blue postocular spots. Thorax blue, black anterior surface with paired blue dots representing ends of antehumeral stripes. Abdomen black, S8–9 blue, each with incomplete black stripe low on sides. *Female*: Polymorphic. Eyes greenish with black cap when mature. Heteromorph originally with head spots and legs pale orange. Thorax pale pinkish-orange with rather narrow black median stripe and very narrow black humeral stripe. Abdomen black above, with sides of S1–2 paler pinkish-orange, S8 and sides of S9 blue with variable black markings low on side of S8. With age, head and legs  become mostly black, and pale markings on thorax become brown, eventually obscured even more by dark gray pruinosity; blue disappears from abdomen tip, so overall a plain

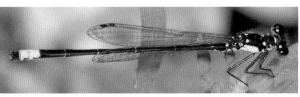

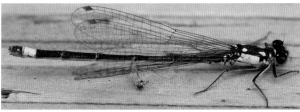

57.1
Pacific Forktail
male—Grant Co.,
WA, August 2007;
andromorph
female—Marion Co.,
OR, June 2005,
Jay Withgott

57.2
Pacific Forktail
immature
heteromorph
female—Kittitas Co.,
WA, July 2003;
immature
andromorph
female—Teton Co.,
WY, September 2007;
heteromorph
female—Kitsap Co.,
WA, September 2005

dark damselfly with greenish face and lower part of eyes. Andromorph similarly patterned to heteromorph at first, but pale colors on head and thorax blue. This also becomes obscured by pruinosity with age, but blue on face, wing bases, and abdomen tip remains; sides of thorax blue-gray.

Identification Only other species with similar male pattern of four pale dots on thorax is **Plains Forktail**. Distinguishable in close view where they overlap, mostly in Rockies, by distinctly elevated "fork" at abdomen tip in **Pacific**, lacking in **Plains**. Young female like same stage of **Plains Forktail**, as mature female identical except prothorax structure. Andromorph of **Plains** more likely than **Pacific** to have dots on thorax lengthened, but not definitive. Mature females differ from female **Mexican** and **Western Forktails** by much darker color, no pale pruinosity. Immature females of latter species have entirely orange thorax and abdomen base, not pinkish.

Natural History Males perch in and around dense beds of aquatic vegetation, and both sexes in herbaceous vegetation near water. Males display at one another with abdomen tip raised. Copulation lengthy, lasting up to several hours and usually occurring in afternoon. Male thus prevents other males from access to female until too late in day. Note both male and female hold onto substrate during copulation, unlike dancers and bluets, in which female usually suspended in midair supported by male. Females then oviposit during following morning, often active before males. Males then dominate later in day. Females oviposit in all kinds of floating vegetation, even on undersides of leaves but submerge only abdomen. Earliest odonate to fly in northern part of range. Length of adult life up to 15 days.

Habitat Any sort of lake, pond, marsh, or ditch with abundant vegetation, much less often at edge of running water. Common in midst of beds of tall cattails and bulrushes where few other odonates are found. In Northwest, more likely to colonize backyard ponds than other damselflies. At hot springs at Banff, Alberta.

Flight Season BC Mar–Sep, AB May–Sep, WA Mar–Oct, OR Mar–Oct, CA all year, MT May–Oct, AZ Mar–Dec.

Distribution Ranges south in Mexico to Baja California Sur and Sonora.

Forktails - female prothorax

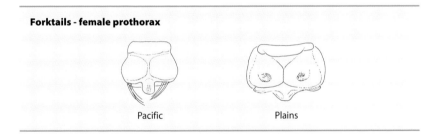

Pacific   Plains

58 Plains Forktail *Ischnura damula*   TL 23–34, HW 11–19

Description Small, slender forktail of Great Plains and Rockies with pale dots on thorax. *Male*: Eyes green with black cap. Small blue postocular spots. Thorax blue, black anterior surface with paired blue dots representing ends of antehumeral stripes. Anterior pair of spots sometimes absent. Abdomen black, S8–9 blue, each with incomplete black stripe low on sides. *Female*: Polymorphic. Heteromorph originally with eyes tan, darker above; head spots and legs pale orange. Thorax with moderate median stripe and quite narrow humeral stripe, pale color pinkish-orange. Abdomen black, base (S1, much of S2) pinkish-orange becoming blue; S8 and sides of S9 blue with variable black markings low on side of S8. With age, head and legs become mostly black, and pale markings on thorax become tan, eventually obscured even more by dark gray pruinosity. Pink postocular spots and blue abdomen tip retained into maturity. Andromorph colored exactly as mature male but less blue on abdomen tip (only S8, sometimes on S9), and lower spots may be extended toward upper ones as exclamation point.

Identification Males similar only to **Pacific Forktail**, overlapping widely in Rockies. Not distinguishable except at close range in side view, when elevated process on end of abdomen characteristic of **Pacific** could be seen. Projection on S10 barely above abdomen and scarcely forked in **Plains**, raised well above abdomen and distinctly forked in **Pacific**. Also, in **Plains** paraprocts strongly curved and rather short, in **Pacific** longer and straighter. Females also similar in all stages and morphs, can be distinguished only by prothorax; **Pacific** has long hairs projecting backward and toward midline, **Plains** lacks them. As in **Pacific**, **Plains** females quite blackish rather than the gray of mature heteromorph female **Eastern**, **Mexican**, and **Western Forktails**. Immature females of those species have entirely orange thorax and abdomen base, not pinkish. Note that very rare individuals of **Eastern Forktail** may have antehumeral stripe represented as spots, and such an individual well outside range of **Plains** could cause some excitement.

Natural History Both sexes, mature and immature, may be abundant in marshy vegetation at edge of water, not flying over open water. Often seen to hold wings partially open after

58.1
Plains Forktail male—Apache Co., AZ, July 2007; andromorph female—Catron Co., NM, July 2003

58.2
Plains Forktail immature heteromorph female, heteromorph female—Apache Co., AZ, July 2007

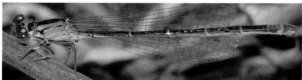

landing, then close them. Females oviposit at surface in floating vegetation, including willow catkins. Adults become mature in a few days.

Habitat Dense vegetation beds at lake margins, ponds, and slow streams and ditches, and hot springs.

Flight Season YT Jun, BC May–Jul, AZ Mar–Oct, NM Feb–Oct, NE May–Aug, TX May–Sep.

Distribution Highly disjunct range puzzling but may indicate special habitat needs. Only at hot springs in isolated populations in British Columbia and Yukon.

59 Mexican Forktail *Ischnura demorsa* TL 21–26, HW 11–15

Description Small, common forktail of southwestern streams with green-striped thorax and blue abdomen tip. *Male*: Eyes green with black cap; small postocular spots also green. Thorax green with black median and rather wide humeral stripes; antehumeral stripe narrower than humeral. Abdomen black above, S1–2 and part of S3 green on sides, middle segments tan on sides; S8–9 blue, usu-

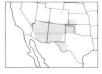

ally with incomplete and tapering black stripe low on side of both segments. *Female*: Polymorphic. Heteromorph with eyes greenish-tan, brown above; moderate-sized separated postocular spots and stripe on face orange. Thorax orange with wide median and fairly narrow humeral stripes, antehumeral about as wide as humeral. Abdomen shiny black above,

59.1
Mexican Forktail
male, female—
Chihuahua, Mexico,
September 2005

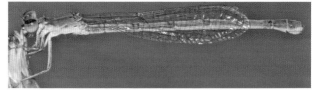

59.2
Mexican Forktail
immature
heteromorph
female, immature
andromorph
female—Chihuahua,
Mexico, September
2005

with variable amounts of orange at base (at least S1 and half of S2, at most all of S1–2 and most of S3); S3–7 may have prominent pale basal rings. Tip also variable in amount of orange, as much as S9–10 and bit of S8. Darkens with maturity until entirely dark and covered with light to dark gray pruinosity, striped pattern of thorax vaguely visible. Less common andromorph also appears to have orange or blue thorax at first, but postocular spots and dorsal spots on abdomen blue (tip of S8, all of S9, and base of 10 blue at greatest extreme); indistinguishable from heteromorph when mature.

Identification No other forktail with green striped thorax and blue-tipped abdomen in much of range. Overlaps with **Western** sparingly in Colorado, Kansas, and New Mexico and **Eastern** in eastern Colorado and New Mexico and western Kansas and Oklahoma. In those areas, males distinguished in hand by side view of appendages, in particular forked paraprocts. Those in **Mexican** with upper part of fork longest and projecting well behind cerci, in **Western** with both forks about equal and not much longer than cerci, in **Eastern** with lower part of fork distinctly longest and projecting. Also, male **Eastern** lacks tiny green spots on either side of prothorax present in **Mexican** and **Western**. **Eastern** and **Western** both average wide, rectangular black markings on S8–9, whereas markings in **Mexican** more likely to be narrower and tapering, but much overlap. Females also impossible to distinguish in field, must be examined in hand. Mesostigmal plates in **Mexican** have flange lifted above only outer two-thirds of plate, whereas other two have this flange extending entire length of plate. **Western** has rear of prothorax rounded, **Eastern** has it angulate, with shelf projecting out from under it. Some immature females with blue thorax could be mistaken for female BLUET with dark abdomen, but forktails would have conspicuous blue at abdomen tip. Also, black on each abdominal segment in most forktails is even-sided, not the tapered condition of BLUETS.

Natural History Males and immature and mature females often common in dense herbaceous vegetation at water's edge. Beds of grass around ponds can be full of them. Most females mate only once, then use that male's sperm to fertilize all eggs they lay. On some occasions, females common with no males in sight. Oviposit in floating vegetation of all sorts. Females mature in just a few days.

Habitat Slow-moving streams and ditches, more common on streams than most other forktails, but also weedy shores of lakes and ponds.

Flight Season AZ Mar–Dec, NM Feb–Nov, TX Apr–Nov.

Distribution Ranges south in uplands of Mexico to Oaxaca.

Forktails - female mesostigmal plates

Mexican   Western   Eastern

60 Western Forktail *Ischnura perparva*   TL 23–30, HW 11–17

Description Typical small forktail with striped thorax and blue-tipped abdomen. *Male*: Eyes green with black cap. Face bright green, postocular spots bluish-green. Thorax bright green, with wide black median and humeral stripes; antehumeral distinctly narrower than humeral. Abdomen black except sides of S1–3 green; S8–9 blue, each with a short but thick black stripe low on sides. *Female*: Polymorphic. Heteromorph when young, black and orange. Eyes greenish-tan with brown cap. Postocular spots large, orange, joined by bar. Thorax with wide black median and narrow black humeral stripes. Abdomen orange on S1–2 and much of S3, in some individuals extending well onto S4, remainder shiny black above (rarely with blue at tip). Some females orange at rear of S8 and S9, and some orange on sides of S8–10; at most extreme much of S8–10 orange. Mature females become entirely gray or even white pruinose with black median and humeral stripes obscured but still visible, face and eyes green. Intermediates so rarely seen in this and other forktails that color change must happen very rapidly. Very rarely seen immature andromorph colored much like male but more black on base of S8. None has been seen at maturity, presumably becomes pruinose and looks like mature heteromorph.

Identification Occurs mostly with **Pacific Forktail**, from which males distinguished by green thorax with stripes and fairly obvious pale rings on some abdominal segments. No trace of rings on **Pacific**, which has blue thorax with four dots on black front. Overlaps broadly with **Eastern** and **Mexican Forktails** on western Great Plains, distinguishable in hand and conceivably in close-focusing binoculars by shape of paraprocts (see **Mexican Forktail**). Also, male **Eastern** lacks tiny green spots on either side of prothorax present in **Mexican** and **Western**. Females of these three species distinguishable only by structure of prothorax and mesostigmal plates.

Natural History Males in small numbers at most locations, usually within dense vegetation. Vastly outnumbered by mature females busily ovipositing. At some times and places, females common at water and no males seen! Immature (orange) and mature (pruinose) females both seen in mating pairs, but pairs rarely seen considering abundance of females. Most females probably mate only once, then use that male's sperm to fertilize all eggs they lay. Females oviposit solo, horizontal on surface of floating vegetation, making short flights between bouts of egg deposition and often quite conspicuous. Very effective at repelling attention of male forktails and bluets by fluttering wings and curling abdomen tip down.

Habitat Marshy edges of lakes, ponds, and slow streams, especially in sedge and grass beds. More common on streams than most other forktails.

Flight Season BC Apr–Sep, WA Apr–Oct, OR Apr–Oct, CA Mar–Nov, MT May–Sep, NM May–Jul, NE Apr–Jun.

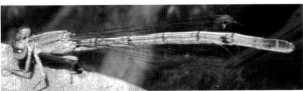

60
Western Forktail male—Kittitas Co., WA, August 2006; immature heteromorph female—Grant Co., WA, June 2003, Netta Smith; female—King Co., WA, June 2004

Description Typical small forktail with black-striped green thorax and blue abdomen tip. *Male:* Eyes and face bright green, postocular spots bluish. Thorax bright green, with black median and humeral stripes. Abdomen black except S1 and sides of S2 green, S8–9 blue, each with short but thick black stripe low on sides. Achieves mature color in 1 day. *Female:* Polymorphic. Heteromorph when young with orange postocular spots, thorax, and abdomen base. Thorax with wide black median and narrow black humeral stripes. Abdomen orange on S1–2 and much of S3, remainder shiny

black above. Mature females become entirely gray pruinose with black median and humeral stripes still visible, face and eyes green. Females take almost a week to achieve mature color. Immature andromorph very rarely seen, colored as male but thorax bluer and more black on base of S8.

Identification Overlaps with **Mexican** and **Western Forktails** (see former species) widely in Great Plains as a major identification challenge, overcome by a close look at paraprocts. Fortunately, there is another close-range field mark for males: **Eastern** lacks tiny green spots on either side of prothorax present in **Mexican** and **Western**. Females also indistinguishable except by structure. See also **Fragile Forktail**, in which mature females can be similar.

Natural History Males perch in dense vegetation at waterside or cruise through it looking for prey and females. Females also present at water and in nearby herbaceous vegetation. Both orange and pruinose females mate. Copulation lasts about 40 min. Most females mate only once, then use male's sperm to fertilize all eggs they lay. Females repel other damselflies,

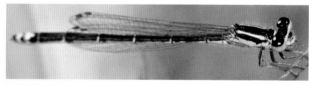

61.1
Eastern Forktail male—Buchanan Co., IA, July 2004; andromorph female—Dallam Co., TX, August 2004, Greg Lasley

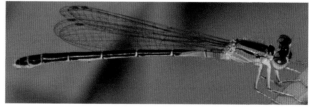

61.2
Eastern Forktail immature heteromorph female—Fayette Co., IA, July 2004; heteromorph female—Buchanan Co., IA, July 2004

including males of their own and other species, by fluttering wings and curving abdomen downward. Oviposition solo on floating or emergent stems of grasses, sedges, and other marsh plants. Females much more likely than males to eat other damselflies, teneral and mature. Unlike Western and Mexican Forktails, males usually as common as ovipositing females.

Habitat Ponds, lakes, marshes, and slow streams, even edges of large rivers, as long as beds of vegetation in quiet water are present. Very wide habitat choice, but not in bogs or fens. Not always obvious in dense grass and sedges but usually quite visible at edges where they can be very common.

Flight Season MT May–Aug, NM May–Oct, NE May–Oct, TX Apr–Oct.

Distribution Also east across southern Canada to Newfoundland and south to Arkansas and South Carolina.

62 Black-fronted Forktail *Ischnura denticollis* TL 22–26, HW 11–15

Description Tiny and shiny blue-tipped forktail of western marshes. *Male*: Eyes green with black cap, postocular spots blue. Thorax unmarked shiny black in front, blue-green on sides. Abdomen black above, blue-green on sides of S1–3, upper surfaces of S8–9 mostly but not entirely blue. *Female*: Polymorphic. Heteromorph with green eyes, wide median and narrow humeral stripe black. Immature with head markings, thorax, and sides of abdomen base pink. Abdomen black above, with blue spots on S8–9 as in male. With maturity, postocular spots, thorax, and abdomen base become blue or green. Still later, thorax gets darker and duller and blue on abdomen tip replaced by black. Rarer andromorph much like heteromorph but pale color always blue; antehumeral stripe may be interrupted, represented by pair of spots, or absent.

Identification Male only damselfly in most of range with combination of unmarked black front of thorax and blue at abdomen tip restricted to dorsal spot rather than filling entire segments. Thorax a bit greener than in **Pacific Forktail**. On central California coast, may occur

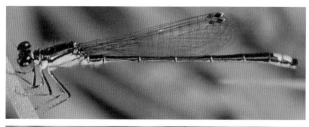

62.1
Black-fronted Forktail male—Harney Co., OR, July 2006; heteromorph female—Chihuahua, Mexico, September 2005; andromorph female—Inyo Co., CA, June 2003

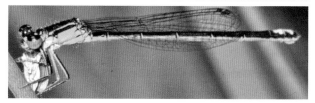

62.2
Black-fronted Forktail
pair—Harney Co.,
OR, July 2005

with extremely similar **San Francisco Forktail**, males distinguished only in hand by append-ages. Female differs from female **San Francisco** structurally; also averages more blue at abdomen tip. Arrangement of that blue, as dorsal spots surrounded by black, different from all other damsels in range except **San Francisco**. **Pacific** and **Plains Forktail** females have blue only on top of S8 (**Rambur's** with blue on S9 have S8–9 entirely blue, not blue-spotted on top). Mature females lacking blue on abdomen may have to have thoracic structures examined for certain identification, although **Black-fronted** distinctly smaller than **Pacific** and **Plains** and darker than mature females of pruinose group (**Citrine**, **Mexican**, **Western**).

Natural History Males can be abundant in dense marsh vegetation, like similarly colored sprites rather hard to see. Often challenge each other head to head. Copulation lasts about 20 min, tandem 23 min before oviposition, then oviposition 24 min. Pairs oviposit in tandem flat on floating vegetation or at moist base of sedge and grass stems. One of only two North American forktails to oviposit in tandem (San Francisco the other). A few eggs are laid in one spot, then the pair moves and rests briefly before another few are laid, and same behavior repeated. Maximum longevity 42 days, quite high for damselfly; sexual maturation in 6–9 days.

Habitat Dense marshes, typically with low sedges but sometimes in cattails, in still or slow-flowing waters, and sometimes on more open streams or ponds wherever marsh plants grow densely. Often associated with springs, both hot and cold.

Flight Season OR Apr–Oct, CA Feb–Dec, AZ Mar–Nov, NM all year, TX Mar–Aug.

Distribution Ranges south in uplands to Guatemala.

Comments Where they overlap in range, this species and San Francisco Forktail hybridize. Hybrids will defy identification.

Forktails - female prothorax

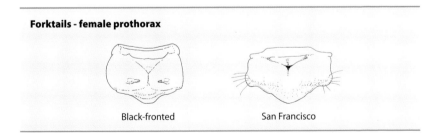

Black-fronted   San Francisco

---

63 San Francisco Forktail *Ischnura gemina*   TL 23–28, HW 12–17

Description Small forktail of central California coast with glossy black front of thorax. *Male*: Eyes black over green, face with transverse green stripes, postocular spots blue. Thorax shiny black in front, blue-green on sides. Abdomen shiny black above, blue-green on sides

of S1–2 and part of 3; S8–9 light blue above. *Female*: Polymorphic. Heteromorph when young with eyes brown, face and head spots orange-pink. Thorax orange-pink with wide black median stripe and narrow black humeral stripes. Abdomen black above, faint blue line at base of S2 and middorsal pale streaks on S8–10. Gets darker, duller with age. Uncommon andromorph colored as male, but blue on abdomen tip varying from like that of male to almost absent.

Identification Black-fronted Forktail   only similar species in range, looks identical in field. Distinguish in hand by male appendages (simple in **San Francisco**, more elaborate in **Black-fronted**, with obvious spine projecting rearward from paraprocts), female prothorax (**San Francisco** lacks pair of projections on top characteristic of **Black-fronted**). Identification may be confounded by hybridization. Color pattern of both sexes different from **Pacific** and **Western Forktails** that may occur with it.

Natural History   Both sexes together in dense vegetation, rarely seen out of restricted breeding habitat. Active at breeding sites in poor weather. Males more likely at edge and over water, females often in nearby grasses and shrubs. Copulation averages 44 min, tandem 36 min, during which oviposition occurs. Some solo oviposition, usually by older females. Maturation time around a week, reproductive adults live 1–3 weeks (maximum 41 days).

Habitat   Still, dense sedgy habitat, including small marshes, some at foot of seepage slopes; also slow-flowing streams and canals. Most are in urban areas.

Flight Season   CA Mar–Nov.

Distribution   Very locally distributed in the San Francisco Bay region. May have disappeared from locations in the southern part of its former range, perhaps from hybridization with Black-fronted.

Comments   Hybridizes with Black-fronted Forktail at few places on south side of bay, hybrids with intermediate structure.

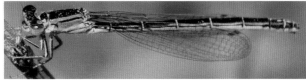

63
San Francisco Forktail male, heteromorph female—San Francisco Co., CA, June 2004

64 Fragile Forktail   *Ischnura posita*   TL 22–29, HW 10–16

Description   Tiny dark forktail with bright exclamation points on thorax. *Male*: Eyes green, face yellow-green; mostly shiny black, thorax with interrupted antehumeral stripe and sides light green, becoming slightly pruinose with age. Abdomen black, sides of S1–2 light green. Very rarely with bit of blue on top of S9 (Mexican subspecies with more). All pale markings yellow-green or even yellow in western part of range. *Female*: Immature as male but postocular spots and thoracic color pale to bright blue. Very rarely spot of blue on S9. Pattern becomes obscure with age, may be almost covered by gray pruinosity on thorax and abdomen but thoracic pattern usually remains at least barely visible.

Identification In flight, especially in sun, males look like tiny spot of golden-green moving through vegetation because bright thoracic pattern is the only feature visible. Often found with equally small **Citrine Forktail**, mature female similar but a bit more robust and with pruinosity usually obscuring black median thoracic stripe that is visible in **Citrine**. Mature female might be mistaken for same of **Eastern Forktail**, but a bit smaller and interrupted humeral stripe often visible even through pruinosity. **Fragile** females more likely to have narrow lateral thoracic stripe than **Eastern**, and line of demarcation between dark dorsal and pale ventral color on S8 often more irregular in **Eastern**. Note **Fragile** females lack vulvar spine on S8 present in **Eastern**.

Natural History Usually remains within dense vegetation, both herbaceous and shrubby, but also on beds of floating plants such as water lilies, pondweeds, and mare's tail. Seems likely to come out from vegetation during cooler, cloudy weather, perhaps because predation from larger odonates unlikely then. Males perch from water surface up to waist high. Females oviposit solo on duckweed and other floating vegetation and emergent stems, even a few inches above water level. Night roost sites average higher than daytime perches, and abdomen more elevated, perhaps to catch morning sun. As other forktails, females persistent predators on other damselflies.

Habitat Ponds, lake shores, swamps, ditches, and slow streams with abundant herbaceous vegetation, including spring-fed. Often in dense grass or sedge beds as other forktails but also common in other herbaceous plant beds in shady wooded situations.

Flight Season NE May–Oct, TX all year.

Distribution Also widespread across all of eastern United States and southern Canada to Newfoundland and south into eastern Mexico.

Comments Northern subspecies *I. p. posita* not found south of border, but additional very local subspecies (*I. p. atezca*) in northeastern Mexico (Tamaulipas, Hidalgo) and another (*I. p. acicularis*) from Yucatán Peninsula to Guatemala.

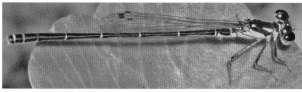

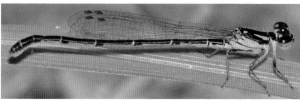

64
Fragile Forktail
male—Alachua
Co., FL, April 2005;
immature
female—Early
Co., GA, July 2007;
female—Alachua
Co., FL, April 2005

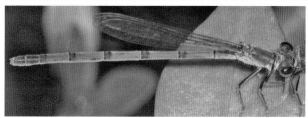

Description Tiny mostly yellow damselfly, unique in that male fore-wing stigmas are not at wing edge. *Male*: Eyes yellow in front, green behind, with large black cap. Thorax striped green and black, with narrow antehumeral and wide humeral stripe. Abdomen yellow, green at extreme base; S1–2 black above, S3–6 with black basal spearpoint and apical ring, may be joined on S6; S7–10 all yellow. *Female*: Eyes dull greenish with brown cap, prominently banded with black. Thorax black in front, gray on upper sides, white on lower sides. Abdomen black above, eventually becoming pruinose gray, often faintly banded with darker gray. Immature with face, postocular spots, pale areas on thorax and abdomen orange. During maturation, S6–8 and parts of S5 and S9 become black above. Gradually becomes greenish and then increasingly pruinose when mature, heavier (paler) in dry climates.

Identification Yellow males unmistakable. Orange immature females differ from orange immatures of other forktails by lacking dark humeral stripe. Mature females in wetter areas more likely to be dull blackish, in drier regions can be quite pale gray pruinose. In either case, very similar to mature females of other pruinose species, **Eastern**, **Fragile**, **Mexican**, and **Western Forktails**, but distinguished by shiny black median stripe on front of thorax (pruinosity obscures stripe in other species) with no trace of distinct humeral stripe (sometimes obvious in other species).

Natural History Both sexes can be found at very high densities in appropriate habitat, sometimes both matures and immatures rising in clouds when disturbed from dense vegetation at shallow grassy or sedgy ponds. Mating seldom seen, perhaps because most females mate only once, then use sperm of that male to fertilize all eggs they lay. Copulation

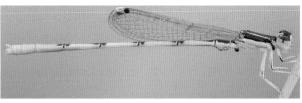

65.1
Citrine Forktail
male—Travis Co.,
TX, June 2007, Eric
Isley; immature
female—Cochise Co.,
AZ, September 2004

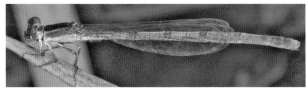

65.2
Citrine Forktail
female—Clay Co., FL,
April 2005; very
pruinose female—
Sonora, Mexico,
August 2006, Netta
Smith

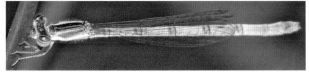

throughout day, with both immature and mature females. Oviposition much less often seen than in other forktails.

Habitat Ponds and lake margins densely vegetated with grasses and/or sedges, also temporary ponds with similar vegetation, at least in southern part of range. Good disperser, taken by winds far above ground, so may turn up in inappropriate habitats.

Flight Season CA Mar–Nov, AZ Feb–Dec, NM Jan–Oct, NE Jun–Sep, TX all year.

Distribution Sparsely distributed in Southwest and on southern Great Plains. Ranges south in lowlands to Venezuela; isolated populations on Galapagos and Azores. Also widespread in East north to Wisconsin, southern Ontario, and Maine.

Forktails - male appendages

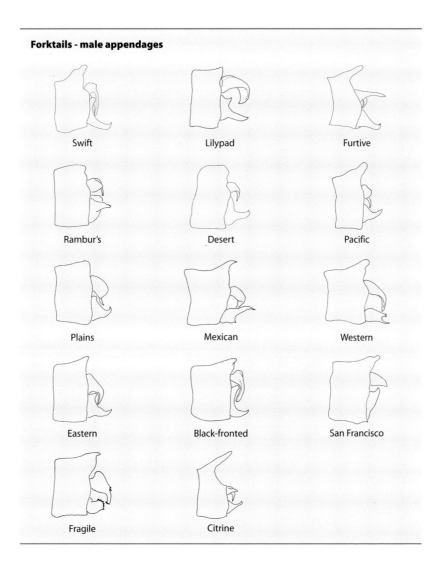

Swift

Lilypad

Furtive

Rambur's

Desert

Pacific

Plains

Mexican

Western

Eastern

Black-fronted

San Francisco

Fragile

Citrine

This is a dominant genus at ponds and lakes in the New World tropics, with up to a dozen species at some South American localities. A single species makes it into our geographic area. Males have blue abdomen tip; those of open ponds and lakes throughout the range have a blue thorax, whereas those of wooded swamps in South America, orange thorax; all have blue S9. The common name for this group comes from the elevated abdomen tip, found in many but not all species and prominent in some of them even more so than ours. World 42, NA 1, West 1.

66 Mexican Wedgetail *Acanthagrion quadratum*　　　TL 29–33, HW 16–18

Description Bluet-like Texas border species, males with distinctive abdomen tip. *Male*: Eyes mostly black, green below; large postocular spots blue. Thorax blue with black median and humeral stripes, fine black line low on sides. Abdomen black with sides of S1–2, end of S7, and all of S8–9 blue. *Female*: Colored as male but eyes tan below, blue of thorax paler, abdomen with side markings and tip

and often base of S8, S9 blue with jagged black marks on either side of base, S10 all blue. Quite variable and less blue on abdomen tip farther south in Mexico.

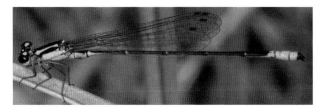

66.1
Mexican Wedgetail
male—Kleberg Co.,
TX, October 2004,
John C. Abbott;
female　San Luis
Potosí, Mexico,
November 2006,
Marion Dobbs

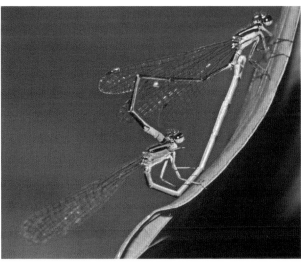

66.2
Mexican Wedgetail
pair—Hidalgo,
Mexico, October
2003, Robert A.
Behrstock

Identification Male superficially like several BLUETS except for triangular side view of end of abdomen but not colored exactly like any of the few BLUETS in south Texas. In that area, similar only to **Stream Bluet,** but that species has only S9 entirely blue (sides of S8 and S10 blue, however) and has obvious blue rings on middle segments lacking in **Wedgetail**. Females surprisingly similar, **Wedgetail** with complexly patterned S8, black low on sides of S9 (on top in **Stream Bluet**). Humeral stripe of **Stream Bluet** often split lengthwise, and some females with green thorax; never that way in **Wedgetail**. In hand, cerci of female distinctive, almost touching in dorsal view, whereas those of BLUETS are well separated. Female also has small pits on either side of middorsal carina of thorax, not found in any other North American damselfly.

Natural History Males and pairs perched low at edge of water or scattered through beds of herbaceous vegetation. Males rise up in air while facing one another, displaying with raised abdomen tip. Pairs oviposit in herbaceous vegetation.

Habitat Well-vegetated ponds and slow streams and open marshes, also swampy areas under tree cover.

Flight Season TX May–Oct.

Distribution Ranges south in lowlands to Costa Rica.

Black-and-white Damsel *Apanisagrion*

This monotypic genus is confined to Middle American plateaus and is distinguished by wing venation and genital morphology. Males have a dusky spot at tip of hindwing, which is actually a patch of very dense veins. The genus is most closely related to *Hesperagrion* (Painted Damsel) and *Anisagrion* of farther south, all noteworthy for dramatic color changes with age. World 1, NA 1, West 1.

67 Black-and-white Damsel *Apanisagrion lais* TL 32–41, HW 18–23

Description Upland pond damsel that looks black and white in its shady habitat. Stigmas in mature individuals with contrasty white borders. *Male*: Eyes and face green. Thorax black in front, with fine median line, narrow antehumeral line, and diagonal line behind that light green. With maturity, sides pruinose white, eventually covering lower pale line. Abdomen black, white pruinosity on sides  of S1–2, all of S8–9, and base of S10. *Female*: Colored as male but eyes duller green. Abdomen black with white pruinosity on top of S8. Immature head tan with narrow dark stripe between eyes. Thorax tan with fine pale striping evident. Abdomen orange, becoming

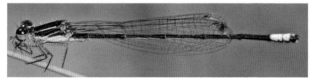

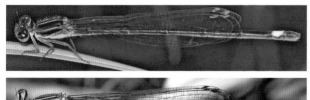

67
Black-and-white
Damsel male,
female—Sonora,
Mexico, September
2005; immature
male—Santa Cruz
Co., AZ, September
2006, Robert A.
Behrstock

black on top from rear forward, tip remaining pale until mature pattern developed; S8 blue in females before becoming pruinose.

Identification No other pond damsel with pruinose white sides of thorax and tip of abdomen, good mark for both sexes at maturity. Pair of fine light stripes on either side of thorax also distinctive. Immatures much like immature **Painted Damsel** but distinguished by dark stripe between eyes and quite different pattern on thorax, with fine lines usually visible.

Natural History Both sexes usually perch in dense tall grass and sedges, flying through surprisingly dense stands while foraging. Immature and mature adults often seen in same area, as in forktails and Painted Damsels; immature phase perhaps lengthy. Mating behavior unknown, oviposition by female alone in plant stems at or just below water level.

Habitat Slow-flowing streams in woodland, localized at extensive patches of grass and sedges, also in watercress.

Flight Season AZ Apr–Nov.

Distribution Ranges south in uplands to Honduras.

Painted Damsel *Hesperagrion*

This genus includes one beautifully colored species of Mexico and the southwestern United States. Because of color changes in females and male appendages, it is considered close to Middle American *Apanisagrion* and *Anisagrion* with which it shares some attributes, for example, brightly colored forewing stigmas, as well as to forktails. World 1, NA 1, West 1.

68 Painted Damsel *Hesperagrion heterodoxum*   TL 28–35, HW 16–21

Description Southwestern pond damsel, dazzlingly colored at any age.  *Male:* Eyes when mature green with black cap as in forktails. Shows dramatic color change with age, starts out orange with red eyespots and light brown eyes, then develops black on top of abdomen and front of thorax with eyespots red and base and tip of abdomen red-orange. Base of abdomen then becomes black, and pale markings on thorax, including four spots on front, become blue. Typically these are spots but may extend toward one another or even be connected to form an antehumeral stripe separated from blue of sides by very narrow humeral. Stigma black, bordered with bright blue at distal end and blue reflection in membrane beyond stigma. *Female:* Eyes green to tan with black cap, as in male. Immature also orange with red eyespots, may vary to orange thorax and red abdomen. Abdomen becomes increasingly black on top, and much of S7 becomes blue; then thorax becomes blue and black exactly as in male, but no trace of red remains.

Identification Nothing else in our area looks like mature red, black, and blue male, and nothing else that occurs with it looks much like any age stage of **Painted Damsel**, including various intermediate color combinations. Combination of pale spots on front of thorax and

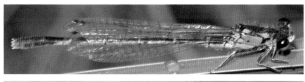

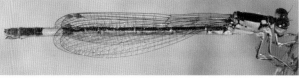

68.1
Painted Damsel
male—Catron
Co., NM, July 2007;
female—Chihuahua,
Mexico, September
2005

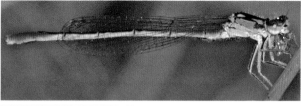

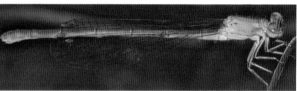

68.2
Painted Damsel
immature
male—Santa Cruz
Co., AZ, September
2004; immature
female—Sonora,
Mexico, September
2005

much red on abdomen of transition male unique, as is extensive blue only on S7 of transitional and mature females. **Pacific** and **Plains Forktails** with spots on thorax but smaller than **Painted** with no red on head or abdomen. Other red pond damsels smaller than **Painted** and almost entirely red. Most similar is female **Western Red Damsel**, orange as immature **Painted** but considerably shorter. Numerous other female pond damsels have blue on S8 or beyond but restricted to S7 in **Painted**.

Natural History Immatures and adults of both sexes may be common at streamside, perching in and flying through dense beds of aquatic vegetation and overhanging grasses and sedges. Like forktails, they spend little time in open. Males mate with females of varying ages (and coloration). Copulating pairs may fly up into nearby trees. Female in mature color oviposits solo in herbaceous vegetation in water, often in watercress. While laying eggs, behaves much as female forktail, flying slowly through vegetation from one stem to another and flashing wings, with conspicuous stigmas, to repel other Painted Damsels.

Habitat Slow-flowing small streams and ditches with much streamside herbaceous vegetation. May become established in garden ponds.

Flight Season AZ Mar–Nov, NM Mar–Nov, TX Apr–Nov.

Distribution Ranges south in uplands of Mexico to Oaxaca.

Red Damsels *Amphiagrion*

This distinctive genus, with two closely related species, is confined to North America. Its nearest relatives are not obvious. Males are bright red, often with much black; females are duller. Thorax is bulky and quite hairy, with a hairy tubercle (bump) beneath it exaggerated in the western species. World 2, NA 2, West 1.

69 Western Red Damsel *Amphiagrion abbreviatum*   TL 24–28, HW 15–19

Description Small, somewhat chunky red or red and black to red-brown or orange damselfly with tubercle (bump) under thorax behind legs. *Male*: Red all over with black on sides, less black on top, of S7–10, sometimes black markings on basal and middle segments, even forming regular black rings around abdomen. Thorax varies from some black in front to entirely black. At darkest, head black and eyes dark brown; paler individuals have red on head and reddish eyes. *Female*: Head, thorax, and abdomen dull brown to orange-brown to bright red-orange. Abdomen often redder than thorax, with black markings on top of S8–9 and sometimes smaller black markings on other segments, forming rings as in male. Some with black on thorax, almost as dark as male. In at least some populations, females entirely dark brown or black. Surprisingly variable.

Identification Only other small red damselfly in range **Western Firetail**, a more slender species with very little black anywhere on body. Bright red eyes and head of male **Firetail** easily distinguish it from **Red Damsel**, with reddish-brown eyes and black head. Females more similar, but female **Firetail** has distinctive (and consistent) pattern of black on front of thorax. In most areas of overlap, **Red Damsel** would occur at higher elevations than **Firetail**, but found together at some places. No other female damselflies in range plain unmarked reddish to reddish-brown.

Natural History Both sexes tend to stay within low and often dense vegetation, although at times feed in more open meadows; night roosts in reeds. Flights short but rapid and more often across open areas than forktails that occur with them. Males and females often found together, as in forktails. Pairs in tandem become common by midday, then decline; oviposit on floating herbaceous vegetation of all kinds, sometimes with male supported upright. Tandem pairs often very elusive, flying some distance when flushed. Tends to be most common in spring.

Habitat Typically marshes associated with springs (both cold and hot) and seepage areas, even in quite arid regions. Also beds of short sedges at borders of lakes, ponds, and slow streams. May be very local in apparently suitable habitat.

Flight Season BC Apr–Sep, WA Apr–Sep, OR Apr–Aug, CA Apr–Sep, MT May–Sep, AZ Apr–Aug, NM May–Sep, NE May–Aug.

Distribution Sparse on northern plains. Ranges south in Mexico to northern Baja California. Extends eastward to eastern Plains, perhaps to Great Lakes.

Comments In eastern part of region, east of Rockies, individuals more slender, with lower bump under thorax, and may be gradual change still farther east to small, slender, relatively

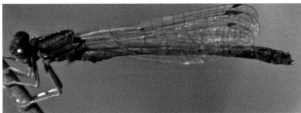

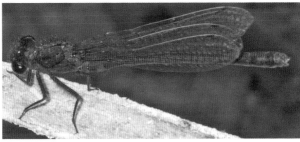

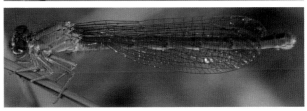

69
Western Red Damsel
male—Harney Co.,
OR, July 2005;
male—Kittitas Co.,
WA, August 1999,
Netta Smith;
female—Grant Co.,
WA, May 2005

long-abdomened Eastern Red Damsel, *Amphiagrion saucium*. The two may be one, but more study is needed.

Aurora Damsel   *Chromagrion*

This monotypic genus is confined to eastern North America, barely extending into the area of this book. Half-open wings, unique color of thorax, and long appendages are unique among North American pond damsels. The anterior side of quadrangle in wings is exceptionally long. It is most closely related to *Pyrrhosoma*, red damselflies with three species in Eurasia. World 1, NA 1, West 1.

70 Aurora Damsel   *Chromagrion conditum*   TL 31–39, HW 20–26

Description   Like large bluet with partially open wings, easily recognized by lack of humeral stripes and yellow on sides. Black front of thorax with wavy edges and no postocular spots. *Male*: Eyes blue, with or without one or two fine black stripes around them. Face blue, most of head black with no postocular spots. Middorsal black stripe on thorax with wavy edges; sides of thorax blue with yellow spot low on each side. Abdomen black with S1, sides of S2, basal rings on S3–7, and S8–9 bright blue. *Female*: Polymorphic, similar to male or with entirely yellow thorax sides. Eyes blue or brown above, gray below, with two black stripes around upper half.

Identification   Some stream-dwelling AMERICAN BLUETS somewhat similarly colored, but all have stripes on thorax, median stripe straight-edged, and of course no trace of yellow.

70
Aurora Damsel
male—Murray Co.,
GA, June 2003, Giff
Beaton; yellow
female—Murray Co.,
GA, May 2006; blue
female—Harvard Co.,
MA, May 2004,
Tom Murray

Long appendages differentiate male from other blue pond damsels. Half-open wings and long appendages may cause confusion with POND SPREADWINGS, but different thoracic pattern and blue color instead of pruinosity make distinction easy.

Natural History Often perches with wings half-open, good field mark at a distance. Males perch on vegetation in open but spend much time in flight over water, presumably looking for stationary females. Copulation for 24–54 min, tandem exploration 12–52 min, oviposition 24–51 min. Pairs oviposit at water surface, male supported upright by female and may be carried at least partially underwater by her. Males and pairs most common at water at midday. Look for it in spring.

Habitat Vegetated edges of clear ponds such as beaver ponds and slow streams, very often spring-fed. Also in wooded swamps with cold water flowing through. Usually in low numbers, even in optimal habitat, and not found far from water.

Flight Season MO May–Jun.

Distribution Widespread in East from Minnesota east to Quebec and Nova Scotia, south to Arkansas and Georgia.

Exclamation Damsel *Zoniagrion*

This monotypic and perhaps primitive genus is superficially forktail-like and may be related to forktails, but its relationships are by no means clear. It is surprisingly restricted in range as the only odonate endemic to California, and this may represent a relict distribution. The eyes are horizontally striped, and postocular spots are well developed as in typical pond damsels. There is a bump under the thorax, as in red damsels, and a slight indication of the paired elevations on S10 of males that are typical of forktails and wedgetails. World 1, NA 1, West 1.

71 Exclamation Damsel *Zoniagrion exclamationis*   TL 33–35, HW 19–22

Description Pond damsel of California that looks like large bluet with blue at base and near tip of abdomen. *Male*: Eyes dark brown or black over light brown or greenish. Postocular spots blue, large, and dumbbell-shaped. Thorax blue and black, antehumeral stripes divided into exclamation point, rarely constricted rather than completely divided; narrow black line low on sides. Abdomen black, blue on S1 and all but lower sides of S2; most of S8 blue, extending onto S7 and S9 but falling well short of abdomen tip. *Female*: Colored as male, antehumeral stripes undivided but slightly pinched;
blue patch on abdomen shorter, on tip of S7 and base of S8. May be polymorphic, with either blue or green thorax, rarely brown. Very old individuals can become pruinose gray on thorax and somewhat on abdomen tip.

71
Exclamation Damsel
male—Sacramento
Co., CA, May 2007,
Jim Johnson;
female—Sonoma
Co., CA, June 2004

Identification Should not be mistaken for BLUET, most like FORKTAIL because of abdominal color pattern. **Swift Forktail**, only large species with which it occurs, has blue extending through S9 and antehumeral stripes undivided and straight-sided. **Swift** has S2 mostly black, **Exclamation Damsel** mostly blue. Bright blue abdomen base distinguishes pruinose female from pruinose FORKTAILS.

Natural History Both sexes forage in sunny clearings near water. Males perch at edge of vegetation over stream, often up to chest height, and both sexes may be present in dense streamside vegetation. Females oviposit alone on vegetation at water surface, including bur-reed leaves hanging into stream. Moves up leaf as she progresses, unlike many other pond damsels that move down.

Habitat Pools in slow small to moderate streams, usually with wooded banks.

Flight Season CA Mar–Aug.

Firetails *Telebasis*

These neotropical damselflies are predominantly red, but there are also a few blue species in South America. Unlike most other North American pond damsels, they lack postocular spots. They are commonly associated with floating vegetation, from tiny duckweeds to water lettuce. Two regional species are distinguished only in hand but may not occur together. World 43, NA 2, West 2.

72 Desert Firetail *Telebasis salva*   TL 24–29, HW 12–16

Description Small red pond damsel of southwestern wetlands. *Male*: Eyes red above, yellowish below. Head black with no postocular spots, face red. Thorax red to red-orange on sides, paler below; wide black median stripe with irregular outer edge and reddish carina; prominent black streak on sides corresponding to humeral stripe, and small black markings closer to wing bases. Abdomen entirely red. *Female*: Thoracic markings as in male. Abdomen brown, with faintly indicated paler narrow basal rings and darker apical rings on S3–8. Abdomen can be rather bright, even reddish like male.

Identification See **Duckweed Firetail**. Narrowly overlaps with **Western Red Damsel**, which is more robust, with relatively shorter abdomen, darker head and thorax, and much black at end of abdomen. Females more similar, but female **Firetails** always have narrow black line down front of thorax, female **Red Damsels** more black.

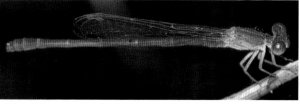

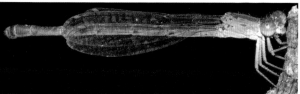

72
Desert Firetail
male—Sonora,
Mexico, August 2006;
female—Sonora,
Mexico, September
2006

Natural History Males become common at water in late morning and early afternoon, followed by buildup in numbers of tandem pairs, but disappear from water later in day. Copulation lasts an hour or more, oviposition a half-hour or more. Pairs oviposit flat on floating vegetation or in grass and other upright emergents; females submerge no more than abdomen. When away from water, may roost up to head height in woody vegetation and probably mate away from water, then fly to it in tandem. Reproductive life up to 22 days.

Habitat Ponds and slow streams, typically in open country but may be bordered by trees and shrubs as long as sun penetrates. Usually associated with floating vegetation such as duckweed or beds of algae, often called "pond scum." Also inhabit grass and sedge beds where floating vegetation is lacking.

Flight Season CA Apr–Dec, AZ Mar–Dec, NM May–Oct, TX Mar–Jan.

Distribution Ranges south to Venezuela.

Comments Another species, Oasis Firetail (*Telebasis incolumis*), occurs not far south of border, in wooded watercourses in Baja California. Extremely similar to Desert Firetail but slightly larger and with slightly different male appendages and female mesostigmal plates. It could well occur in far southern California, and observers should watch for firetails larger than usual.

---

73 Duckweed Firetail *Telebasis byersi*   TL 25–31, HW 13–17

Description Small red pond damsel of duckweed carpets in wooded southeastern wetlands. *Male*: Eyes red above, yellowish below. Head black with no postocular spots, face red. Thorax reddish on sides, paler below; wide black median stripe with irregular outer edge and reddish carina; prominent black streak on sides corresponding to humeral stripe, and small black markings closer to wing  bases. Abdomen entirely red. *Female*: Eyes brown to reddish brown above, tan below. Head brown with black markings. Thoracic markings as in male but light brown on sides, paling to whitish below. Abdomen brown, upper surface darker and may be almost blackish, with narrow pale basal rings on S3–8.

Identification Ranges of two firetail species approach in eastern Texas, but so far not found together. Cerci in this species obviously more than half length of paraprocts, about half in **Desert Firetail.** Hind lobe of prothorax with two small processes pointing up in **Desert**, no such processes in **Duckweed**. Female abdomen tends to be less marked, more reddish brown in **Desert**, darker and more patterned in **Duckweed**, with little hint of reddish.

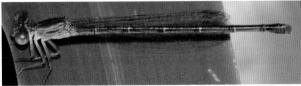

73
Duckweed Firetail
male, female—
Hillsborough Co.,
FL, June 2004

Habitat choice sufficiently different that they may not be found together, but if so, they would have to be distinguished in hand.

Natural History Males fly low over and rest on carpet of duckweed on water surface: Pairs oviposit on same carpet.

Habitat Swamps, usually under canopy and typically with an abundance of duckweed or water lettuce, less commonly in water lilies or grass beds.

Flight Season TX May–Sep.

Distribution Widespread in Southeast, north to southern Illinois and Delaware, mostly in Coastal Plain.

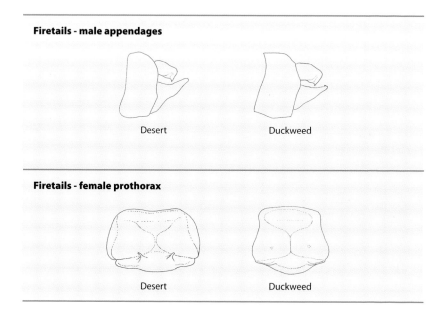

Firetails - male appendages

Desert   Duckweed

Firetails - female prothorax

Desert   Duckweed

Swampdamsels   *Leptobasis*

This is a small neotropical group of slender, swamp-dwelling damselflies, one of which was recently and surprisingly found in Texas. Most have dramatic color change with age, although not closely related to other genera such as *Hesperagrion* that do the same. Females have exceptionally long ovipositors. At least some species spend long dry season as immature-colored adults; this may may be universal in the group. World 8, NA 1, West 1.

74 Cream-tipped Swampdamsel   *Leptobasis melinogaster*   TL 38–40, HW 17–21

Description Very elongate pond damsel of wooded wetlands with cream-colored abdomen tip. *Male:* Eyes green, black above; postocular spots green. Thorax green, black-striped; abdomen with black stripe above, sides of S1–6 pale gray to greenish, with conspicuous black rings at ends of segments; and S7–10 cream-colored, slightly expanded. *Female:* Colored rather like male, with eyes greenish above, yellow below; thorax more obscurely patterned. Abdomen black above, sides green at base, blue on S3–10; S9 with paired black basal triangles, S10 with no black;

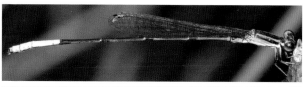

74
Cream-tipped
Swampdamsel
male—Hidalgo Co.,
TX, November 2006,
Robert A. Behrstock;
immature female—
Hidalgo Co., TX, June
2005, John C. Abbott

vivid black rings at ends of S3–7. Ovipositor prominent, with valves extending beyond tip of cerci. Immatures of both sexes overall pale orange-brown with faintly indicated pattern, dominated by black apical rings on S3–6 or 7; abdomen tip pale brown.

Identification Nothing else in its range has glowing cream-tipped abdomen. That color at end of the abdomen more extensive than in other damselflies with contrasty tips of any color, thus recognizable even in deep shade. No other female damselfly in its range has such conspicuous black rings around all middle segments. Otherwise somewhat similar to various BLUETS, but thoracic pattern more obscure. Might be mistaken for POND SPREAD-WING because of size but more delicate, stigmas and legs shorter, and wings always closed.

Natural History Males perch from near ground to head height in shrubs, vine tangles, and tall grass, over or near water, and appear to be present at water during midday hours. Both sexes bob abdomen up and down after landing. Seem less likely to be seen when conditions are hot and dry, even in their shaded habitat. Females much less often seen but may dominate locally toward end of season. May spend dry season (winter) as adults, as related species does to south, or may lay eggs that overwinter. Oviposition has not been observed.

Habitat Shaded pools in slow-flowing streams or forested ponds, all with herbaceous vegetation, especially tall grasses. Not obviously present when these areas dry up.

Flight Season TX Jun–Nov.

Distribution So far known from populations at Santa Ana National Wildlife Refuge, McAllen Nature Center, and Santa Gertrudis Creek on King Ranch. Otherwise known only from southern Mexico (Jalisco and Oaxaca).

Sprites *Nehalennia*

These are very small damselflies of North and South America, with one species also widely distributed in Eurasia. Top of head, front of thorax, and abdomen are metallic green in four temperate species, black in two tropical ones; tropical species have postocular spots and pale antehumeral stripes, temperate species lack them when mature. All have blue on the sides of thorax and tip of the abdomen. Typically they inhabit dense sedge and grass and are less conspicuous than most damselflies but may be very common. Long tibial spines and wings held above abdomen when perched may indicate they are flycatchers. Wing position is also an identification clue, shared only with dancers in North America. Pairs oviposit with the male held up like a little blue-tipped stick. World 6, NA 5, West 3.

Description Tiny dark damselfly extremely restricted in range if still extant in West. Bright blue colors visible at close range. *Male*: Eyes blue, blackish above and paling to whitish below. Narrow pale postocular spots joined by bar. Very narrow pale blue antehumeral stripe at outer edge of front of thorax, with black humeral stripe varying in thickness. Sides of thorax blue, brownish in younger indi-  viduals. Abdomen black, sides of S1–2, S8–9, and all of S10 blue; also narrow blue basal rings on S3–9, scarcely visible from above. *Female*: Colored as male but duller. Immature patterned as mature but all pale areas tan rather than blue; sides of thorax with brown stripe.

Identification Nothing in its range in West is much like it. **Southern Sprite**, similarly small and slender, is metallic green above with a bit more blue visible on abdomen tip. Mature **Fragile Forktails** show no trace of blue and have divided antehumeral stripes closer to middle of thorax. Immatures can have bluish thorax but still show green eyes and typical striping. No females of similarly small dark damselflies have S10 blue.

Natural History Adults apparently spend much time away from water and are often seen in immature coloration, perhaps a lengthy maturation period waiting for rains. Flies throughout year in Florida.

Habitat Grass and sedge marshes, may retire to woodland when not breeding.

Flight Season TX Oct.

Distribution Specimens collected on upper Texas coast in 1918, not found in our region since and may not be here now. Otherwise restricted to Florida.

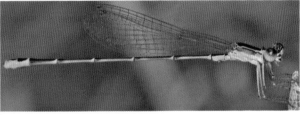

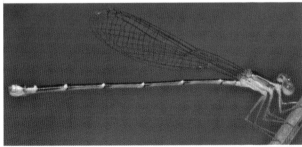

75
Everglades Sprite
male, female—Dade
Co., FL, April 2005

Description Tiny metallic green and blue southeastern damselfly. *Male*: Eyes blue with dark brown to black cap; head metallic green. Thorax metallic green in front, blue on sides. Abdomen metallic green above, blue on sides of S1–2, sides of S8, and all of S9–10 except for paired black streaks on top of S9. *Female*: Eyes blue with brown cap, usually bit duller than male. Color as in male, but blue at abdomen confined to S10 and spot at tip of S9, sometimes sides of S9 as well. Immature duller, with narrow postocular spots joined by fine line and very fine pale antehumeral stripe.

Identification Nothing else like this species in its range in West, not even any very small damselfly with thorax all dark in front and limited blue abdomen tip. **Sedge Sprite** very similar but has more blue on abdomen tip, including tip of S8; known range separated by whole tier of states. Immature might be confused with BLUET or FORKTAIL, but pattern of pale color on abdomen tip as well as very small size should be distinctive.

Natural History Can be very common in dense vegetation, so close search is always advised when looking for sprites. Often detected only by sweeping net through grass beds. Pairs remain in tandem for oviposition in floating vegetation and grass and sedge stems at water level. While laying eggs, female holds abdomen between wings like female threadtail.

Habitat Ponds and lakes with dense vegetation, usually sedges or grasses, at shore or throughout.

Flight Season TX Apr–Jul.

Distribution Widespread in the Southeast from Arkansas to Rhode Island.

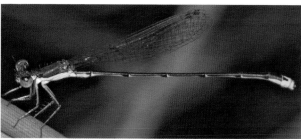

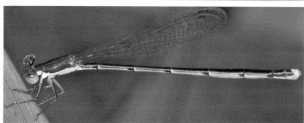

76
Southern Sprite male, female—San Jacinto Co., TX, May 2005

77 Sedge Sprite *Nehalennia irene*   TL 24–29, HW 13–17

Description Tiny metallic green and blue northern damselfly. *Male*: Eyes blue with black cap; head metallic green. Thorax metallic green in front, blue on sides. Abdomen metallic green above, tip of S8–10 blue with pair of black spots on top of S9; also sides of S1–2, S7–8 blue. *Female*: Polymorphic in overall range, but western populations all seem to be andromorphs, colored exactly as male but less blue on S8, just small spot at tip. Situation at east end of region not known, but farther east most females heteromorphs, with blue only at tip of S10.

Identification Nothing else like this species in its range in West. Other small damselflies with blue abdomen tips are FORKTAILS, all

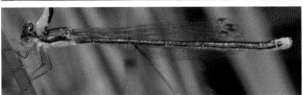

77
Sedge Sprite male,
female—Chelan Co.,
WA, June 2004

of which are black rather than metallic green on thorax and have green rather than blue eyes, obvious postocular spots, and either striped or dotted thorax. **Black-fronted Forktail** most similar, with unmarked thoracic dorsum and blue abdomen tip, approaches range of **Sedge Sprite** in California and Oregon mountains. Equally similar **San Francisco Forktail** only on California coast.

Natural History Abundant at many localities. Mostly associated with dense beds of sedge and often much more common than first apparent, as many individuals deep in vegetation, both short and tall. Sometimes seen in more open vegetation, however, and can be common on open sphagnum mats. Pairs oviposit in upright or floating stems, mosses, and detritus just below water surface.

Habitat Wide variety of habitats, from sedgy lake and pond shores to sedge meadows, bogs, and fens.

Flight Season YT Jun–Jul, BC May–Aug, AB Jun–Aug, WA Jun–Sep, OR Jun–Aug, CA Jun–Aug, MT May–Aug, NE Jun–Aug.

Distribution Also across southern Canada to Newfoundland and south to Iowa and South Carolina.

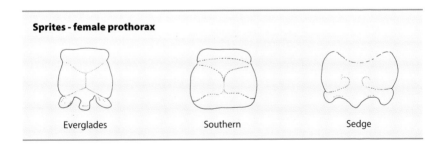

Sprites - female prothorax

Everglades   Southern   Sedge

Dancers *Argia*

The flight style of dancers gives them their name, a bouncy or jerky movement through the air quite different from the smooth forward motion of bluets, forktails, and other pond damsels and more like JEWELWING flight. They are much more likely to land flat on rocks, logs, or soil than other types of pond damsels, and they stay in the open, not moving slowly through vegetation as do many other pond damsels. Males show varying proportions of blue and black, not so different from bluets in coloration but quite different in behavior. A few lack blue, and a few add red eyes and copper thorax to the basic dancer color pattern. They usually perch with wings raised above the abdomen (all other North American pond damsels except sprites hold them alongside the abdomen) and sally forth after flying insects (others glean from vegetation). The wings have a slightly shorter petiole (they become broader closer to the base), probably an adaptation for quick flight to catch flying insects, and the long tibial spines that characterize the genus are surely effective in that activity. Many, perhaps all, species open and then close their wings in what has been called "wing-clapping."

Blue dancers are usually distinguished from bluets by humeral stripe either notched at end, clearly forked into two branches, or distinctly narrowed in the middle and expanded to a spot at the upper end (bluets are never forked, are same width throughout or smoothly narrowed to upper end). Seepage Dancer, an exception, looks much like bluet but holds wings above the abdomen. Dancers with blue-tipped abdomen have blue S10, whereas bluets with blue tips have S10 black on top (except long, slender Attenuated Bluet and very local Baja Bluet). Also, black markings are usually on sides of S2 in dancers but on top in bluets. In female dancers, dark markings on abdomen are paired dorsolateral spots or stripes; in other pond damsels, dark markings are central, either covering entire top of the segment or, in bluets, torpedo-shaped. Finally, female dancers lack vulvar spine present on American bluets.

Although most species perch on rocks or ground in open areas, some prefer vegetation. At water, they typically perch along overhanging stem or leaf facing stream rather than hanging at tip facing shore, as do bluets; this is an excellent clue. Copulating pairs are infrequently seen (copulation must be brief), but tandem pairs are commonly seen moving about habitat and ovipositing. Typically females insert eggs in plant tissues at and just below water surface, the males either grasping plant stem or, more commonly, held vertically in air (not all pond damsels do this). Pairs aggregate at good oviposition sites, perhaps attracted because a pair already present indicates no nearby predators. Exposed rootlets of streamside trees in swift currents are preferred by many species for egg laying. Some, especially females, will have to be captured for identification. This largest genus of Odonata in the New World is well represented in North America, although most species are southwestern. A half-dozen species can easily be found at some streams in Arizona. Diversity increases southward (over 50 species exist in Mexico), and additional Mexican species might be found north of the border.

Most dancers have similar head patterns, with front of head mostly pale, then transverse black occipital bar between eyes surrounding ocelli, with one or two pairs of forward projections from bar. Behind that are two moderate to large postocular spots almost connected by a rather thick pale bar. When the black head bar is narrow, entire head looks pale; when broader, the head looks dark above, and postocular spots are prominent as in other damselflies. Descriptions of spots are included for species that differ from this general pattern. Important field marks in both sexes include width of black median and humeral stripes on thorax and whether humeral is forked or not. The abdomen usually has S8–10 blue, middle segments blue at front and black at rear, either blue or black taking up most but not all of segment. On the abdomen, look for presence or absence of spots on sides of middle abdominal segments in males; presence or absence of black low on sides of terminal segments of males; basal spots or continuous stripes on middle abdominal segments in females; and unmarked, spotted, or striped S8–10 in females. Extent of black markings is at least somewhat variable on all species—but fortunately not infinitely variable! One venation character is important in this group: the number of cells between the quadrangle and the vein that runs back from the nodus. Generally the larger species have more of these cells, but this character may be used to distinguish some similar species. World 111, NA 32, West 32.

Table 4   Dancer (*Argia*) Identification

	A	B	C	D	E	F
Seepage	1	3	1	3	15	5
Blue-fronted	2	4	6	4	1	1
Powdered	6	4	1	34	1	1
Blue-ringed	2	4	1	3	234	7
Paiute	2	3	1	1	2	1
Leonora's	12	3	1	2	2	1
Golden-winged	2	4	1	23	1	7
Variable	4	4	2	24	1	1
Lavender	4	3	1	24	1	1
Amethyst	5	4	2	4	1	1
Spine-tipped	1	4	1	4	124	3
Aztec	1	4	1	24	2	1
California	1	4	1	24	2	1
Yaqui	1	3	1	1	24	7
Kiowa	4	3	1	12	24	7
Harkness's	1	5	2	4	1	1
Tarascan	1	4	1	1	321	7
Sierra Madre	1	5	1	24	1	2
Pima	1	5	1	1	1	2
Sabino	1	5	1	1	1	4
Springwater	14	4	1	4	24	73
Vivid	1	4	1	42	4	73
Apache	1	5	2	4	21	7
Comanche	1	5	1	4	1	1
Tonto	4	5	2	4	1	2
Emma's	4	4	2	4	26	7
Sooty	3	5	3	1	1	1
Blue-tipped	2	4	1	3	15	6
Dusky	23	5	4	1	1	1
Tezpi	3	5	4	1	15	1
Coppery	2	4	5	5	1	1
Fiery-eyed	14	4	5	5	1	1

A, male abdomen: 1, mostly blue; 2, mostly black, blue tip; 3, all black; 4, mostly violet; 5, red-violet; 6, black with pruinose tip.

B, modal postquadrangular cells in forewings

C, median thoracic stripe in male: 1, wide; 2, narrow; 3, wide and split; 4, very wide; 5, entire front metallic; 6, none.

D, humeral stripe in male: 1, wide, forked; 2, narrow, forked; 3, wide, unforked; 4, narrow, unforked; 5, entire front metallic.

E, black markings on S4–5 in female: 1, stripe; 2, stripe plus spot; 3, spot; 4, stripe + double spot; 5, all black; 6, none.

F, black markings on S8–9 in female: 1, stripe on both or stripe on S8, basal spot on S9; 2, basal spot on both; 3, small spot on S9; 4, tiny apical spot low on S8; 5, S9 all black; 6, all black; 7, no black.

78 Seepage Dancer   *Argia bipunctulata*   TL 23–30, HW 13–18

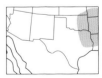

Description Small bluet-like dancer of vegetation beds. Head dark above, no postocular spots, but pale connecting bar present. *Male*: Eyes blue. Entirely blue except for these black markings: wide median and humeral stripes on thorax; black blotch at end of S2; rings at end of S3–5; mostly black S6, blue on dorsal base; and all of S7. *Female*: Eyes tan with dark brown cap. Thorax patterned as male but light areas light tan. Abdomen mostly pale brown on S1–3, S4–5 mostly black above and at end and pale on sides, S6–7 and S9–10 black, S8 entirely pale tan above. Apparently no blue females.

Identification Most similar dancer **Aztec Dancer**, with which it overlaps in eastern Oklahoma and Texas. **Seepage** has much wider black humeral stripe and top of head mostly black, **Aztec** narrow humeral stripe and top of head mostly blue. Typically **Seepage** has more black on S6 than **Aztec**, which has only apical ring. Easily mistaken for BLUET but distinguished by wings held above abdomen and very wide humeral stripe, much wider than blue antehumeral, so thorax looks mostly black in front. Small BLUETS such as **Atlan-**

78
Seepage Dancer
male—Calhoun Co.,
FL, April 2005;
female—Clay Co.,
FL, April 2005

tic Bluet also have black on top of S10, lacking in blue dancers. No other female dancer has mostly dark abdomen, black at tip but with pale S8. Rather similar female **Skimming Bluet** has black markings on blue S8.

Natural History Populations are scattered because of special habitat needs and not very dense where they occur. Males perch in low and often dense vegetation, sometimes at edge of open water, and seem rather sedentary. Females seldom seen except in pairs.

Habitat Boggy areas with abundant sedges and weedy flowing ditches.

Flight Season TX May–Jun.

Distribution Widespread in Southeast, north to Missouri, Ohio, and New Jersey.

79 Blue-fronted Dancer *Argia apicalis* TL 33–40, HW 20–25

Description Widespread eastern dancer with distinctive unpatterned thorax. Front and top of head blue, black at rear with small separated postocular spots. *Male*: Eyes blue in front, brown behind, look all blue in face view. Thorax blue in front, pale whitish-tan on sides, with conspicuous black polygon above each middle leg. Abdomen mostly black above, pale tan on sides; S8–10 blue, with black low on sides, rarely extending up onto top of S8. Males often have gray-violet thorax, much duller than usual, this color more likely at low temperatures or in tandem. *Female*: Polymorphic.
Eyes brown, darker above; lack of blue in eyes in andromorph good distinction from male. Black blotch at base of middle leg smaller or even lacking. Andromorph with all-blue thorax as in male, blue extending back along sides of basal abdominal segments; heteromorph entirely brown. Abdomen black above, dorsolateral markings often split by fine pale line but continuous from S2 to S9. Ovipositing blue females also become duller, blue-gray. Some females green and there is known to be color change with age in individual females.

Identification No other species colored just like male, with almost entirely blue thorax and blue abdomen tip. Unique even with gray thorax. Females much like female **Powdered Dancer**, with unmarked thorax, and can be difficult. **Blue-fronted** has more vividly striped abdomen tip, with conspicuous ventrolateral stripe on S9 that is mostly lacking in **Powdered**; thus, **Powdered** shows large pale area on side of abdomen tip. **Blue-fronted** also more likely to have partially developed black humeral stripe.

79.1
Blue-fronted Dancer
male—Jackson Co.,
FL, June 2004; dull
male—Caldwell Co.,
TX, July 2004

79.2
Blue-fronted Dancer
brown female, blue
female—Johnson
Co., IA, July 2004

Natural History Commonly perches on ground but also low in vegetation. At one pond, males arrived early and became spaced at 6-foot intervals, typically at water for about 3 hr but not retaining territories from day to day. Often aggressive toward other males, rarely to pairs. Females arrive about 2 hr after males, maximum mating just past midday. Copulation lasts 10–27 min, tandem exploration 10–50 min, and tandem oviposition 53–115 min. About half of females continue ovipositing alone for 20 min after release by males. Horizontal floating substrates typically used for egg laying, with pairs concentrating in small areas. Both sexes reproductively active for about 1 week, averaging about one and one-half matings; maximum age about 1 month after maturity.

Habitat Very wide habitat choice includes rivers and streams, less often lakes and ponds. More typical of larger, muddier rivers than smaller streams.

Flight Season NM Mar–Oct, NE May–Oct, TX Mar–Dec.

Distribution Ranges south in lowlands of Mexico to Nuevo León, also east to Ontario and New Hampshire and south to northern Florida.

Description Large dull brown to black dancer with areas of whitish pruinosity ("powder"). Head almost entirely pale in pale western populations, pale postocular line in dark eastern populations. *Male*: Eyes brown, in some populations darker on top, or pale gray-brown. Heavily marked immatures with wide median stripe and wide humeral stripes, some populations with humeral stripe forked or entirely divided with outer fork wider. Wide and complete black stripes on abdomen make it mostly black above with contrasty white rings at base of S3–7 (at most extreme these may
show tinge of blue); S9–10 pale except for dorsolateral stripes. Pruinosity varies with population, in wetter regions (eastern edge of West) may be only on front of head, front of thorax, and S9–10 at most, in drier regions covering head, all of thorax, and S8–10, and in arid Southwest may extend to cover much of abdomen as well, producing a whitish to gray damselfly. Thoracic pattern at least faintly visible no matter how pruinose. Very young males in West can virtually lack thoracic markings, like females, and have S8–10 conspicuously pale as if abdomen tip were to become blue; but it does not! Range of variation can be seen across width of Texas. *Female*: Polymorphic, either entirely light brown, only head and thorax blue, or largely blue (latter in arid regions); usually brown more common than blue. Blue individuals may be green when younger. Most western populations have relatively lightly marked females with only fine median and humeral stripes, thorax essentially unmarked. Abdomen with wide complete dorsolateral stripes on S2–9 enclosing brown middorsal stripe; obvious ventrolateral stripe on S8, but S9–10 pale from side. In wetter eastern part of region, thorax obviously striped, with wide median stripe and narrow, either unforked or widely forked humeral stripe; never broad humeral stripe as in males. Dorsolateral stripes may almost fuse, so abdomen looks entirely dark above.

Identification Typical heavily pruinose mature male like nothing else, as pruinosity rare in male pond damsels. Male POND SPREADWINGS have pruinose thorax and abdomen tip but

80.1
Powdered Dancer very pruinose male—Hidalgo Co., TX, May 2005; lightly pruinose male— Jasper Co., TX, May 2005; immature male—Fayette Co., IA, July 2004

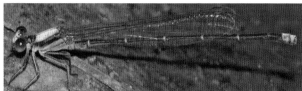

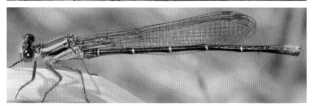

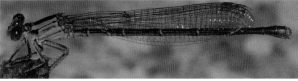

80.2
Powdered Dancer
brown female,
striped blue
female—Jasper Co.,
TX, May 2005;
unstriped blue
female—Llano Co.,
TX, July 2004

are otherwise very different from **Powdered Dancer**. Immature males quite variable but could be distinguished by outer fork of humeral stripe much wider than inner and entirely black abdomen with S9–10 pale, with or without stripes. Female distinguished from female **Blue-fronted Dancer** by almost entirely lacking ventrolateral stripe on S9, so abdomen tip looks pale from side rather than striped. In hand, helpful mark is relatively long stigma, often with crossvein contacting it in middle; in all other dancers but **Sooty**, crossveins at ends of stigmas. Interestingly, this species perches more often than other dancers with wings alongside abdomen.

Natural History Males typically perch on rocks at shore or out in water, also on streamside vegetation. May hover over water for lengthy periods, much more likely to do so than other dancers; typically face upstream but perhaps merely into wind. Males can be abundant in late morning, tandem pairs not until midafternoon, but copulating pairs much less often seen. Copulation lasts 14–31 min, tandem exploration 3–49 min, oviposition 37–67 min. Females oviposit at surface or submerged, either solo or in tandem, to several feet below surface; may stay under for 30 min. Pairs often aggregate in large numbers, even landing on each other while female probing with abdomen. Eggs laid in almost any substrate in live or dead plant tissue, including algal films on rocks. Sexual maturation takes about 2 weeks, average adult longevity 3 weeks, maximum 4 weeks.

Habitat Streams and rivers, from muddy to sandy but often rocky, with wooded or open banks; also on irrigation canals. Typical of largest rivers inhabited by dancers but may also be on quite small streams. Also on large lakes in northern parts of range.

Flight Season CA Apr–Nov, AZ Apr–Nov, NM Mar–Dec, NE May–Oct, TX Jan–Dec.

Distribution Ranges south in Mexico to Jalisco and Tabasco and in East from Ontario and Nova Scotia south.

81 Blue-ringed Dancer *Argia sedula* TL 30–34, HW 18–21

Description Common small, mostly black dancer with bright blue rings and blue abdomen tip. Sexual dimorphism more extreme than in most dancers. *Male*: Eyes blue. Postocular spots small, separated. Median and humeral stripes very wide, so thorax mostly black in front with narrow blue antehumeral stripe. S2 with black side stripes widely fused above,

S3–7 black with conspicuous blue basal rings, S8–10 blue with black lower side stripes. Black on middle segments slightly constricted toward rear, so small amount of tan or blue shows from sides. Males in tandem often have postocular spots and front of thorax dull blue-gray, much duller than usual, and this color also characteristic of individuals at low temperatures. *Female*: Eyes brown; head en-  tirely pale above. Body color light brown. Thoracic markings dull and obscure; wide brown median stripe usually faintly evident with median carina a black hairline, and wide brown forked humeral stripe may also be visible. Mature females show touch of pale blue-green on tiny plates at wing bases. Abdominal markings brown, consisting of apical spot on S2–4 becoming more obscure on rearward segments, dorsolateral spots or stripes on S3–5 or S3–6 similarly disappearing into dull brown patternless S7–10, becoming paler toward end. Pale bluish-green basal rings and sides evident on S3–6 in some, perhaps sign of maturity. Wings distinctly amber-tinted.

Identification In this region, male most like **Paiute Dancer** but only slight range overlap. **Blue-ringed** has thorax mostly black in front, with relatively narrow antehumeral stripes, whereas **Paiute** shows more blue, with those stripes wide. **Paiute** has narrower blue rings on abdomen but two per segment in middle. Female **Blue-ringed** less patterned than most other brown female damselflies, in some almost patternless and only black markings, upright mesostigmal plates and hairline median carina. Also more likely to have wings brownish. Most similar to female **Tarascan**; see that species. Also similar to female **Golden-winged**, which has more deeply tinted amber wings, broad dark median stripe, and black markings on head (scarcely visible if present on **Blue-ringed**). **Blue-ringed** also like **Paiute**, but latter has prominent black markings on all segments back to S9.

Natural History Males perch on herbaceous vegetation at streamside, much more likely on vegetation than rocks. Females often common in open areas near water, perched in low vegetation and on ground. Pairs oviposit in live herbaceous stems or floating dead vegetation at water surface. Reproductive life averages 4–5 days, maximum 2 weeks.

Habitat Small to large streams and rivers, often most common where much herbaceous vegetation along shore but also found in open, rocky stretches. Wide habitat tolerance probably contributes to general abundance and wide range.

Flight Season CA Feb–Dec, AZ Mar–Dec, NM Mar–Nov, NE Jun–Sep, TX Jan–Dec.

Distribution Ranges south in Mexico to Baja California, Querétaro, and Veracruz, and in East from Nebraska, Michigan, and New Jersey south.

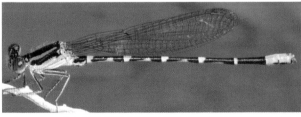

81
Blue-ringed Dancer
male—Caldwell Co.,
TX, July 2004;
female—Upson Co.,
GA, May 2006

Description Small blue-and-black dancer of sedgy habitats. Head with much black above, postocular spots quite narrow. *Male*: Black of top of head extends out as short stripe onto each eye, as if eye has black pupil. Thorax blue with black median and fairly wide forked humeral stripes. Fork varies from rarely absent to one-third of stripe length. S1–2 blue, S2 with narrow black dorsolateral stripes ending in bulge and turning inward to point, like neck, head, and bill of bird. Middle segments of abdomen with black dorsolateral stripes almost filling space available (rarely leaving fine middorsal line), so looks dark with narrow blue rings and light brown to blue on sides. At darkest, entire middle of abdomen black; at lightest, stripes may end so second blue ring visible at two-thirds length of middle segments. S7 entirely black above, S8–10 blue with black low on sides (little on S8, extensive on S10). *Female*: Polymorphic, entirely brown or thorax light blue, abdomen light brown to blue. Eyes tan with limited dark brown cap. Color pattern as in male but humeral stripe often narrower, abdomen with narrow to broad incomplete dark-pointed dorsolateral stripes and dorsal blotches at ends of segments; stripes complete and vivid on S7–9.

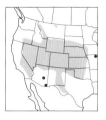

Identification Male most like **Blue-ringed Dancer**, with which it occasionally occurs, but **Paiute** has distinctly wider antehumeral stripes (thorax looks as much blue as black in front) and narrower, forked humeral stripes. **Blue-ringed** entirely black above on middle segments with wider basal rings that make its name appropriate. In many male **Paiute**, pale color of sides visible from above at two points on each middle segment; thus, abdomen looks more patterned. No pale color visible on sides of **Blue-ringed** abdomen.

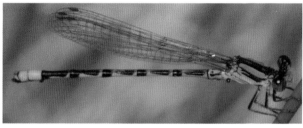

82
Paiute Dancer
male—Harney Co.,
OR, July 2005; brown
female—Inyo Co.,
CA, June 2003; blue
female—Harney Co.,
OR, July 2005

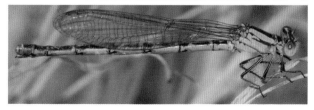

Heteromorph female **Paiute** not too dissimilar from **Blue-ringed**, but paired abdominal stripes usually also evident, not so in **Blue-ringed**. Tip of abdomen entirely pale in female **Blue-ringed**, with black markings in **Paiute**. Female occurs with female **Aztec Dancer** (and, more rarely, **California Dancer**) in marshy wetlands, on average more vividly striped on abdomen but distinguished for certain only by examination of mesostigmal plates. Male **Paiute** more purplish-blue than most other dancers with which it might occur.

Natural History Males can be common in and at edge of dense marsh vegetation, especially tall bulrushes, also out in low sedges, especially when at water. Pairs become more common in midafternoon.

Habitat Shallow sedge marshes, often associated with springs, hot springs in the northern part of the range. Also small sandy streams that flow out of such springs, with slight current and abundant sedges.

Flight Season OR Apr–Sep, CA Apr–Nov, MT May–Oct, AZ Jun–Sep, NM Mar–Nov, NE Jun–Aug, TX Apr–May.

Distribution Ranges south in Mexico to Chihuahua, also east through Iowa to northern Missouri.

83 Leonora's Dancer *Argia leonorae* TL 28–32, HW 15–19

Description Small Texas dancer of sedgy areas, more blue than black. Postocular spots small. *Male:* Eyes blue. All blue except for black markings, which are surprisingly variable. Thorax with wide median stripe and humeral stripe varying from wide and shallowly forked to narrow and deeply forked. Black markings on side of S2 variable, either spot at rear or stripe along side, may or may not  meet across rear of segment. S3–6 with posterior black ring prolonged forward into dorsolateral points or median point extending from a bit less than halfway to as much as almost entire segment; abdomen in most heavily marked individuals looks black above with blue rings but extensively blue from side. S7 mostly black above but still with prominent blue sides, S8–10 blue with black stripe low along sides of all segments. *Female:* Eyes brown. Entirely brown except for black markings. Humeral stripe narrow, unforked or showing faint fork. S2–5 with small basal and large apical dorsolateral spots, in some individuals

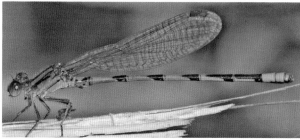

83
Leonora's Dancer
male, female—Jim
Hogg Co., TX, June
2005

fused on most segments; S6–8 with stripes large enough so most of upper surface black; S9 with paired incomplete stripes.

Identification No other blue dancer is so variable in amount of black. Those with maximum amount look something like **Blue-ringed Dancer** from above but show much more blue from the side. In **Blue-ringed**, little or no blue is visible from the side except on rings. Also, **Blue-ringed** has wider median and humeral stripes, thus much narrower blue antehumeral stripe. **Leonora's** less like **Paiute Dancer** because **Paiute** has black markings extending from front of segment pinched off before posterior black ring, whereas black of **Leonora's** widens to rear. These two species not known to overlap in range. Striped abdominal pattern of female **Leonora's** shared with several other dancers of similar size that overlap in range: **Amethyst**, **Aztec**, **Lavender**, and **Variable**. Unfortunately, all are similar enough that capture and examination of mesostigmal plates are necessary to distinguish them, although **Amethyst** probably distinguishable by narrower median stripe. Female **Comanche** has similar pattern but is considerably larger and has complete stripes along middle abdominal segments.

Natural History Males perch in streamside sedge beds and are rather sedentary. At times both sexes common in herbaceous vegetation near breeding sites. Both males and females watched at one weedy site regularly opened and closed wings, seemingly more than in other damselflies. More observations needed.

Habitat Slow streams and seeps and small ponds, usually associated with clumps of very fine, long-stemmed sedges that fall over in tangles rather than standing upright. Often associated with Blue-ringed Dancers.

Flight Season TX May–Nov.

Distribution Ranges south in Mexico to Nuevo León.

84 Golden-winged Dancer *Argia rhoadsi* — TL 34–37, HW 18–21

Description Bright blue and black dancer of south Texas with golden-brown wings. *Male*: Eyes blue. Thorax with moderate median stripe, humeral stripe seemingly either wide and shallowly notched or much narrower and more deeply forked, intermediates not seen. S2 with black side stripes joined across middle. S3–7 black with narrow blue basal ring, S3 with narrow blue midline; S8–10 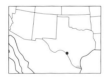 entirely blue. Thorax and even abdomen become darker and duller in tandem pairs. *Female*: Polymorphic, entirely brown or with pale blue-green wash mixed with brown on

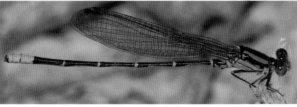

84
Golden-winged
Dancer male—
Hidalgo Co., TX,
November 2005;
female—Kinney Co.,
TX, August 2004,
Martin Reid

head, thorax, and abdomen. Brown eyes. Thoracic pattern as male but stripes duller, brown instead of black. Abdomen with dark brown dorsolateral stripes on S2–3, fused across middle so S4–7 entirely dark above with pale basal rings; S8–10 pale.

Identification Male somewhat similar to male **Blue-ringed Dancer** but a bit larger and paler, more sky-blue, with narrower, forked humeral stripe (the best mark). Blue ring on S3 prolonged backward, not so in **Blue-ringed**. **Blue-ringed** in Texas with wings only lightly tinted with color at most. Females also similar, and with tinted wings, but wings of female **Golden-winged** darker yet. Also has black markings on head, darker stripes on thorax, more distinct dark stripes on abdomen than **Blue-ringed**. **Golden-winged** with blue thorax easily distinguished, as **Blue-ringed** has no andromorph female with that coloration.

Natural History Males perch in vegetation along streams and at pond edges. Males and pairs also seen at outer edge of riparian belt above stream in morning, become more common at water in afternoon. Rather elusive, often fly back into vegetation when disturbed. Both sexes watched at one site regularly opened and closed their wings, more apparent than other damselflies in same area. Pairs oviposited in grass at rain pools in Mexico.

Habitat Dense herbaceous vegetation at border of slow streams, often associated with spring runs. Also found during rainy season in Mexico at isolated rain pools, unusual habitat for a dancer.

Flight Season TX Mar–Dec.

Distribution Ranges south in eastern Mexico to Puebla and Veracruz.

85 Variable Dancer *Argia fumipennis*   TL 29–34, HW 18–23

Description Small violet dancer with two-colored abdomen tip. *Male*: Eyes brown above, violet below. General coloration violet. Thorax with narrow median stripe and narrow humeral stripe with long and divergent fork; lower sides white. Black markings on abdomen include narrow dorsolateral line ending in triangle on S2, complete or almost-complete rings on S3–6, almost all of S7, and conspicuous lower edges of S8–10. S8 violet-blue, S9 blue, S10 blue tinged slightly with violet, unique color pattern among North American damselflies. *Female*: Eyes brown. Coloration brown or reddish-brown with thoracic markings as male. Abdomen with dorsolateral black stripes continuous on all middle segments, extending back to S8 and faintly on S9; most

85
Variable Dancer
male, female—
Tyler Co., TX, May
2005

of last three segments pale but also indistinct black ventrolateral stripe on S8. No blue female.

Identification Male likely to be mistaken for other violet dancers, especially **Lavender Dancer**. Differs from that species by typically having S8 violet in contrast with blue S9–10 (but violet blue enough that confusion possible), also having conspicuous black lower edges of S8–10, whereas **Lavender** has short and not very conspicuous stripes low on S8–9 (good side view necessary to see this). Mature males often but not always show white dot on side of thorax behind humeral stripe that seems distinctive when present, but not seen in **Lavender**. Male **Lavender** also has very white sides, more contrasty white than **Variable**. No other violet species, including **Kiowa** and **Springwater** that occur with **Variable**, shows contrast between purple S8 and blue S9–10. Female in group with narrow forked or unforked humeral stripe and dark abdominal stripes extending length of S8 and onto S9, much as do female **Amethyst**, **Aztec**, **Lavender**, **Leonora's**, and **Paiute**. **Amethyst** has very narrow median stripe, but all others could be confused with **Variable**, so either capture them and look closely at their mesostigmal plates or associate them with male in mating pair.

Natural History Common over riffles and in vegetation along pools. Males rest on rocks, ground, or vegetation at water. Often slowly open and close wings. Pairs oviposit on live and dead plant stems and detritus at surface, occasionally submerging.

Habitat Wide habitat choice, from small streams and ditches with much vegetation to open sandy lake shores. More often at ponds than most other dancers.

Flight Season MT Jun–Aug, AZ Apr–Oct, NM Apr–Nov, NE May–Sep, TX Mar–Nov.

Distribution Sparsely distributed in western part of range. Ranges south in uplands of Mexico to Durango and Hidalgo, throughout East from southern Ontario and Nova Scotia south.

Comments All western populations are subspecies *Argia fumipennis violacea*, which has been called Violet Dancer. Two other subspecies in Southeast have dark wings.

86 Lavender Dancer   *Argia hinei*   TL 30–35, HW 17–21

Description Slender violet dancer of the Southwest. *Male*: Eyes dark brown above, tan below, with bluish-green tinge around dark cap. Thorax violet with moderate median stripe and narrow, forked humeral stripe; lower sides become whitish pruinose and often blue-tinged. Abdomen mostly violet, dark markings include narrow dorsolateral line ending in triangle on S2, partial or complete  rings on S3–6, spots with pointed tails on S4–6, almost all of S7, and indistinct stripes on lower edges of S8–9; S8–10 blue. *Female*: Eyes brown, body brown with reddish tinge. Color

86
Lavender Dancer
male—Travis Co.,
TX, June 2005;
female—Presidio Co.,
TX, February 2000,
Robert A. Behrstock

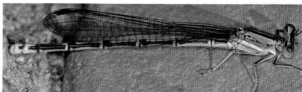

pattern as male but prominent black dorsolateral stripes extend from basal spots on S2–7. S8 with broad but incomplete stripe, S9 with shorter one, faint ventrolateral stripe on S8–9. No blue female.

Identification Delicate and slender damselfly, most like **Variable Dancer,** but S8 blue rather than violet, stripes on sides of S8–9 short and sometimes faint. Eyes not violet as in that species. Other violet dancers even less similar. **Kiowa** duller, darker purple with big spots on bases of middle abdominal segments. **Springwater** a bit larger with narrow, unforked humeral stripe. Females usually with forked humeral stripe, abdominal stripes continuous on middle segments and extending onto S8–9, most like **Variable Dancer** but distinguished by association with male or examination of mesostigmal plates. **Aztec, California**, and **Paiute** have middle segments with anterior stripe and posterior spot not connected. **Amethyst** has narrow median stripe on thorax.

Natural History Males perch on vegetation or rocks, usually at riffles. Pairs oviposit in shallow riffles in upright or trailing sedges, submerged roots, or floating leaves.

Habitat Small to medium streams, usually with much emergent vegetation; often at spring runs. Occasionally at ponds with open sandy edges.

Flight Season CA May–Oct, AZ Mar–Nov, NM May–Nov, TX Apr–Oct.

Distribution Ranges south in Mexico to Sinaloa and Coahuila.

---

87 Amethyst Dancer   *Argia pallens*   TL 32–35, HW 20–22

Description Slender red-violet dancer of southwestern streams. *Male*: Eyes reddish-brown. Body red-violet, becoming increasingly blue-violet toward abdomen tip, but not blue. Thorax with narrow median and unforked humeral stripes, becomes pruinose white below with age. Abdomen with black lateral spots on S2–7, fusing into rings on S6–7; S7 also has black lateral stripe; S8–9 with black  stripe low on sides. *Female*: Eyes brown. Thorax patterned as male. Abdomen with narrow dorsolateral stripes on S2–9, may be interrupted on more basal segments and reduced on S8–9.

Identification Male unmistakable, as no other dancer in its range is such a reddish shade of violet (the color of amethyst, in fact). Could be confused with females, as it looks brown at a distance. Even tip of abdomen not blue as in other dancers. Female in group with spotted abdomen with striped tip. Differs from others of group (**Aztec, California, Lavender, Variable**)

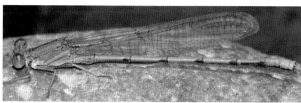

87
Amethyst Dancer
male—Graham Co.,
AZ, August 2007;
female—Cochise Co.,
AZ, June 2004,
Robert A. Behrstock

in narrow median stripe and unforked humeral stripe (forked stripe typical of others, but some individuals unforked).

Natural History Males perch on sun-bathed rocks, most common in open, rocky parts of streams but may also be found at other extreme, on vegetation in shaded pools. Pairs or lone females oviposit in emergent or floating herbaceous stems and leaves, including sago pondweed, at water surface.

Habitat Prefers open, rocky parts of shallow streams and rivers.

Flight Season AZ Mar–Nov, NM Feb–Oct.

Distribution Ranges south in uplands to Guatemala.

88 Spine-tipped Dancer *Argia extranea*   TL 32–37, HW 18–23

Description Heavily marked blue dancer of southwestern streams. *Male*: Eyes blue, uppermost part black. Entirely blue to purplish-blue with moderate median stripe and narrow humeral stripe, latter widened in lower half and with large rectangle at lower end. Black irregular polygon on side of S2 notched at rear; pointed anterior spot or stripe on each side and apical ring on S3–6, the two joined  on S6; S7 almost entirely black, S8–10 blue with black stripe low on sides of S8–9. *Female*: Polymorphic, either entirely brown or with blue thorax and abdomen tip (rarely blue on rest of abdomen). Color pattern as male, black on abdomen a bit more extensive with dorsolateral stripes often extending full length of segments; thus, abdomen can look either spotted or striped; S8–10 usually unmarked except for pair of tiny basal black spots at base of S9.

Identification More purplish males, especially immature ones, could be mistaken for **Spring-water Dancer**, but that species has paraprocts shorter without obvious long point and lower expanded part of humeral stripe with smaller rectangle. Also, **Springwater** shows no black on lower edge of S8–9 and usually no anterior dark spot on S3. **Spine-tipped** not known from range of blue eastern populations of **Springwater**. **Kiowa Dancer**, somewhat similarly marked, is duller, more purplish, and with larger stripes on middle segments joined across segment, producing two rings per segment. Also similar to rare **Yaqui Dancer**, which has rather broad forked or unforked humeral stripe. Black marking on S2 of male, with notch at rear end, distinguishes from **Apache** and **Springwater** of same area, but not **Yaqui**. Female in group of southwestern species with spotted abdomen and mostly unmarked tip (S8–10). Very much like female **Apache, Springwater,** and **Tarascan Dancers**, also with narrow humeral stripe, but black at lower end of humeral stripe more ex-

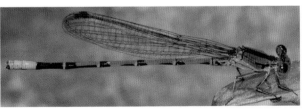

88
Spine-tipped Dancer male—Sonora, Mexico, August 2006, Netta Smith; blue female—Sonora, Mexico, August 2003, Doug Danforth

panded in **Spine-tipped** than in **Apache** or **Tarascan**. **Apache** also has very narrow median thoracic stripe, and **Tarascan** has only sparse spotting on middle abdominal segments. Extremely like female **Springwater** but almost always with black spots on S9, lacking in all but about 10% of **Springwater**. In **Spine-tipped**, posterior edge of mesostigmal plate is straight; in **Springwater**, it bulges to rear on either side of median carina. Others of that type that might be confused with **Spine-tipped** include **Kiowa**, **Sabino**, and **Yaqui Dancers**, all with split or forked humeral stripe.

Natural History Males perch on rocks or vegetation, usually at riffles. Pairs oviposit in herbaceous plants such as watercress in riffles.

Habitat Small to medium rocky and sandy streams with moderate current and usually much emergent vegetation; in open or with shrubby or riparian borders. Also seepage areas above streams.

Flight Season AZ Apr–Nov.

Distribution Ranges south in uplands to Colombia.

89 Aztec Dancer *Argia nahuana* TL 28–35, HW 18–23

Description Small sky-blue dancer of southwestern weedy streams. *Male*: Bright blue all over, thorax with narrow black median and humeral (usually forked) stripes. S2 with pair of subapical black spots, S3–6 with black ring at rear, S7 mostly black, S8–10 blue with black stripe low on sides. Rare lightly marked individuals have median stripe narrower, humeral stripe unforked, and/or S7 all blue. *Female*: Polymorphic, brown or blue (andromorph rare). Eyes brown in brown female, blue in blue female. Color pattern as male but black dorsolateral stripes ending in points along each abdominal segment, continuous on S7, usually incomplete on S8–9, lacking on S10.

89
Aztec Dancer
male—Cochise Co.,
AZ, September 2005;
brown female, blue
female—Chihuahua,
Mexico, September
2005

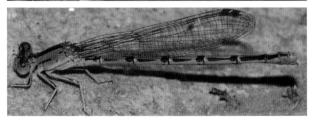

Identification Virtually identical to **California Dancer** in field and overlaps with it locally in Nevada, California, and Arizona; note differences in pattern of dark markings on S2 under that species. Females will have to be captured and scrutinized with at least a hand lens if not a microscope, although close-up photo of male appendages from above could distinguish them. See **California Dancer** for details. Other than this difficult pair, male **Aztec** most likely to be confused with **Vivid**, which is slightly larger and darker blue with black basal spots on middle abdominal segments. Both sexes of **Aztec** likely to have forked humeral stripe, uncommon variant in **Vivid Dancer**, but anterior spots on middle segments should still distinguish male **Vivid**. In females, note larger mesostigmal plates of **Vivid** that are entirely flat and project toward the rear as a flange rather than having tubercle at inner margin. Female **Aztec** quite similar to female **Amethyst**, **Lavender**, **Leonora's**, and **Variable**; capture will be necessary to distinguish them. Andromorph females have blue eyes, unusual in dancers (condition in **California Dancer** not known). Finally, note small percentage of **Aztec** males with S7 largely blue. Could be confused with **Pima** and **Sierra Madre Dancers** because of this, but both of the latter much larger, with longer black marking on S2.

Natural History Typically perches in vegetation at riffles. Pairs oviposit in herbaceous vegetation, including grass leaves, at water level.

Dancers - male appendages

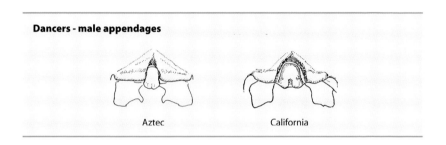

Aztec   California

Habitat Small streams with open banks but abundant emergent vegetation, both sedges and watercress. Usually at riffles rather than pools.

Flight Season OR Aug–Sep, CA Apr–Nov, AZ Mar–Nov, NM Mar–Nov, NE Jun–Aug, TX Mar–Nov.

Distribution Sparsely distributed in Great Basin and eastern Great Plains. Ranges south in Mexico to Jalisco.

---

90 California Dancer *Argia agrioides* TL 30–34, HW 18–20

Description Small sky-blue dancer of west coast streams. *Male*: Bright blue all over, thorax with black median and narrow humeral stripes; humeral stripe usually but not always forked. S2 with black stripe on each side, stripe broken into anterior and posterior spots, or only posterior spot on either side. S3–6 with black ring at rear, S7 mostly black, S8–10 blue with black stripe low on sides. *Female*: Polymorphic, brown or blue, but blue females must be quite rare. Color pattern as in male but black dorsolateral stripes ending in points along each abdominal segment, continuous on S7, usually lacking on S8–9 but present in small proportion of individuals.

Identification Big identification problem here is with identical-looking **Aztec Dancer**. One difference in males involves black markings on S2. **Aztec** west of plains always has apical spot on either side, **California** either spot or stripe, often wider at ends (front and

90
California Dancer
male—Shasta Co.,
CA, July 2004, Ray
Bruun; brown
female—Maricopa
Co., AZ, June 2007,
Pierre Deviche

back end of stripe not joined in some). This means that in areas of overlap, individuals with stripes should be **California**, those with spots must be captured or at least photographed close-up. With a good hand lens or sharp photo from above, tip of abdomen can be examined and species easily distinguished, **California** with gap between paired tori (see Glossary) about width of each torus, **Aztec** with gap narrower than tori. Females more difficult, but in **California**, middorsal carina splits toward front well before mesostigmal plates, whereas in **Aztec**, it remains narrow and diverges only after passing between mesostigmal plates. Female **California** typically lacks markings on S8–9, **Aztec** typically has them, but both vary so that this character is not definitive. Other than this difficult pair, male **California** most likely to be confused with **Vivid**, which is slightly larger and darker blue with black basal spots on middle abdominal segments. Both sexes of **California** likely to have forked humeral stripe, uncommon variant in **Vivid**. Abdominal pattern still distinguishes males. Females distinguished by larger mesostigmal plates of **Vivid** entirely flat, projecting toward rear as flange rather than having tubercle at inner margin. **California** also looks distinctly smaller than **Vivid**, once you gain familiarity with both.

Natural History Males at outer edge of and in emergent vegetation over stream, less often on rocks. Pairs oviposit in tandem, usually flat on plants at water surface.

Habitat Shallow sandy and rocky streams in open with moderate current and beds of vegetation such as watercress. More often at larger streams than Aztec Dancer.

Flight Season OR Jun–Sep, CA Apr–Nov, AZ Apr–Nov.

Distribution Ranges south in Mexico to Baja California Sur.

91 Yaqui Dancer *Argia carlcooki*   TL 29–32, HW 17–20

Description Blue dancer barely present in United States. Postocular spots large, two dashes between them. *Male:* Eyes black over blue. All blue except for following black markings: wide median stripe and moderate humeral stripe, either deeply forked, outer fork wider than inner, or wide with only shallow fork; irregular polygon on side of S2, low basal spot on each side and apical ring on S3–6, all of S7 except narrow basal ring, and lower sides of S8–10. *Female:* Polymorphic. Eyes brown, body blue or brown. Thorax patterned as in male but humeral stripe often partially obscured by brown. Abdomen pale, middle segments with two pairs each of dorsolateral and ventrolateral long oval spots on each; spots joined to make most of S7 black; S8–10 mostly pale, black low on sides of S8. Unusually, extent of black markings may vary seasonally, less extensive early in flight season and more so later. More observations needed.

Identification Only other male dancers along southern Arizona border with same abdominal pattern of dark markings both fore and aft on middle segments are **Apache**, **Kiowa**, **Spine-tipped**, and **Springwater Dancers**. **Kiowa** and **Springwater** both violet. **Spine-tipped** differs from **Yaqui** in prominent square lower end of relatively narrow and unforked humeral stripe, also long tip of paraprocts often visible at close range. **Apache** differs from **Yaqui** in reduced anterior spots on middle abdominal segments and very small black spot on S2. Female as females of many other dancers in size and geographic range but most like those with spotted S4–6 and unmarked S8–10. Distinguished from others of this group as follows. From **Apache**, **Spine-tipped**, **Springwater**, and **Tarascan** by wider, forked humeral stripe, and from **Sabino** by less developed fork (narrow and parallel to stripe in **Yaqui**, wide and diverging from stripe in **Sabino**). **Yaqui** also averages slightly smaller than those other species. **Yaqui** seems more likely than these or other species to have both anterior and posterior spots on sides of middle segments doubled, a larger upper one on top of a smaller lower one, although this also in **Spine-tipped** and **Springwater**. See diagrams of mesostigmal plates.

Natural History Males perch on leaves and rocks, usually at riffles. Pairs oviposit in emergent vegetation.

Habitat Shallow, slow-flowing sandy and rocky streams, small to medium in size, with shrubby or open borders and often abundant emergent vegetation such as grasses and watercress. Often at spring runs.

Flight Season AZ Sep.

Distribution Ranges south in Mexico to Oaxaca.

Comments Tandem pair photographed at spring in San Bernardino National Wildlife Refuge, southeastern Arizona, but no further record. Dancers of this type should be closely

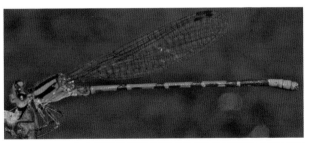

91
Yaqui Dancer
male—Sonora,
Mexico, September
2004; blue
female—Sonora,
Mexico, June 2004,
Doug Danforth

scrutinized in that area. Kiowa Dancer, although rather local in Arizona, found at same spring another year.

92 Kiowa Dancer *Argia immunda*   TL 33–38, HW 19–25

Description Robust dull purple dancer, heavily marked with black. *Male*: Eyes violet, paler beneath. Head, thorax, and much of abdomen violet. Thorax with wide median stripe, moderately wide forked humeral stripe. Black abdominal markings include large polygon on S2, wide black rings and large black basal spots on S3–6, black spots larger toward rear, often fused with ring on S6 and rarely reaches ring on S3–5; S7 almost entirely black; S8–10 blue with continuous black stripe low along sides. *Female*: Polymorphic. Eyes brown. All brown or head, thorax, and S8–10 blue. Patterned as male, but dorsolateral abdominal spots usually longer, in some forming stripes. S8–10 usually plain but may have indistinct ventrolateral stripe, especially on S8.

Identification Male distinctive by dull violet coloration with large basal spots on middle segments, making abdomen look heavily spotted. Other violet dancers with basal markings on middle segments (**Lavender**, **Springwater**, **Variable**) all have markings much smaller, so middle segments largely unspotted. Still other violet species (**Emma's**, **Tonto**) have no anterior markings on middle segments. Exceptionally dark individuals, with only pale rings on middle segments, recall **Blue-ringed Dancer** but of course are violet. Heavily marked middle segments of female contrast with unmarked S8–10, but a few other species similar. Female **Yaqui** most similar, examination of mesostigmal plates necessary to distinguish them. Female **Spine-tipped** distinguished by narrow humeral stripe expanded at lower end and pair of dark spots on S9. **Apache**, **Springwater**, and **Vivid** also similar, but all have narrow, unforked humeral stripe, and **Apache** has very narrow median thoracic stripe abruptly widened at front end.

Natural History Males and pairs common in vegetation in riffles, where they oviposit in green stems. Females often on ground away from water.

Habitat Small to medium rocky streams in open; may have shrubby or riparian borders.

Flight Season AZ Mar–Nov, NM Mar–Nov, TX Jan–Dec.

Distribution Ranges south to Belize, also in northwestern Arkansas.

92
Kiowa Dancer
male—Graham Co.,
AZ, August 2007;
blue female—
Brewster Co., TX,
November 2007,
Netta Smith

Pond Damsel Family   **159**

Description Bright blue, fairly large dancer rare in Southwest. *Male*: Eyes black above, blue below. Thorax blue with rather narrow median stripe and narrow, unforked humeral stripe. Abdomen blue, dark rings at rear of S3–6 with points extending forward on either side; almost all of S7 black, S8–10 blue bordered by black at lower edge. Some with black rings more extensive, S6 may be mostly black. When in tandem, color of thorax shifts from blue to dull purple. *Female*: Eyes brown above, tan below. Polymorphic, thorax brown with blue tinge or entirely blue with narrow dark stripes of male often somewhat obscured by brownish. Abdomen mostly dull blue, middle segments with dark brown posterior ring and stripe extending forward on either side almost to front of segment; thus, prominently striped. S7–9 with dark dorsolateral stripes, extending full or not quite full length of S8, only two-thirds length of S9, but sides of segments completely striped to S9; S10 entirely blue.

Identification Male must be distinguished from other mostly blue, black-ringed (black at rear of segments, no markings at front) dancers in its range. Most similar is **Tarascan**, about same size, bright blue, and with S7 mostly black. Differs from it in narrower thoracic stripes, humeral unforked; also **Tarascan** has simple black rings, **Harkness's** rings with points on either side and often more black. **Sabino** has widely forked humeral stripe, usually more purplish-blue. **Aztec** and **California** both obviously smaller, usually with forked humeral stripe. **Pima** and **Sierra Madre** slightly larger, with mostly blue S7. Female **Harkness's** distinguished by moderate-width unforked humeral stripe, entirely striped abdomen including tip. Most like **Sierra Madre**, which is larger with narrower humeral stripe; also like **Pima**, which has widely forked humeral stripe. Somewhat similar **Springwater** and **Tarascan** have tip of abdomen unstriped.

Natural History Males perch on vegetation or rocks, usually at riffles. Ovipositing pairs collect in shallow riffles with abundant rootlets.

Habitat Fair-sized rivers with moderate current.

Flight Season AZ Jun.

Distribution So far known in United States only from San Francisco River, Greenlee County, Arizona. Ranges south in western Mexico to Guerrero.

93
Harkness's Dancer
male—Sonora,
Mexico, August 2006,
Netta Smith; blue
female—Sonora,
Mexico, August 2006

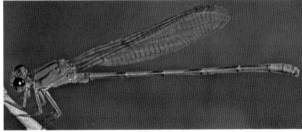

Description Bright blue dancer of southwestern forested streams. *Male*: Eyes dark blue. Wide median and humeral stripes, latter with short to long fork, forks even or outer fork thicker. Side stripe of S2 expanded at rear. Black rings on S3–6, S7 mostly black, S8–10 entirely blue. Front of thorax becomes dull purplish when in tandem. *Female*: Polymorphic, either brown or with much blue. Eyes brown. Median stripe dark brown, borders sometimes obscure, with black median carina; humeral stripe scarcely evident but slightly enlarged at both ends, rarely forked (quite unlike that of male). Abdomen pale with obscure darker dorsolateral stripes on S2–7, more distinct dark dorsal apical spots on S2–6.

Identification Male most like **Pima** and **Sierra Madre Dancers** because of lack of basal markings on middle segments, but black S7 distinctive in Arizona. Males with forked humeral stripes patterned much as **Sabino Dancer** but brighter blue and with less than half of S6 black. Female very plain, black apical spots on midabdomen and darker S7 most conspicuous part of pattern in most and only like one other dancer, **Blue-ringed**, which is a bit smaller, with strongly amber-tinted wings, black median carina only a hairline if visible at all (more conspicuous in **Tarascan**), and prominent upright black mesostigmal plates (plates present but not conspicuous in **Tarascan**). When midabdomen stripes more evident, could be confused with other dancers with narrow humeral stripes such as **Spine-tipped** and **Springwater**, but humeral stripe not ending in bold black area at lower end. Median stripe wider than in **Apache Dancer**.

Natural History Males perch mostly on woody vegetation overhanging pools, but both sexes also on rocks.

Habitat Stream and river banks, wooded or open, with overhanging vegetation; more likely to be at pools than riffles. Irrigation ditches farther south in Mexico.

Flight Season AZ May–Nov.

Distribution Ranges south in uplands of Mexico to Oaxaca.

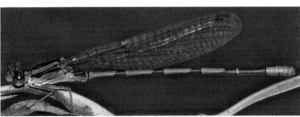

94
Tarascan Dancer
male—Sonora,
Mexico, September
2005; blue
female—Sonora,
Mexico, July 2003

97
Sabino Dancer
male—Santa Cruz
Co., AZ, September
2004; blue
female—Sonora,
Mexico, September
2004, Netta Smith

mostly blue dancers that may occur with it (**Apache, Aztec, Spine-tipped, Tarascan, Yaqui**). Especially look at S6, more than half covered by black; a good distant field mark. Bit of black at rear ends of S8–10 also distinctive but best seen in hand and not present on all. Female distinguished from all other dancers by unique combination of wide, conspicuously forked humeral stripe and much black on abdomen before entirely pale tip. Black spot low on S8 may also be good mark. In both sexes of **Sabino**, large pale postocular spot separated from eye by black line; in similar blue to violet species, pale area extends to eye.

Natural History Males defend small territories on rocks near pools, even on sides of big boulders; move closer to stream by midday, as females come increasingly to water and mating occurs. Pairs oviposit in tandem by slowly backing underwater to several inches deep (may be up to a foot or more) and laying eggs among fine algal growth on rock faces. Pair floats to surface and separates after oviposition, may be under water for up to 37 min.

Habitat Open, rocky streams, most often encountered where stream runs over huge boulders, forming deep plunge pools.

Flight Season AZ May–Oct.

Distribution Ranges south in uplands of Mexico to Jalisco.

98 Springwater Dancer *Argia plana* TL 33–40, HW 22–25

Description Common southwestern darner with blue and violet populations. *Male*: Eastern populations blue, with blue eyes, western populations (west of Texas) violet, with brown to violet to greenish eyes. Thorax with moderate median stripe and narrow unforked humeral stripe (rarely forked) widening at lower end. Lower part of thorax often bluish in violet populations. Black on abdomen: S2 with narrow (sometimes interrupted) stripe ending in triangle on each side; rings on S3–6, becoming wider on posterior segments; long pointed spot on S4–6, (rarely S3–6) joined to  ring on S6; S7 mostly black with slight extension backward from pale basal ring; S8–10 entirely pale to bluish in violet populations. Immature males may be mixed blue and pale violet. *Female*: Polymorphic, either brown or, depending on population, blue or violet, with eyes usually brown but blue in eastern populations. Blue/violet females can be those colors all over or just on thorax and/or abdomen tip. Thoracic markings as in male. Pointed narrow dorsolateral stripes on S2–7, fused with apical ring on S7; S8–10 entirely pale or rarely with small pair of basal spots on S9.

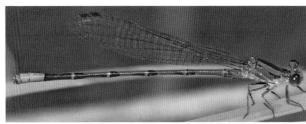

98.1
Springwater Dancer
violet male—Cochise
Co., AZ, September
2005; brown
female—Cochise Co.,
AZ, August 2006;
blue female—
Gila Co., AZ, July 2007

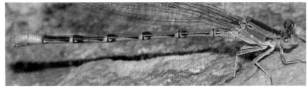

98.2
Springwater Dancer
blue male—Travis
Co., TX, June 2005

Identification Male must be distinguished from species of similar color in range of both blue and violet types. Blue type overlaps with **Vivid Dancer** locally in western plains. Where they occur together, might be distinguishable by **Vivid** being slightly heavier-bodied than **Springwater** and usually having black spot at base of S3, usually lacking in **Springwater** (exceptions in both species). Otherwise, check cerci in top view: inner side distinctly longer than outer in **Vivid**, about same in **Springwater**. Other mostly blue species in range of blue males are **Aztec** and **Comanche Dancers**. **Aztec** smaller and lacks black basal markings at midabdomen. **Comanche** larger, with narrower median stripe and black lines along all middle segments (only on S6 if at all in **Springwater**). Other violet species in range of violet males are **Apache**, **Kiowa**, and **Variable**. **Variable** smaller, with forked humeral stripe and blue S9–10 (entire abdomen violet in **Springwater**). **Kiowa** has much more extensive black basal markings on mid abdomen. **Apache** most similar but usually blue, with very narrow median stripe. See also male **Spine-tipped**, which when immature looks much like **Springwater**. Some female **Springwater** have prominent basal spots on S9 and might be mistaken for female **Sierra Madre** or **Tonto Dancer**, with their narrow humeral stripes, but those two larger species always have obvious spots on S8–9. Female **Springwater** generally similar to two other species with unmarked abdomen tip and narrow humeral stripe. Of these, differs from **Apache** by broader median stripe. **Spine-tipped** most similar, always has black spots on S9 and further distinguishable by examination of mesostigmal plates (see **Spine-tipped Dancer**). Female **Springwater** differs from **Kiowa**, **Sabino**, **Tarascan**, and **Yaqui Dancers** by narrow, usually unforked humeral stripe.

Natural History Males typically perch on vegetation at shore or in water. Only eastern populations studied quantitatively. Males live more than 11 days, females more than 8 days after maturation, and both sexes mate one or two times during their lives. Males not territorial but may be spaced about 3 feet apart along streams because of aggressive behavior toward intruders. Copulation takes 19–40 min, tandem exploration 16–35 min, and oviposition 38–56 min. Pairs oviposit at surface in aquatic vegetation or floating detritus, also in damp clay adjacent to spring; separate quickly after oviposition.

Habitat Small to medium shallow streams, rocky and sandy, in or out of woodland; may be on quite small spring runs and hillside trickles, often at tiny seeps with no other species of dancers. Considered indicative of springs in some parts of range.

Flight Season AZ Jan–Nov, NM Apr–Nov, NE Jun–Sep, TX Jan–Dec.

Distribution Blue populations mostly east of New Mexico and south into northeastern Mexico, east to Wisconsin, Missouri, and Arkansas. Violet populations mostly west of Texas and south in uplands to Guatemala.

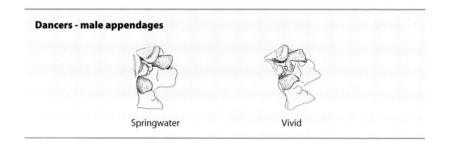

Dancers - male appendages

Springwater   Vivid

99 Vivid Dancer *Argia vivida*   TL 29–38, HW 18–26

Description Large blue dancer common to north of related species. *Male*: Eyes blue. Bright blue with moderate median stripe and narrow, unforked (rarely forked) humeral stripe. S2 with lateral stripe expanded at rear. S3–6 with black apical ring and teardrop-shaped dorsolateral spot; S7 almost completely black, S8–10 entirely blue. Males in tandem or at low temperature may have duller, purplish thorax. *Female*: Polymorphic, some blue females as bright as any male and probably brightest female dancers (but also duller at low temperature). May be all brown, all blue, or blue thorax and abdomen tip, brown in between (least common). Eyes appropriately colored, blue in andromorph, unusual as most andromorph dancers have brown eyes. Thorax marked as in male; abdomen with subapical dorsal spots on S2–6, becoming larger toward rear but not fused with smaller ventrolateral spots; similarly, basal teardrop-shaped spots on S3–6 becoming stripes toward rear and fused with apical spots on more posterior segments; S7 mostly black, S8–10 pale, with small pair of basal spots on S9 (may be lacking). Immatures tan to cream-colored. Some populations in southeastern New Mexico with golden-tinted wings.

Identification Only blue dancer in northern part of its range, where it occurs with **Emma's Dancer**, a violet species with narrow median stripe. Overlaps extensively with **Aztec** and **California Dancers**, from which it differs in having basal spots on middle segments and usually unforked humeral stripe. Also slightly larger and usually darker blue in color than those two species. Much like eastern blue populations of **Springwater Dancer**, with which it overlaps in western Great Plains and Rockies, but more robust with relatively shorter abdomen (typically wings in **Springwater** extend to end of S6, in **Vivid** to middle or even end of S7). Female **Vivid** usually has subapical spot on side of S2, whereas **Springwater** has stripe. See **Springwater** for other differences.

Natural History Males perch on rocks or vegetation adjacent to stream, but most mating occurs away from water, where males bask in morning sun and wait for females moving streamward. Copulation lasts about 30–40 min, followed by an hour of tandem flight before

99.1
Vivid Dancer
male—Grant Co.,
WA, May 2005;
immature
male—Grant Co.,
WA, August 2007

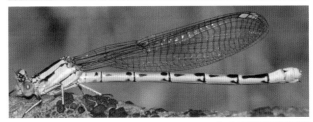

99.2
Vivid Dancer
immature
female—Grant Co.,
WA, May 2005; blue
female—Inyo Co.,
CA, August 1973

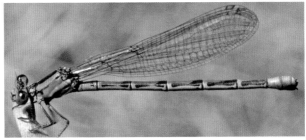

oviposition. Tandem pairs appear at water in late morning, oviposition most common at midday. Single females oviposit into later afternoon. Eggs laid just below water level in stems of aquatic plants. Immatures can be very common near stream, probably do not disperse very far from water in arid country they inhabit. Night roost sites in shrubs or trees, even up in canopy.

Habitat Spring runs and small to medium streams with good current and much emergent vegetation. In open or with associated riparian belt. Both dense sedge beds and rocks can be common, and woody vegetation for roost sites is important. Hot springs in northern part of range. Also at flowing irrigation canals in open country. Typically at smaller and more heavily vegetated streams than Emma's Dancer where they coexist.

Flight Season BC May–Sep, AB Jun–Sep, WA Mar–Oct, OR May–Oct, CA all year, MT Jun–Oct, AZ Mar–Nov, NM Mar–Oct, NE May–Oct.

Distribution Ranges south in Mexico to Baja California Sur.

---

100 Apache Dancer *Argia munda* TL 36–40, HW 22–27

Description Medium-sized bright blue dancer of southwestern mountain streams with early flight season. *Male*: Eyes dark blue over light blue. Body blue (Arizona) to bluish-violet (west Texas) except for black markings. Median stripe narrow but abruptly widened at lower end; humeral stripe narrow but expanded into small rectangle at lower end. Obvious pale line between darker front  and lighter sides of thorax. S2 with small black dorsolateral spot, S3–6 with apical ring and small basal dorsolateral spot, S7 mostly black, S8–10 entirely blue, sometimes bluer than rest of body color. In tandem, color of front of thorax and much of abdomen shifts from blue to dull purple. *Female*: Polymorphic, either brown or blue/violet. Eyes brown in brown female, with distinct blue tinge in blue/violet female. Pattern on thorax and abdomen surprisingly like male in every way, including white markings on thorax, small spots on S3–6, and entirely pale S8–10.

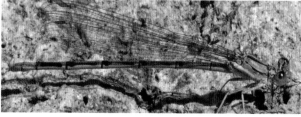

100
Apache Dancer
male—Cochise Co.,
AZ, July 2003; brown
female—Santa Cruz
Co., AZ, April 2007,
Doug Danforth; blue
female—Jeff Davis
Co., TX, May 2005,
Martin Reid

Identification Males slightly on darker or violet side of blue (except for abdomen tip), in comparison with species such as **Aztec Dancer**. **Aztec** lacks dark basal spots of **Apache** on middle segments, but **Springwater** has them. **Springwater** quite similar but darker violet where they occur together and with wider median stripe on thorax and larger black spots on S2. **Springwater** usually has black basal spot and apical ring connected by black stripe on S7, not connected on **Apache**. White line down side of thorax and white at base of first two legs (mostly on the basal leg segment, the *coxa*) seems to be distinctive for this species, although other species may be paler in that area. See also **Spine-tipped** and **Yaqui Dancers**. Female distinguished from similar species by these markings as well as male-like pattern of minimal markings on S3–7 and entirely pale S8–10. Generally similar to **Springwater**, however, with minor differences: one apical spot on each side of S3 (two in **Springwater**) and five antenodal cells (four in **Springwater**).

Natural History Males perch on rocks along stream bed, often at isolated pools when stream low. While awaiting summer rains, pairs may oviposit in low herbaceous plants in dry stream bed, quite unusual for dancers and most other pond damsels. Life history in Arizona seems to be different from those of many other dancers, with peak populations earlier in summer, then disappearing with appearance of monsoon rains. However, later records in New Mexico and Texas indicate some variation.

Habitat Rocky shallow streams with moderate current, drying into pools by midsummer.

Flight Season AZ May–Sep, NM Jul–Oct, TX May–Oct.

Distribution Ranges south in uplands of Mexico to Durango and San Luis Potosí.

---

101 Comanche Dancer *Argia barretti*   TL 38–43, HW 22–25

Description Large blue dancer of rocky rivers in Texas. Large postocular spots with two dashes between them rather than solid stripe. *Male*: Entirely blue, black thoracic markings include somewhat narrow median and narrow humeral stripes, the latter widening into rectangle at lower end. Black abdominal markings include distinctively jagged blotch on sides of S2, narrow ventrolateral lines and black apical rings on S3–6, much of S7 (except for basal blue ring or more extensive pointed dorsal stripe), and ventrolateral lines on S8–10. *Female*: Eyes brown. Polymorphic, either all brown or mostly blue, varying much in intensity. Color pattern as male on

101
Comanche Dancer
male, blue
female—Uvalde Co.,
TX, June 2005

thorax but humeral stripe with only poorly defined enlargement at lower end; somewhat similar on abdomen, but stripes on sides of middle segments broader, leaving pale dorsal stripe on S2–6; also dorsolateral and ventrolateral stripes on S8–9 varying from faint (abdomen tip largely pale) to quite distinct (abdomen tip striped).

Identification This species larger than any other all-blue dancer in its range and with characteristic fine black lines extending forward from black rings on middle abdominal segments; lines visible from side, not distinctive from above. Variation in amount of black on S7 thus should pose no problem. **Aztec Dancer** much smaller with rings but not lines on middle segments and usually with forked humeral stripe. **Springwater Dancer** most similar but has black markings on sides of middle abdominal segments rather than complete lines, smaller black blotch on S2, and S9–10 entirely blue with no black on sides. Females very different looking from other large species with which they might occur: **Dusky**, **Powdered**, and **Sooty Dancers**. Colored most like female **Springwater Dancer** but continuous dark dorsolateral stripes enclosing central pale stripe from base almost to tip of abdomen; **Springwater** has two spots on each middle segment. Considerably larger than female **Aztec Dancer**, with unforked humeral stripe; that species also with spotted abdomen.

Natural History Males perch on rocks and in willows over stream. Pairs oviposit in vegetation in riffles.

Habitat Small to medium rocky streams, usually with swift current and often where herbaceous terrestrial vegetation projects from bare rock substrate.

Flight Season TX May–Nov.

Distribution Ranges south in uplands of Mexico to Puebla.

---

102 Tonto Dancer *Argia tonto* **TL 38–44, HW 25–29**

Description Large violet dancer of southwestern mountains. *Male*: Eyes dark brown. Body violet, slightly reddish on thorax and becoming bluer toward abdomen tip. Narrow median stripe and narrow humeral stripe that widens in lower half and ends in large square. S2 with wide lateral stripe that extends across rear of segment; S3–6 with black apical rings; S7 almost entirely black; S8–10  violet with faint black stripes low on sides. *Female*: Eyes brown. Brown overall with moderate median stripe (distinctly wider than in male), humeral stripe as in male. Complete dor-

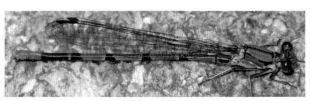

102
Tonto Dancer
male—Cochise Co.,
AZ, August 2007;
female—Cochise Co.,
AZ, August 1999,
Robert A. Behrstock

solateral stripes on S2–6, S7 mostly black, S8–10 pale with paired black spots at base of S8–9 (that on S9 or both lacking in small percentage of individuals).

Identification Male only robust violet damselfly with narrow median stripe and unspotted middle abdominal segments in range; larger than others that might be called violet (**Lavender**, **Springwater**, **Variable**), narrow median stripe diagnostic. No overlap known with **Emma's**, perhaps most similar. Smaller **Amethyst Dancer** red-violet, not blue-violet. Female as other species with spotted S8–9. Unforked humeral stripe distinguishes from **Pima** and some **Sierra Madre**, but other **Sierra Madre** with unforked stripe might be distinguished only by mesostigmal plates.

Natural History Males typically on boulders at riffles. Both sexes feed in sunny clearings in nearby woodland. Pairs oviposit in submerged detritus and twigs along edge of stream.

Habitat Small rocky streams in or out of woodland in mountains.

Flight Season AZ May–Oct, NM Jun–Aug.

Distribution Ranges south in uplands of Mexico to Morelos.

103 Emma's Dancer *Argia emma* TL 33–40, HW 20–25

Description Violet dancer of western rivers. *Male*: Eyes violet, darker above. Basic body color violet, thorax with fine black median and humeral lines and whitish sides; black smudge on side of S2, black rings on S3–6, wider on S5 and much wider on S6; S7 mostly black, and S8–10 blue, with very little black low on sides. *Female*: Eyes brown. Polymorphic, either rich tan all over or thorax and abdomen base pale blue-green. Thorax marked as male, abdomen either entirely pale with black blotches at ends of S3–6 or in addition narrow black dorsolateral stripes from S3–7; S8–10 always entirely pale.

Identification No other violet/purple damselfly in its range. Most like **Lavender Dancer** to south but distinctly more robust; not known to overlap but might at south edge of **Emma's** range. Female palest and least-marked dancer in range; very narrow median, and humeral stripes distinctive. Almost lightly enough marked to be confused with light-colored female **Western Red Damsel**, but **Emma's** always with some black dancer markings on abdomen.

Natural History Males can be very common at breeding habitat, typically associated with no other species of dancers except, at some streams, **Vivid**. Females in nearby uplands, and

103.1
Emma's Dancer male—Benton Co., WA, August 2007; brown female—Yakima Co., WA, June 2005

103.2
Emma's Dancer pairs
with brown and
violet females—
Gilliam Co., OR,
August 2004

pairs become common in afternoon, sometimes in clusters at appropriate oviposition sites, including floating vegetation of all sorts, rootlets of streamside trees, and emergent shore vegetation. Tandem pairs also common away from water, perhaps form there, and spend hours paired. More likely to perch on rocks and ground than vegetation.

Habitat Moderate to large streams and rivers, typically rocky but also with sand or mud substrates and with open banks or shrubby or riparian borders. Typical of larger rivers than most other western dancers but occurs also on small, swift, rocky streams, also large lakes with open, rocky shores in both forested and desert regions.

Flight Season BC Jun–Aug, WA Jun–Sep, OR May–Sep, CA Mar–Sep, MT Jun–Aug, NE Jun–Sep.

Distribution Also in western Iowa.

104 Sooty Dancer *Argia lugens* TL 40–50, HW 25–25

Description Large dark dancer, often with brown suffusion on wings. *Male*: Eyes dark brown. Thorax entirely black or black in front, gray below. Abdomen dark brown to black with tan low on sides of middle segments and narrow tan ring at base of S4–7, at least extreme ring barely visible; S8–10 black. In some populations, perhaps only along southern border, gray pruinosity overlays basic black of head and thorax. When in tandem, thoracic stripes usually become evident. *Female*: Polymorphic, all brown or with head and thorax light blue. Eyes brown. Thorax with split median stripe forming narrow dark stripe on each side of carina, fairly broad but almost completely divided humeral stripe. S2–7 with pale ring at base, dark dorsolateral stripes fusing at posterior end to dark ring; S8–9 with stripes usually incomplete, extending full length of segments in some. Immature male brown, with same pattern as female on thorax and abdomen, gets progressively darker with age until mostly black.

104

Sooty Dancer
male—Santa Cruz
Co., AZ, September
2005; brown
female—Josephine
Co., OR, July 2005,
Jay Withgott; blue
female—Chihuahua,
Mexico, September
2005

Identification Largest dancer in region, followed by **Comanche** and **Powdered**, both of which occur with it but look very different. Nothing else in region looks like mostly blackish male except **Tezpi Dancer**, somewhat smaller and with white instead of tan rings on middle abdominal segments. Male **Dusky Dancer** has purple eyes and, usually, some blue at abdomen tip. Immature male and female **Sooty** have unique thoracic pattern with both median and humeral stripes split. In hand, note that stigma rather large, so crossvein below it contacts middle of stigma rather than only at either end as in all other western dancers but **Powdered**.

Natural History Males perch on large rocks mostly, also on sand, but much less often on plants. Females also perch on rocks when at water, tandem pairs as well. Females more often seen than those of many other dancers, perhaps merely because of larger size. Most mating in early afternoon, when tandem pairs frequently seen. Males usually not held erect in ovipositing pairs. Remain active in shade late in day, probably adaptive in its very hot habitat; also, males on rocks frequently fly out and touch water, presumably for cooling.

Habitat Small to medium open rocky streams with good current; may be bordered by riparian zone.

Flight Season OR Aug–Oct, CA Apr–Oct, AZ Apr–Nov, NM May–Oct, TX May–Oct

Distribution Ranges south in uplands of Mexico to Chiapas.

---

105 Blue-tipped Dancer *Argia tibialis*   TL 30–38, HW 18–24

Description Dark but vividly marked dancer of eastern forests. Postocular spots small and separated to nonexistent, may be dashed line in younger individuals. *Male*: Eyes dark brown over tan, with tinge of blue. Thorax dark violet in front, whitish on sides, with rather narrow median stripe but quite broad humeral stripes. Abdomen entirely black with very narrow pale rings on S5–7, S9–10 blue. Males in tandem often have front of thorax dull red-violet, duller than usual, and this color also characteristic of individuals at low temperatures.

Female: Polymorphic, entirely brown or head, thorax and abdomen base bright blue. Eyes dark brown over tan or greenish gray. Thorax with narrow median stripe and fairly broad forked humeral stripe, outer fork widest. Abdomen mostly black above, S10 and sometimes tip of S9 pale.

Identification Only dancer with its combination of violet, black, and white thorax and blue only on extreme abdomen tip. Thorax color distinguishes from blue-tipped stream-dwelling BLUETS such as **Stream** and **Turquoise**. Other female dancers in range and habitat with mostly dark abdomen are **Blue-fronted**, **Dusky**, and **Powdered Dancers**. Differs from all by its combination of prominently striped thorax and mostly black abdomen tip. Female **Dusky** has fully forked humeral stripe and striped abdomen tip. Females with bright blue, vividly striped thorax could also be mistaken for stream BLUETS, distinguished by lack of blue abdomen tip.

Natural History Males perch on streamside vegetation and rocks, often in shade. Females typically on ground in sunny patches in woodland, but both sexes prefer vegetation to rocks. Pairs oviposit on floating leaves, twigs, and rootlets, even on wet wood above water, sometimes in large aggregations.

Habitat Small wooded sandy streams with slow to moderate current and with or without riffles, less often larger rivers. More tied to forest streams than other dancers in its range.

Flight Season NE May–Sep, TX Apr–Sep.

Distribution Also in East from Minnesota, southern Ontario, and New York south.

105
Blue-tipped Dancer male—Jackson Co., FL, June 2004; brown female—Putnam Co., FL, April 2005; blue female—Jackson Co., FL, June 2004

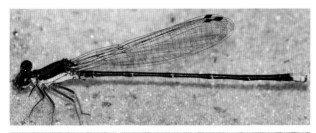

Description Slender dark dancer with limited blue. Postocular spots small, separated; younger individuals with line between them. *Male*: Eyes bright blue-violet. Thorax when immature shows moderate median stripe and fairly broad but almost completely split humeral stripes. Humeral stripe area, then entire thorax, becomes entirely dark gray to blackish with pruinosity during maturation.

Abdomen black with narrow bluish basal rings on S3–6 or S3–7, most also with blue rings on S8–9, blue extending backward on sides of segments to produce somewhat blue-tipped abdomen. Considerable variation in this, from no blue to almost all of S9–10 blue above. Males in tandem undergo dramatic transformation, brightly marked thorax with pale tan to whitish sides and antehumeral stripes contrasting with black humeral stripes; single

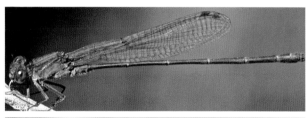

106.1
Dusky Dancer
male—Edwards Co.,
TX, July 2004;
male—Tamaulipas,
Mexico, November
2006, Marion Dobbs;
blue female—Sonora,
Mexico, August 2006

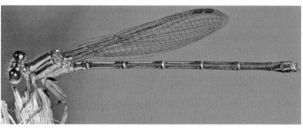

106.2
Dusky Dancer pair
with striped
male—Edwards Co.,
TX, July 2004

males sometimes seen like this, presumably after mating. *Female*: Polymorphic, light areas usually brown or with blue on head and dorsal stripe on abdomen. Thorax as immature male with deeply divided humeral stripe, outer fork wider. Black dorsolateral stripe on S2–10, fused with black apical ring top on S3–6; S7 mostly black. Contrasty ventrolateral stripe on S8–9 and base of S10, so abdomen tip vividly striped in side view.

Identification No other dancer, or pond damsel for that matter, has blue on abdomen tip exactly as male of this species. Quite dark **Tezpi Dancer** superficially most similar, males easily distinguished by blackish eyes, brownish wings, and usual lack of blue color at abdomen tip. **Tezpi** often has darker thorax with metallic purple overtones lacking in **Dusky**. **Sooty Dancer** also has brown eyes, brownish wings, and no blue anywhere on abdomen. Female **Dusky** most like other species with heavily striped abdomen tip. See **Coppery** and **Fiery-eyed Dancers** for differences from those species. Female **Tezpi** similar but usually has humeral stripe less deeply forked, abdominal stripes not extending onto S10, brownish wings.

Natural History Males perch on rocks or vegetation adjacent to water. Pairs oviposit in herbaceous vegetation and rootlets at riffles. In woodland away from water, perch in trees.

Habitat Small to medium slow-flowing sandy or rocky streams or larger rivers, in quite open areas or with wooded banks.

Flight Season AZ May–Nov, NM Jun–Aug, TX Feb–Jan.

Distribution Sparsely distributed in southern Great Plains. Ranges south in lowlands to Argentina, widest distribution by far of any dancer species. Also in band across East from Missouri, southern Ontario, and Maine south to Arkansas and South Carolina.

Description Slender dark dancer of open rocky areas with brown-tinted wings. Postocular spots small, separated, become obscured with maturity. *Male*: Eyes dark brown, almost black, sharply paler beneath. Thorax metallic purple-black in front, sometimes showing narrow brown to reddish-purple antehumeral stripe, dark gray on sides. Abdomen black with narrow white rings at base of S4–7;  S9–10 may be largely brown above, so tip looks a bit paler than rest of abdomen. Males in tandem with thorax more strongly patterned, pale yellowish antehumeral stripe and white sides. *Female*: Polymorphic, either all brown or with blue thorax. Eyes dark brown above, tan below. Thorax with moderate median stripe and broad humeral stripes that vary from undivided to deeply forked with outer fork much wider than inner; many forked about halfway. Abdomen largely dark above, black dorsolateral stripes extending to base of S9; also dark ventrolateral stripes on S8–9 varying from obscure to conspicuous.

Identification Overall dark body with four narrow white rings on middle abdominal segments distinctive. Male **Sooty Dancer** quite similar but clear-winged, a bit larger, and with rings tan rather than white. **Sooty** also usually shows more pale coloration on each middle segment and never shows slightly paler antehumeral stripe except when immature. **Dusky Dancer** also similar but has purple to purple-blue eyes, clear wings, and usually at least a bit of blue on abdomen tip. Pattern of pale markings on end of abdomen different, distal on S9–10 in **Tezpi** and basal on S8–9 on **Dusky**. Both species have contrasty black and white thorax when in tandem, sometimes in lone males. Female with end of abdomen striped, similar to female **Dusky,** but dorsolateral stripes at end of abdomen extend onto S10 in **Dusky**, not so in **Tezpi**. For some reason, striped pattern becomes obscure more often in **Tezpi** than in **Dusky**. **Dusky** always has deeply forked humeral stripe, **Tezpi** rarely so. Humeral stripe and abdomen striping of **Tezpi** more like female **Fiery-eyed**, but tinted wings should distinguish it from **Fiery-eyed** as well as **Dusky**.

107
Tezpi Dancer
male—Sonora,
Mexico, August 2006,
Netta Smith; brown
female—Sonora,
Mexico, September
2004; blue
female—Sonora,
Mexico, July 2005,
Doug Danforth

Natural History Males usually on rocks in open. Both sexes in shady woodland when not at water, often perching in shrubs.

Habitat Open, rocky parts of shallow streams and rivers. Very often with Fiery-eyed Dancer.

Flight Season AZ May–Nov, TX May.

Distribution Ranges south on Mexican Plateau and in Pacific lowlands to Costa Rica.

---

108 Coppery Dancer *Argia cuprea*　　　　　　　　**TL 38–42, HW 22–25**

Description Red-eyed, copper-backed, mostly black dancer of Texas Hill Country. Postocular spots tiny, well separated. *Male*: Eyes bright red in front, dark brown over pale blue or greenish at rear. Thorax coppery-red in front and blue on sides; abdomen black with blue rings at anterior end of each segment (wider on S8), S9–10 entirely blue. *Female*: Eyes brown over pale blue or green. Polymorphic,  pale color mostly light tan or light blue to gray. Front of eyes reddish-tinged, thorax striped with metallic red-black; abdomen prominently black-striped above, also side stripes on S8–9.

Identification Only species in its North American range in which male has bright red eyes and copper and blue thorax. **Fiery-eyed Dancer** of farther west superficially similar but with extensive blue or pale violet on abdomen (**Coppery** mostly black), S8–10 blue (S9–10 in **Coppery**). Females much more similar, but humeral stripe wide and undivided in **Coppery**, largely divided in **Fiery-eyed**. No known locations of co-occurrence in North America.

108
Coppery Dancer
pair—Tamaulipas,
Mexico, November
2006, Marion Dobbs

Only other female with strongly striped abdomen and small postocular spots in range is **Dusky Dancer**, which also differs by divided humeral stripe. Front edge of eyes reddish in female **Coppery**, quite distinctive.

Natural History Pairs oviposit in live vegetation or dead leaves in riffles.

Habitat Open rocky streams with good current; also on forested streams in tropical part of range.

Flight Season TX Apr–Nov.

Distribution Ranges south in lowlands of Mexico to Chiapas.

109 Fiery-eyed Dancer *Argia oenea*   TL 33–39, HW 20–24

Description Red-eyed, copper-backed, mostly blue dancer of rocky southwestern streams. Postocular spots small, well separated, and often obscure. *Male:* Eyes bright red in front, blackish over pale greenish behind. Thorax coppery-red on front and blue to violet on sides. S2 with side stripe expanded at rear; S3–6 blue to violet with black at rear of segments increasing rearward and filling most  of S6 and almost all of S7; S8–10 entirely blue/violet with black stripe low down on side. *Female:* Polymorphic, pale color mostly light tan or light blue to gray. Eyes brown, reddish in

109
Fiery-eyed Dancer
male—Sonora,
Mexico, September
2006, Netta Smith;
blue female—
Sonora, Mexico,
August 2006, Netta
Smith

front. Thorax with wide median and wide, deeply forked humeral stripes metallic dark red. S2–9 black-striped, stripes fusing across middle on S7; also ventrolateral stripes on S8–9, creating striped look from sides.

Identification Although varying from blue to violet, nothing else in range like male. Might be found to overlap with **Coppery Dancer**, another red-eyed species with copper thorax, which occurs to east in Texas Hill Country and has abdomen mostly black with blue tip. Female **Fiery-eyed** differs from **Coppery** in having largely divided humeral stripe. Female much as female **Dusky** and **Tezpi Dancers**, with small postocular spots, divided humeral stripe, and conspicuously striped abdomen tip. Blue female **Fiery-eyed** easily distinguished by coppery thorax and reddish eyes, but brown females more similar to other two. Stripes on abdomen tip extend to S10 in **Dusky**, not in **Fiery-eyed** or **Tezpi**. Also, **Dusky** has humeral stripe split for almost entire length (**Fiery-eyed** slightly more than half length) and narrow but obvious dark stripe behind humeral stripe (lacking in **Fiery-eyed**). Andromorph **Tezpi** lacks reddish head and thorax of andromorph **Fiery-eyed**, but brown heteromorphs quite similar; brownish wings of **Tezpi** then distinctive.

Natural History Males typically perch on rocks in riffles, where tandem pairs also often seen. Pairs oviposit in live (herbaceous plants) or dead (floating leaves and stems) vegetation at water level, female sometimes submerging entirely.

Habitat Shallow, rocky streams and rivers. Very often with Tezpi Dancer.

Flight Season AZ May–Nov, NM May, TX May–Sep.

Distribution Ranges south in lowlands to Colombia.

Dancers - male appendages

Seepage

Blue-fronted

Powdered

Blue-ringed

Paiute

Leonora's

Golden-winged

Variable

Lavender

Amethyst

Spine-tipped

Aztec

California

Yaqui

Kiowa

Harkness's

Tarascan

Sierra Madre

Dancers - male appendages (*continued*)

Pima

Sabino

Springwater

Vivid

Apache

Comanche

Tonto

Emma's

Sooty

Blue-tipped

Dusky

Tezpi

Coppery

Fiery-eyed

Dancers - female mesostigmal plates

Seepage

Blue-fronted

Powdered

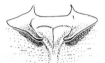

Blue-ringed

Paiute

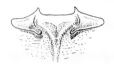

Leonora's

Golden-winged

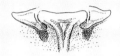

Variable

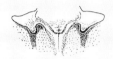

Lavender

Amethyst

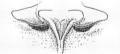

Spine-tipped

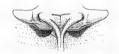

Aztec

California

Yaqui

Kiowa

Harkness's

Tarascan

Sierra Madre

Dancers - female mesostigmal plates (*continued*)

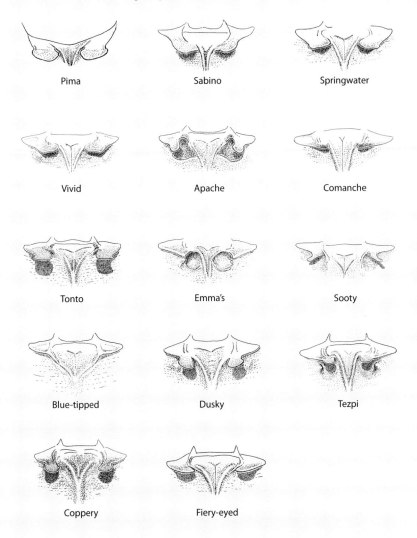

Pima

Sabino

Springwater

Vivid

Apache

Comanche

Tonto

Emma's

Sooty

Blue-tipped

Dusky

Tezpi

Coppery

Fiery-eyed

Shadowdamsel Family *Platystictidae*

This family of forest-stream-dwelling damselflies occurs in the American and Asian tropics. Wings are narrow as in pond damsels and threadtails, and abdomen is usually long and slender. Most species live in dense forest and probably have poor powers of dispersal, as there are many species with small ranges and many known from only a single locality. Many undescribed species are in collections. World 199, NA 1, West 1.

Shadowdamsels *Palaemnema*

Ranging from Arizona to Peru and French Guiana, this genus is most diverse in southern Central America. Often difficult to find in deep forest shade, some species are easiest to locate when one gets down on hands and knees and crawls through the shrubbery along a tiny trickle of water. Often they are found in places where nets cannot be swung but are captured by hand easily. They are the size of large pond damsels, with blue on thorax and usually abdomen tip; a few are entirely black. Prothorax is prominently large, legs long; male cerci are forceps-like, unlike most pond damsels. World 42, NA 1, West 1.

110 Desert Shadowdamsel *Palaemnema domina*   TL 35–44, HW 19–24

Description Moderate-sized slender damselfly of shaded areas with black and blue thorax and blue abdomen tip. *Male*: Eyes dark brown above, dull yellowish below. Prothorax entirely blue, conspicuous. Thorax blue with black stripes, a wide median stripe, narrower humeral stripe, and lower side stripe. Abdomen brown with black terminal ring on S2–7, white to pale blue basal ring on

110

Desert Shadowdamsel male—Graham Co., AZ, August 2007; female—Sonora, Mexico, July 2003, Netta Smith

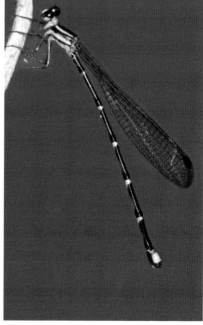

S3–7; S8–9 blue, S10 black. *Female*: Colored as male, but blue on thorax duller and slightly less extensive; only S9 blue, abdomen tip abruptly enlarged.

Identification Easily distinguished from similarly colored POND DAMSELS by relatively longer abdomen with narrow pale ring at base of each abdominal segment. POND DAMSELS with similar coloration all have some blue at abdomen base, **Desert Shadowdamsel** only on thorax and abdomen tip. POND DAMSELS with mostly black abdomens have either no rings or bright blue rings. Females easily distinguished from other similar-sized damselflies by bulbous abdomen tip. Large spot of blue at tip also eliminates female SPREADWINGS that might be perched with wings closed. Other damselflies not likely to hang up in dense woody thickets.

Natural History During sunny periods, adults perch, usually with abdomen hanging down, in dense woody thickets of root masses (often of fallen cottonwoods or sycamores) or branch tangles within a few dozen feet of stream, up to head height. Seem to aggregate in groups of both sexes, easily caught by hand in this situation. Roosting individuals open wings slowly, then clap them shut; significance of this very distinctive behavior unknown. Active at stream when cloudy and cooler, even raining. Pairs have been seen in Mexico on hot midday, however. Move around in tandem for lengthy periods, then female oviposits solo but (at least in closely related species) with male perching adjacent, presumably to guard her from other males. Tandem pairs often locally concentrated, perhaps attract one another. Probably oviposits in herbaceous plants in water as other shadowdamsels do.

Habitat Small, clear, rocky streams bordered by dense riparian vegetation.

Flight Season AZ Jul–Sep.

Distribution Ranges south in lowlands to Nicaragua.

Threadtail Family   *Protoneuridae*

This is a worldwide tropical family of mostly small and quite narrow-winged species. Wing venation is the most simple of all odonates. Generally abdomen is most slender of all odonates, thus the common name; the family is also called "pinflies" elsewhere. Males are often highly colorful, with red, orange, yellow, or blue thorax and/or abdomen. In many species, only the thorax is colored, a slender black abdomen is difficult to see, and males look like spots of color moving over water. Some are entirely dark-colored, almost impossible to see. Note that robust threadtails are rather different from other members of this family. Typically they are found over streams and rivers or edges of large lakes, a very few on ponds. World 259, NA 3, West 3.

Robust Threadtails   *Neoneura*

These are small damselflies, thicker-bodied than slender threadtails *Protoneura*, thus similar in size and shape to typical pond damsels such as bluets and forktails (in fact, it has been suggested that they actually are pond damsels with reduced wing venation). Males of most species are brightly colored, females much duller. Males and tandem pairs hover over quiet water, sometimes well away from shore, or perch on overhanging leaf tips at shoreline. Pairs oviposit on top of floating wood chips or plant stems, with male supported by female. Look for them in eddies behind fallen logs or other objects that interrupt water flow. Females are shorter-bodied than *Protoneura* and do not bend the abdomen sharply to wedge it between wings when ovipositing. They are entirely neotropical, from Texas to Argentina and in Cuba. World 28, NA 2, West 2.

111 Amelia's Threadtail   *Neoneura amelia*   TL 29–35, HW 16–18

Description   Small, slender, red-bodied, mostly dark-tailed damselfly of south Texas. *Male*: Eyes orange-brown, head scarlet. Thorax bright scarlet in front, dull orange on sides, with narrow black median line. Abdomen dark brown with faintly indicated apical black and basal whitish rings; S1–3 mostly red-orange. *Female*: Quite different, pale tan all over with black dots and dashes on head and thorax and darker rings on abdomen.

Identification   Compare with **Coral-fronted** and **Orange-striped Threadtails**. Bright color much more obvious in hovering males than in **Coral-fronted**, which look dark. Females very similar, but **Amelia's** a bit more tinged with orange, especially on head, than **Coral-fronted**; easily seen in comparison. Female differs from most other brown female damselflies by lack of stripes on thorax.

Natural History   See genus account. Males more often seen hovering out over water than perched but often perch in shrubs, sheltered from above by foliage. Females rarely seen except in pairs. Pairs oviposit in floating detritus, sometimes several species of threadtails together with various dancers. Oviposition peaks at midday, then declines.

Habitat   Quiet backwaters at edges of rivers, pools in streams, and large ponds, often muddy. Together with Coral-fronted on Lower Rio Grande.

Flight Season   TX Apr–Dec.

Distribution   Ranges south in lowlands to Panama.

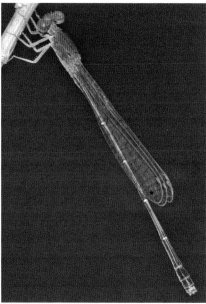

111

Amelia's Threadtail pair—Zapata Co., TX, June 2005; male—Sonora, Mexico August 2006

Description Small, slender Texas damselfly with red on head and thorax. *Male*: Eyes dark red-orange, head mostly red. Thorax bright coral-red in front, dark bluish-gray on sides. Abdomen black, but gray pruinosity develops with age and can produce bluish individuals. *Female*: Quite different, eyes light brown, pale tan all over with black dots and dashes on head and thorax and faintly darker rings on abdomen.

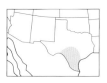

Identification Only species somewhat similar is **Amelia's Threadtail**, which has much more red coloration. **Amelia's** head and thorax look red at a distance, **Coral-fronted** black or bluish-gray with bit of red. Females very similar, distinguished best by structure of prothorax in hand, but **Coral-fronted** does not show slight orange tinge of **Amelia's**. Female **Amelia's** has lobes projecting at either end of rear margin of prothorax that **Coral-fronted** lacks, should be visible under magnification. Other red damselflies (**Desert Firetail**, **Orange-striped Threadtail**, **Western Red Damsel**) are all red. **Western Red Damsel** not in range of threadtails. Female **Desert Firetail**, likely to occur with threadtails, has narrow black stripes between eyes and down front of thorax, no dusky rings on abdomen, shorter body, and not usually in same habitat. Female *Neoneura* easily distinguished from all other female damselflies by black dashed line down either side of carina of thorax.

Natural History See genus account. Males have been seen hovering as high as 15 feet above stream, even with top of high bank. Large numbers sometimes seen with all individuals facing light breeze, whether upstream or down. Can be very difficult to distinguish from

112
Coral-fronted
Threadtail
pairs—Zapata Co.,
TX, June 2005

water surface, so watch for motion. Pairs or sometimes lone females oviposit in floating wood, often in dense groups.

Habitat Slow-flowing hill streams and rivers, more rarely lowland rivers, even Rio Grande.

Flight Season TX May–Sep.

Distribution Also known from Nuevo León in Mexico; Guatemalan literature record probably in error.

Robust Threadtails - female prothorax

Amelia's   Coral-fronted

Slender Threadtails *Protoneura*

These neotropical damselflies are much more slender and elongate than robust threadtails. Males often are brilliantly colored with red, orange, yellow, purple, or blue; females are duller than males but brighter than female *Neoneura*. Pairs oviposit in floating detritus, and abundance of some species may be related to the presence of such detritus. They also use live plants, however, and females can also oviposit alone. In all of these very slender species, the female oviposits with her abdomen sharply bent at end of third and fourth segments with sixth segment wedged between wings, bringing ovipositor close behind thorax, where she can guide it more effectively. The male is conspicuous sticking up from female, usually whirring wings, which may help support him. This action may also move the pair and its floating substrate over the water surface! Entirely neotropical, from Texas to Brazil and in West Indies. World 21, NA 1, West 1.

Description Medium-small slender damselfly with tan- or orange-striped thorax, dark abdomen with white rings. *Male*: Eyes dark red, thorax vivid orange and black striped, legs and first two abdominal segments largely orange; abdomen black with well-defined but narrow white rings. *Female*: Eyes and thorax much duller and paler, little or no orange on abdomen, and abdomen conspicuously thickened toward tip.

Identification Nothing else like it in North America. ROBUST THREADTAILS shorter and thicker, **Amelia's** with redder head and thorax with very little black and extensive orange on abdomen base, **Coral-fronted** with red only on front of thorax. Superficially similar **Orange Bluet** has sides of thorax orange without black stripes and S9 entirely orange, more conspicuous than touch of orange in threadtail. Female with overall darker thorax than **Orange Bluet**, little or no orange at end of abdomen. More similar to female **Burgundy Bluet**; latter smaller, probably no overlap in range with threadtail. Female threadtail has pale antehumeral stripe partially split, not the case in BLUETS. Threadtail abdomen slender at base, looks more swollen than other damselflies when full of eggs.

Natural History See genus account. Males spend much of their time hovering just above water but also perch on slender stems; may slowly rise into trees when disturbed. Occasional aggregations of males have been seen hovering well above head height in sunny clearings near water in what may have been mating swarms, as tandem formation was seen when females entered swarm. Flying individuals often very difficult to see against

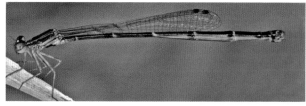

113.1
Orange-striped Threadtail male—Sonora, Mexico, September 2006, Netta Smith; female—Kerr Co., TX, September 2000, Robert A. Behrstock

113.2
Orange-striped Threadtail pair—Sonora, Mexico, September 2006, Netta Smith

dark water, often look like nothing more than orange dot (thus very effectively camouflaged from odonate predators). Pairs may hover for long periods (greater than a minute), then oviposit in any sort of floating detritus, even tiny wood chips barely large enough to hold them. Typically concentrate where currents concentrate detritus but also lay eggs in grass stems and rootlets of woody plants near shore.

Habitat Small, clear streams with sand bottoms in woodland or open.

Flight Season TX Jun–Oct.

Distribution Ranges south in lowlands to Nicaragua.

Dragonflies *Anisoptera*

Petaltail Family *Petaluridae*

These dragonflies are often considered the most primitive living odonates. They are characterized by large size, very long stigmas, somewhat clubtail-like small eyes (but brown, not green or blue), camouflage colors, and semiterrestrial larvae that live in mud or burrows and forage at night on terrestrial insects and spiders. Broad, petal-like cerci of males of the Australian species have given the family its scientific and common names. Females have ovipositors like those of darners but unlike other North American dragonflies. Different authors have placed them at the base of dragonfly evolution or in a slightly more advanced group together with clubtails. Formerly among the dominant odonates and with much greater distribution, there remain relict species in cool, moist climates of Australia, New Zealand, Chile, Japan, and northwestern and eastern North America. World 11, NA 2, West 2.

Gray Petaltail *Tachopteryx*

This southeastern petaltail barely enters our region. Sufficiently distinct in both adult and larval anatomy to warrant a genus of its own, but it is a typical petaltail in behavior. World 1, NA 1, West 1.

114 Gray Petaltail *Tachopteryx thoreyi*   TL 71–80, HW 48–53

Description Large brownish-gray and black dragonfly with separated eyes. *Male*: Eyes brown with gray highlights, becoming grayer with maturity; face gray-brown. Thorax slightly purplish gray with wide black humeral stripe and single black side stripe. Abdomen same color gray, strongly banded with black at rear of S3–7, rest black. *Female*: Similarly colored, readily distinguished by ovipositor and slightly thicker abdomen.

Identification Nothing else like it. Other large dragonflies in range more brightly marked with yellow, usually with eyes blue or green and mostly with eyes larger and touching.

114.1
Gray Petaltail
male—Liberty Co., TX,
April 2005

114.2
Gray Petaltail female—San
Jacinto Co., TX, May 1998,
Robert A. Behrstock

Perching behavior alone is good field mark, although some DARNERS also land on tree trunks. All are colored differently, most much smaller.

Natural History Flight fast and direct. Rests on flat surfaces at any angle—rocks, fallen logs, tree trunks, stone walls—flat against substrate, where sometimes very well camouflaged. Commonly perches vertically on tree trunks from waist height up to 15 feet, also on large stones and rock walls closer to ground. Very likely to land on light-colored net or clothing, a dragonfly you can actually attract by standing in or walking slowly through its habitat! Also quite easily approached, surprisingly tame for large dragonfly. Feeds by sallying out from perch after flying insects, including quite large ones such as butterflies and other dragonflies. Also lazy figure-eight foraging flights to catch small insects in clearings. Sexual maturation in 2–3 weeks, adult life span to 7 weeks. Males not territorial but spend much time at limited breeding sites, also encountered in forest clearings and edge, sometimes at some distance from seeps. Typically search for females on tree trunks. Females oviposit into wet soil or tiny puddles. Larvae in wet mud, usually under leaves.

Habitat Shallow, mucky seeps in forested areas, flat or on hillside and associated with either streams or ponds; also sedge-covered open fens in Ozarks. Ranges from breeding habitat into open woodland, pine flatwoods.

Flight Season TX Apr–Jun.

Distribution Widespread in East from Missouri, Michigan, and New Hampshire south to northern Florida.

These are the smallest and most brightly marked petaltails. Two species of this genus show biogeographic affinity between Japan and the Pacific Northwest. World 2, NA 1, West 1.

Description Moderately large black and yellow dragonfly with separated eyes; black with yellow spots on front and sides of thorax, yellow half-rings on abdomen. *Male*: Eyes dark brown; face cream. Thorax dark brown to black with yellow spots scattered over it. Abdomen black with bright yellow paired and partially joined basal spots on S2–8, becoming smaller to rear. *Female*: Colored similarly but yellow spots on front of thorax may be joined to form pair of stripes, and spots on abdomen more and larger, forming complex pattern. Also distinguished by ovipositor and slightly thicker abdomen.

Identification Nothing else very similar. **Pacific Spiketail** considerably larger with blue eyes, stripes on sides of thorax; flies up and down streams and hangs up to perch. Female DARNERS with big eyes, stripes on thorax; most are larger.

Natural History Flight rather leisurely for dragonfly except when male chases another. Rests against flat surfaces—rocks, fallen logs, tree trunks, stone walls, tops of dried pitcher-plant leaves—at any angle but typically with abdomen inclined downward 20° from horizontal and pressed to substrate. Very likely to land on light-colored net or clothing, and easily approached. Males spend much time at limited breeding sites at midday, defending rather small territories against other male petaltails and chasing most other dragonflies. Frequent moving, time on one territory usually no more than a half-hour. Females rarely seen except when they come to mate and oviposit, more likely later in day after male numbers decline. Copulation at rest, from ground to well up in tree. Females oviposit directly into substrate, including moss, rotting vegetation, and mud, walking slowly and probing for 5–30 min at one site. Larvae in shallow burrows in wet areas, come out to forage for terrestrial prey at night. Burrow openings hard to see. Reproductive adult life span at least 5 weeks.

115.1
Black Petaltail male—
Josephine Co., OR, July 2005,
Jay Withgott

115.2
Black Petaltail female—
Butte Co., CA, June 2003,
Chris Heaivilin

Habitat Seeps at stream edges and shallow boggy areas, usually in hills or mountains but down to sea level in British Columbia, mostly in open or at forest edge. Low sedges typical, also pitcher plants in southern Oregon and northern California.

Flight Season BC Jun–Aug, WA Jun–Sep, OR May–Sep, CA May–Aug.

117.1
Fawn Darner male—
Murray Co., GA,
September 2005,
Giff Beaton

117.2
Fawn Darner female—
Worcester Co., MA,
September 2005,
Tom Murray

Natural History Both sexes cruise around clearings to feed and hang vertically from almost any sort of shaded perch low in woodland, on cliffs, or under bridges. Males patrol streams low over water and usually near shore, typically in late or very late afternoon but also at other times of day, including dawn. May fly when too dark to see them clearly, intermingled with bats. Presumably looking for females, males often spend much time moving slowly along the bank, poking in and out of root tangles and downed branches, circling projecting sticks, and seldom hovering as other darners do. Also forage while doing this, sometimes capturing damselflies they flush. Flight very low and somewhat fluttery, almost butterflylike, distinguishing it from other stream-patrolling darners. Mating occurs at any time, even at dusk; copulation at rest in shrubs or trees. Females oviposit in wet wood at and just below water level, well known for landing on legs of people wading in streams and attempting same (ouch!), surely source of myth that dragonflies sting.

Habitat Wooded streams with some current, from very small to moderate-sized rivers. Sometimes along lake shores.

Flight Season NE Jul–Oct, TX Apr–Jun.

Distribution Ranges throughout East from southern Ontario and Nova Scotia south.

Swamp Darner *Epiaeschna*

A very large darner in a monotypic genus, it is superficially like the pilot darner but with different wing venation and prominent spine in male projecting upward from end of S10. It is related to Cyrano Darner and, as in that genus, there is no hint of constriction ("waist") near base of abdomen. World 1, NA 1, West 1.

118 Swamp Darner *Epiaeschna heros* TL 82–91, HW 52–59

Description Large dark darner with brown, green-striped thorax and green-ringed abdomen. *Male:* Eyes blue, face brown. Thorax brown with green stripes. Abdomen dark brown to black with pale green rings, two to three per segment. *Female:* Colored as male but eyes with more brown, less blue, at maturity. Also distinguished by thicker abdomen and lack of epiproct; cerci about size of male's.

Identification Most similar is **Regal Darner**, less robust but just as long. **Regal** differs in having overall paler look, thorax and abdomen base with more green than brown on both sides and front and more brightly marked abdomen, with green line down center as well as green rings (only ringed with green in **Swamp**). Male **Regals** also have green eyes (blue in **Swamp** and mature female **Regal**) and green face (brown in **Swamp**). Stem of T-spot on frons in **Swamp** very broad, as broad as pale green spots on either side (narrower in **Regal**, bordered by green extending down side of face). Flight style different (see **Regal**). Small **Swamp Darner** might be mistaken for superficially similar **Cyrano Darner**, as they fly in similar ways over swampy pools. Both have blue eyes and green-striped brown thorax, but **Cyrano** has much more extensive green all over slightly thicker abdomen, as well as projecting frons.

Natural History Adults fly back and forth through swamps or forest, sometimes in large numbers, also forage over open areas in woodland habitats. Flight from near ground to well up in trees. Males flying at chest height over ponds and stream pools may be searching for females. Sexes often meet away from water for mating, then female oviposits in woodland pools or extensive swamps, usually in wet dead wood just above water level but also in mud or up on standing trunks.

Habitat Swamps and slow streams for breeding, more confined to woodland than Regal Darner.

Flight Season TX Mar–Oct.

Distribution Ranges throughout East north to Wisconsin, southern Ontario and Quebec, and Maine.

118.1
Swamp Darner
male—Grant Par., LA,
May 2005 (posed)

118.2
Swamp Darner
female—East Baton
Rouge Par., LA, May
2004, Gayle and
Jeanell Strickland

Cyrano Darner *Nasiaeschna*

This is another relatively primitive monotypic genus confined to eastern North America, probably most closely related to Swamp Darner. It is characterized by a rather stocky body, with no waist, and projecting frons. World 1, NA 1, West 1.

Description Stocky, mostly greenish darner of eastern swamps. *Male*: Eyes blue, face green to blue-green above, chartreuse to dull yellow on front. Thorax brown with green stripes: short and L-shaped on front, first side stripe jagged and interrupted, second straight and complete. Abdomen with complex striped pattern, mostly green in middle above brown dorsolateral and green ventrolateral zones; tapers toward tip. *Female*: Colored as male, abdomen thicker with very short cerci.

Identification Looks thick-bodied in comparison with other darners, with no hint of waist. In hand, large, angled frons distinctive, also visible at close range in flight. Jagged thoracic stripes and abdomen with as much green as brown distinguish it from **Regal** and **Swamp Darners** that fly in similar places.

Natural History Males fly regular, often fairly short, beats at waist to head height over open water. Abdomen usually held slightly curved in flight, wings above horizontal and fluttered continuously. Both sexes fly low and slow and pluck prey from herbaceous vegetation, unusual feeding method for flier dragonfly. Prey often large and regularly other odonates up to moderate-sized clubtails, immediately taken to perch in vegetation and eaten. Taking large prey more typical of this species than most other darners, perhaps associated with use of gleaning as foraging method. Also cruise back and forth over clearings in forest at about head height. Females oviposit in sodden logs or stumps at water's edge or just above it, remaining for lengthy periods in one spot.

Habitat Wooded wetlands of all kinds; swamps, lake edges, and slow streams. Border of at least shrubs if not trees seems necessary. Floating and emergent vegetation present or not. Widespread but usually not common.

Flight Season NE Jun–Aug, TX Mar–Sep.

Distribution Ranges throughout East north to Minnesota, southern Ontario, and New Brunswick.

119.1
Cyrano Darner male—Jasper Co., TX, May 2005 (posed)

119.2
Cyrano Darner female—Morris Co., NJ, June 2005, Allen Barlow (posed)

Superficially this is like the larger mosaic darners but a member of a more ancestral group, with more primitive wing venation. Eyes are smaller than those of mosaic darners, the line where they meet no more than twice the length of occiput (three times or more in mosaic and neotropical darners). It also inhabits woodland streams and flies in spring, unlike most mosaic darners. World 1, NA 1, West 1.

120   Springtime Darner   *Basiaeschna janata*   TL 53–64, HW 32–40

Description   Medium-sized stream darner patterned much like mosaic darners but smaller than any of those that occur with it and with small brown spot at base of each wing. *Male*: Eyes brown with blue tinge, face dull yellow-brown. Lateral thoracic stripes bright yellow to whitish, bit of blue at upper ends; frontal stripes narrow and dull greenish, scarcely visible. Appendages simple, quite narrow in top view. *Female*: Eyes brown, face brownish. Polymorphic, abdominal spots green or blue. Cerci very narrow, slightly shorter than S9–10.

Identification   Color contrast between yellow-white thoracic stripes and blue abdominal spots more striking than in most other darners; dull eye color also distinctive. Only other small darner in its western range that flies up and down woodland streams is **Turquoise-tipped Darner** of different genus that overlaps with **Springtime** in central Texas. **Turquoise-tipped** has irregular thoracic side stripes, no trace of brown at wing base, and blue under abdomen tip. Abdomen of **Turquoise-tipped** ringed with green beyond its blue base, whereas **Springtime** has blue spots for entire length. Flight season of **Springtime** separates it in time from most MOSAIC DARNERS and probably from **Turquoise-tipped**. In Canadian part of **Springtime's** range, might be found in same area as **Zigzag Darner**, but latter inhabits different habitats and has different flight season, and **Zigzag**, true to its name, has zigzag thoracic stripes. Also marked only with blue, whereas **Springtime** has yellow to whitish thoracic stripes.

120.1
Springtime Darner male—
Bibb Co., AL, April 2005,
Giff Beaton

120.2
Springtime Darner
female—New Haven Co.,
CT, May 2004, Tom Murray

Natural History Males fly long beats rapidly up and down streams, at knee height or lower and usually near shore when on larger rivers. Often capture damselflies while cruising. Active at water from dawn until dark, one of common dusk-flying species. Both sexes hang up in trees, sometimes in weeds, and often seen in low feeding flights in clearings at forest edge, within few hundred yards of breeding sites. Females oviposit in upright herbaceous plants and floating leaves (for example, bur-reed and spatterdock) out in current, often with abdomen submerged.

Habitat Woodland streams and rivers with some current; also at beaver ponds along stream course, and, sometimes, rock-bordered lakes.

Flight Season TX Feb–May.

Distribution Also throughout East, north to Ontario, Quebec, and Nova Scotia and south to northwestern Florida.

Riffle Darners *Oplonaeschna*

This represents another small genus superficially like mosaic and neotropical darners but with the pointed projection on top of S10 in male distinctive. Otherwise, minor differences involve wing venation and larval morphology. It is thought to be more closely related to two-spined and three-spined darners. Presumably evolved on Mexican Plateau, where it is common. A second species is larger and poorly known. World 2, NA 1, West 1.

121 Riffle Darner *Oplonaeschna armata*   TL 66–75, HW 45–55

Description Medium-sized blue-marked darner of southwestern mountain streams. *Male:* Eyes blue, face yellow. Thorax brown with typical pair of darner stripes, blue above and yellow to greenish below; frontal stripe constricted in middle, often separated. Abdomen brown to black with much blue on S2–3, smaller markings and light blue or greenish on S4–9. Most prominent abdominal mark-ings backward-pointing green triangles above, irregular blue spots on sides. *Female:* Polymorphic, color pattern much like male in andromorph or duller, with eyes brown, pale markings on abdomen light green to yellow in heteromorph. Cerci relatively short (slightly longer than S9, sharply pointed; break off in maturity.

121.1
Riffle Darner male—
Catron Co., NM, July 2007
(posed)

121.2
Riffle Darner female—
Cochise Co., AZ, August
2001, Doug Danforth

Identification Males easily distinguished from NEOTROPICAL DARNERS in its range (**Arroyo** and **Blue-eyed**) by lack of pale areas on either side of black T-spot, yellow face contrasting with dark blue eyes, and prominence of dorsal triangles rather than dorsolateral spots on abdomen. Broken frontal stripe on thorax also good mark when visible. In hand distinguished by projection on S10 and lack of tubercle under S1. Females differ by same color characters. Slightly smaller **Turquoise-tipped Darner** not usually in habitat with **Riffle** but could be distinguished by even finer abdominal markings and S10 mostly blue.

Natural History Both sexes feed over clearings and among trees in open woodland. Males fly up and down stream just above water, mostly on riffles but also check ponds for females, moving slowly along fallen logs and bank. One copulating pair observed flew high into treetops. Females oviposit in moist wood and dead leaves in shallow water.

Habitat Small rocky streams with pools and riffles in mountain woodland.

Flight Season AZ May–Nov, NM May–Sep.

Distribution Ranges south in uplands to El Salvador.

Two-spined Darners *Gynacantha*

The largest genus in this family is well represented in both New and Old World tropics. Most are shades of brown, but many have bright green or blue markings on the thorax and abdomen base. Some have basal wing markings, and wings of mature adults can be quite brown. Females have a prominent two-spined structure projecting downward from S10, probably involved with steadying the abdomen for egg laying. All but a few species are dusk fliers, and their biology is poorly known. Large eyes and broad wings fit them for their daytime flight in forests and dusk flight in clearings. Two North American species are rather similar, but the group includes very small to very large species. World 84, NA 2, West 2.

122 Twilight Darner *Gynacantha nervosa*   TL 75–80, HW 47–56

Description Plain brown dusk-flying forest darner. *Male*: Eyes greenish when fully mature, brown otherwise. Entirely light brown from face to abdomen tip. Small black spots on thorax and faint indications of patterning on abdomen. Mature individuals at brightest with front of thorax greenish, light green sclerites between wings, and fine light green abdominal rings and lines. Abdomen very  slightly constricted at S3. *Female*: Colored exactly as male, with thicker abdomen. With age, wings become increasingly suffused with brown.

Identification Bar-sided Darner, not known to overlap in range in United States, has dark bar on sides of thorax and blue markings on thorax and abdomen base when mature (green when immature, however; from above, note different abdomen shape). Not similar to any other regional species except **Fawn Darner**, which has vivid yellow spots on sides of thorax and never any green markings. Differs from **Pale-green Darner** of similar habits in larger size, mostly brown thorax. Note **Mocha Emerald**, an all-brown emerald that is smaller and darker and with very different terminal appendages.

Natural History Both sexes patrol for small insects in dusk feeding flights, sometimes in large numbers and usually at woodland edge. May establish "beats" along edges a few feet wide and 40–50 feet long. Rarely fly above head height. Commonly enters buildings after dusk flight and may be found next morning. Daytime roosts within forest, also usually below head height, and difficult to see because abdomen usually backed by vertical stem and wings invisible in shade. Mating in forest away from water, apparently at any time of day. Female searches out shallow pools in or at edge of woodland and oviposits in rotten wood or mud. May lay eggs in depressions before they fill with rain, and late-emerging adults may live for months to breed in next rainy season.

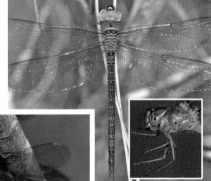

122.1
Twilight Darner male—
Alachua Co., FL,
October 2005, Giff Beaton;
male thorax—Collier Co., FL,
December 2007

122.2
Twilight Darner female—
Highlands Co., FL, August
2002, David McShaffrey

Habitat Wooded swamps and low-lying woodland, even in garden pools; no minimal size for larval habitat.

Flight Season Florida all year.

Distribution Presumably only vagrant to this region, one record in eastern Oklahoma may have been stray from Mexican population. Otherwise mostly in Florida within United States, few records in Alabama and Georgia. Ranges from northern Mexico south in lowlands to Brazil and throughout West Indies.

123 Bar-sided Darner *Gynacantha mexicana* TL 70–76, HW 46–50

Description Large rather plain brown darner with black spots on sides of thorax and dark bar bordering its lower edge. Some individuals with dark bar between second and third longitudinal veins in wings. *Male*: Eyes brown, tinged blue on front with maturity. Very slender abdomen, base expanded and obvious constriction ("wasp waist") at S3. Thorax pale greenish, abdomen brown with  fine green markings across segments. Blue sclerites between wings and on second abdominal segment. Green appears on immatures before blue. *Female*: Colored as male, although not known to develop blue markings (but still poorly known). Abdomen slightly thicker and less constricted near base.

Identification Unmistakable in its range. So far, **Twilight Darner** not found in range of **Bar-sided** in North America. Would be distinguished by lack of markings on sides of thorax and unconstricted abdomen in male. **Fawn Darner** also brown but has bright yellow spots on each side of thorax, no overlap in range with **Bar-sided**. **Pale-green Darner** much like **Bar-sided** in habits and habitat, considerably smaller with no dark markings on side of thorax, no blue color anywhere, and no wing markings. Also typically brighter green thorax.

Natural History Typically seen in dusk feeding flight or hanging from branches or against trunks in shady woodland. Known to gather in communal roosts, several individuals

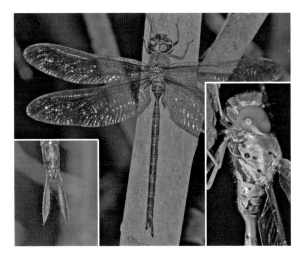

123
Bar-sided Darner male,
male thorax, female
abdomen tip—Hidalgo
Co., TX November 2005

hanging together in small area. As do others of the genus, presumably overwinters in dense woodland, not necessarily associated with water, and female lays eggs on wet ground or in wet wood at beginning of rainy season. One known breeding site in Texas a permanent pond, however. Often enters buildings or attracted to lights at dusk, found hanging there next morning.

Habitat Woodland with low areas that fill with water during wet season. Usually found in dense to open woodland in south Texas.

Flight Season TX Jun–Feb.

Distribution Ranges south in lowlands to Brazil.

Three-spined Darners *Triacanthagyna*

Closely related to two-spined darners, with very large eyes and broad wings, these display three prominent spines under S10 of females rather than two. Most species are smaller than two-spined darners, and all have green and brown thorax and mostly brown abdomen. Forest-based and crepuscular, they make up a prominent component of dusk flight of darners throughout the New World tropics. World 9, NA 2, West 1.

124 Pale-green Darner *Triacanthagyna septima* TL 59–66, HW 34–43

Identification Small, rather drab darner of south Texas. *Male*: Eyes blue-green, face olive. Thorax light green with faintly indicated brownish wash on front. Abdomen mostly brown, sides of S1–2 greenish, also light green dorsal stripe on S2 and fine transverse markings on middle segments becoming obscured toward rear. *Female*: Colored about as male; eyes greenish to brown, perhaps never with blue. Cerci long, narrow, and pointed, broken off in mature individuals. Close look at appendages necessary to determine sex, as females with appendages look much like males.

Similar Species Should not be confused with anything else where it occurs. Conspicuously smaller than **Bar-sided Darner**, with no blue markings on top or dark brown markings on sides of thorax and never dark stripes along front of wings. Both have pale legs. Closest to **Blue-faced Darner** in size and even in color pattern, but that species darker overall with dark legs and usually blue face. Mature **Blue-faced** much more brightly colored than

124
Pale-green Darner female—
Hidalgo Co., TX, October
2006, R. A. Behrstock; male—
Madre de Dios, Peru, July
2002, Netta Smith

Pale-green, also has fine green median line down length of abdomen not present in **Pale-green**.

Behavior Both sexes hang up during day in woodland, usually dense, and forage at forest edge from ground level to well up in trees at dusk; may be second flight at dawn when warm enough. Breeding poorly known, but females probably oviposit in fallen logs and branches in depressions on dry forest floor that fill with water at beginning of rainy season.

Habitat Known from dense woodland adjacent to Rio Grande, where locally common. Breeds in forested swamps in tropics. Some potential breeding habitat in south Texas woodlands.

Flight Season TX Jun–Oct.

Distribution Ranges south to Bolivia and Brazil, also throughout West Indies and recorded from far southern Florida.

Pilot Darners *Coryphaeschna*

These are rather large neotropical darners, the mature males either with conspicuously green thorax and black abdomen or mostly red. Males have simple appendages, females often very long cerci that are easily broken off. Adults are usually associated with forest but may breed in open marshes. They forage well above ground (thus "pilot darners"), even among treetops, sometimes in virtual swarms and often with other species. World 8, NA 3, West 2.

125 Blue-faced Darner *Coryphaeschna adnexa*　　TL 66–69, HW 42–45

Description Small darner of south Texas with blue face, green thorax, slender brown abdomen. *Male*: Eyes green with bright blue margins, face and rear of head blue. Thorax green with fine brown suture lines. Abdomen with fine green line down center and two narrow green rings around each segment (one on S9, none on S10). Teneral with wide brown stripes on thorax, quickly obscured.

125
Blue-faced Darner male,
female—Kleberg Co.,
TX, May 2005

Female: Colored as male but eyes and face green with blue tinge, abdomen often some-what richer brown. Might also be distinguished by intact cerci considerably longer than those of male; cerci lost in older females.

Identification No other darner in range in this book with entirely green thorax, slender brown abdomen except rare **Pale-green Darner**, which is paler, with pale legs and green-ish face. Much larger **Regal Darner** has brown stripes on thorax; so far not known to over-lap in range. Note that thorax in teneral **Blue-faced** striped just like **Regal**, distinguish by size and brown abdomen of **Blue-faced**. GREEN DARNERS all much bulkier, relatively shorter abdomens.

Natural History Males patrol back and forth low over floating vegetation. Females oviposit in floating vegetation, for example water lettuce, just above water. Both sexes in feeding flight, sometimes numerous individuals, from near ground to treetops. Flies within tiny woodland clearings as well as out in open, then hangs up under branch at about head height.

Habitat Ponds, usually covered with floating vegetation such as water lettuce, water hya-cinth, or duckweeds. Usually associated with forest.

Flight Season TX Jun–Nov.

Distribution Ranges south in lowlands to Argentina, also southern Florida and Greater Antilles.

---

126 Regal Darner *Coryphaeschna ingens*   **TL 86–90, HW 54–59**

Description Large darner with green- and brown-striped thorax, green-ringed black abdomen. *Male*: Eyes green, face green with brown crossline. Thorax green with wide brown stripes. Abdomen black with narrow green dorsal line and fine narrow green rings around it, two to three each on middle segments. *Female*: Colored as male but with blue eyes when mature, green when immature.

This must be primary way sexes distinguished away from water. Very long, slender cerci in immature female shed with maturity. Wings in immature female orange at base, clear beyond, this reversed in maturity, when much of wings become brownish-orange.

Identification See **Swamp Darner**, closest in size and color to this species and overlapping with it extensively. **Regal** with more green, less brown on thorax. Flight styles somewhat different, **Swamp** zipping through air like fighter plane, **Regal** with more buoyant flight like glider. Birders should think of difference between merlin and kestrel. When wings suffused with orange, tends to be more toward tip and rear edge in **Regal**, more in center and front edge in **Swamp**. Intact female cerci longer in **Regal** than in **Swamp**.

Natural History Unusual among dragonflies in males not patrolling at wetlands, but pairs meet and mate away from water. Apparently both "immature" and mature females mate. Both sexes in feeding swarms from near ground to high in air at any time, including dawn, but especially late afternoon to dusk; in open or among trees, sometimes in large numbers. May be mixed with other large species such as Common Green Darners or Prince Baskettails. Tends to feed on quite small prey. Females oviposit in vegetation at lake or pond shores, even in fairly dense marshes. Spend much time dropping into vegetation and rising again, as if testing for appropriate and/or safe site, meanwhile with abdomen curled downward to discourage male attention. Both sexes hang up in shrubs and trees.

Habitat Typically wooded country, although may be seen feeding over open areas, larvae common in open ponds as well.

Flight Season TX Apr–Oct.

Distribution Also across Southeast to Virginia, mostly in Coastal Plain.

126.1
Regal Darner male—TX,
Curtis Williams (posed)

126.2
Regal Darner immature female—
Citrus Co., FL, April 2005 (posed);
female—Early Co., GA, September
2002, Giff Beaton

Another small neotropical genus that barely makes it into North America, these are related to pilot darners but with quite different larva, and female cerci are shorter than usual in the family (unusually long in pilot darners). World 4, NA 1, West 1.

127 **Malachite Darner** *Remartinia luteipennis* — TL 76–80, HW 45–49

Description Big green and brown darner of southwestern canyons. *Male*: Eyes and face turquoise. Thorax brown with wide green stripe in front, very wide green stripes on sides so thorax looks green with brown stripes. Abdomen brown, mostly green on S1–2 and base of S3. Small forward-pointing green dorsal triangles on S3–6, fine green lines and partial rings on much of abdomen. *Female*: Colored as male, eyes and face blue (green when immature). Cerci very short and narrow, quite different from other darners (but looks like those with broken-off cerci). Wings become brown with maturity.

Identification Mostly green thorax with vivid brown stripes like no other darner in range. Abdomen more slender than in **Common Green Darner**, longer than in **Blue-eyed Darner**. **Regal** and **Swamp Darners**, somewhat similar, occur well to east. Green of GREEN DARNERS paler and unmarked with stripes.

Natural History Males cruise at waist to above head height over water or in and out of tall emergent marsh plants, hovering at intervals. Cruising beats may be 15–25 feet in length. Male seen to land on perched female in forest, presumably mating attempt. Females oviposit in dense vegetation, often dropping down into vegetation and hovering repeatedly before deciding on appropriate site.

Habitat Marshy or marsh-edged ponds and pools in streams. More common in wooded areas but may be on entirely open wetlands.

Flight Season AZ Jul–Oct.

Distribution Ranges south in uplands to Argentina.

127
Malachite Darner
male, female—Coclé,
Panama, January
2008, Giff Beaton

These bright and contrastingly patterned dragonflies fly over lakes and marshes throughout temperate regions of the northern hemisphere, where they dominate dragonfly faunas. Often seen away from water, sometimes at great distances, they commonly form mixed-species feeding swarms, probably aggregating only because flying prey is also aggregated. Both sexes move from water body to water body, not at all tied to one place (probably true of most if not all mosaic darners). Typical coloration includes a pair of stripes on the front and two slanted stripes on either side of the thorax, often green shading down into blue, and a series of paired blue dots down the entire length of the abdomen. Females in most species are polymorphic, with male-colored andromorph and usually more common heteromorph with yellow stripes and spots; green is also an alternate color to blue in some. Check out enough mating pairs, and you will usually see both morphs. Males either fly regular beats along the shoreline or hover here and there in smaller territory, but some of them circumnavigate a pond or lake and then move on to another one, all to search for females ovipositing in shore vegetation. Females disturbed during oviposition usually fly rapidly up into trees. Males can be identified by combination of thoracic stripe shape, abdominal pattern, and appendage type. With enough practice, some species can be picked out as they fly by, but please believe that this is tricky! Because you do not see these dragonflies perch very often, capturing them and identifying them in the hand is the best way to find out what species are present.

Neotropical darners (*Rhionaeschna*), very similar to but recently separated from mosaic darners, are distinguished by pale borders to T-spot on the frons and tubercle under S1 in both sexes. Some marks to look for to distinguish mosaic and neotropical darners that fly together include (1) shape of male appendages, whether simple, paddle, or forked; (2) eye color bright blue, darker turquoise, or clearly greenish; (3) face unmarked or with obvious dark stripe on frontoclypeal suture; (4) lateral thoracic stripes wide or narrow, straight or notched, extended posteriorly or not at upper ends; (5) presence or absence of markings between lateral thoracic stripes; (6) S10 black or with prominent blue spots; (7) pale spots under most abdominal segments or not; and (8) much hovering or continuous flight along shore. Females are much more difficult to distinguish but share the facial, thoracic, and abdominal pattern-

Table 5   Mosaic/Neotropical Darner (*Aeshna/Rhionaeschna*) Identification

	A	B	C	D	E
Black-tipped	1	1	3	1	1
Sedge	2	1	1	1	2
Subarctic	2	4	1	1	2
Green-striped	1	4	1	1	2
Canada	1	4	1	1	2
Lake	1	4	1	1	1
Variable	1	36	213	1	1
Zigzag	1	5	1	1	1
Azure	1	5	1	1	1
Lance-tipped	1	4	1	2	1
Paddle-tailed	1	2	1	2	1
Shadow	1	2	3	2	2
Walker's	1	2	3	2	1
Persephone's	1	1	3	2	1
Turquoise-tipped	1	4	13	1	1
California	13	3	1	1	1
Arroyo	3	4	12	3	1
Blue-eyed	3	3	1	3	1

A, male eye color: 1, turquoise to blue; 2, more greenish; 3, bright blue.

B, lateral stripes: 1, wide, straight; 2, medium, straight; 3, narrow, straight; 4, wide, flagged at upper end and slightly indented in middle; 5, zigzag; 6, spotted.

C, spots on male S10: 1, blue; 2, white; 3, absent.

D, male appendages: 1, simple; 2, paddle; 3, forked.

E, spots under abdomen: 1, absent; 2, present.

ing (but not necessarily coloration) of males and may have distinctive ovipositors or cerci. Identification problems are greater in the North, where more species occur together. Bear in mind that several other genera of this family can be mistaken for mosaic darners, especially Springtime and Riffle Darners. World 42, NA 15, West 14.

Description Rather dark darner, with limited blue. Lateral thoracic stripes straight (may be slight extension at wing base) and narrow, all blue or blue above, greenish below (no yellow). Frontal stripes very narrow, greenish. *Male*: Eyes turquoise. Simple appendages, cerci with prominent ventral tubercle in side view at about one-fourth length. *Female*: Polymorphic, common andromorph colored exactly as male; heteromorph, with yellow-green spots on abdomen, much less common. Cerci large and conspicuous, obviously longer than S9–10 and acutely pointed. Abdomen shaped as male's.

Identification For both sexes, dark color, with narrow, straight thoracic stripes, limited blue on abdomen and none on S10, distinctive, Superficially most similar **Shadow Darner** has paddle appendages, pale spots under abdomen. Female **Black-tipped** identified by looking amazingly like male, same narrow-waisted shape and bright color, large cerci simulate male appendages.

Natural History Males fly along lake shores or over marshes, not hovering, and often above vegetation at head height or above. Females behave exactly like males, presumably mimicking them to avoid harassment when ovipositing, but often fly lower, below tops of sedges. If a seeming male darner, flying among other males in the same manner, suddenly lands and begins ovipositing, it is probably this species. Abdomen of female more slender than in other darners, enhancing similarity to male. Oviposits both well above water level and at water level, on emergent or floating plants, even mud. Cattails, irises, bur-reeds, and

128.1
Black-tipped Darner male—Skamania Co., WA, August 2005 (posed); blue female— Chelan Co., WA, August 2007 (posed)

128.2
Black-tipped Darner green female— Waushara Co., WI, September 2002, Mike Reese

other plants used. Sometimes in mixed feeding swarms. Usually one of least common darners where it occurs.

Habitat Clear lakes and ponds, often with associated bog vegetation, in forested regions.

Flight Season BC Jun–Sep, WA Jul–Oct, MT Jul–Sep.

Distribution Also in eastern Canada east to Nova Scotia and south to northern Iowa, Ohio, and Maryland, in mountains to North Carolina.

129 Sedge Darner *Aeshna juncea*   TL 61–69, HW 40–46

Description Lateral thoracic stripes wide and straight, may be slight extension; long streak parallel to and between them and often short streak in front of anterior one. Frontal stripes well developed and conspicuous. Fine black line across face. Whitish spots beneath middle abdominal segments (not in all individuals). *Male*: Eyes greener than in most other common species, at least in Northwest. Lateral thoracic stripes blue above and yellow-green below. Very wide in some individuals, with in-between streak especially prominent. Blue abdominal spots relatively small but well developed on S10. Simple appendages. *Female*: Polymorphic, heteromorph with green-tinged eyes and yellow-green markings, andromorph with blue-tinged eyes and blue markings. Cerci rather small for mosaic darner, not as long as S9–10.

Identification Combination of simple appendages and broad, straight thoracic stripes good for this species. Pale streak between stripes more prominent than in other species. Stripes much broader than in **Variable Darner**, lacking prominent indentations and extensions of **Canada**, **Lake**, and **Subarctic Darners**. Most similar to **Subarctic** and often occurs with it. Identification can be tricky and may have to be resolved by scrutinizing hamules of individuals in hand, but careful look at thoracic stripe shape will usually do it. Female also with well-developed and fairly long pale stripe between especially broad and straight

129.1
Sedge Darner male—Fairbanks, AK, August 2002, Wim Arp; green female—Kittitas Co., WA, September 2007

129.2
Sedge Darner blue female—Kittitas Co., WA, August 2006

lateral thoracic stripes. Closest in pattern because of broad, straight lateral stripes would be **Black-tipped Darner**, which has no in-between stripe. Also, female **Sedge** one of few species with obvious pale spots under abdomen (usually); thoracic pattern distinct from that of **Canada**, **Shadow**, and **Subarctic**, the other species. See **Subarctic** for differences from female **Sedge**. In flight, could be confused with **Paddle-tailed Darner**, often found with it.

Natural History Most abundant darner in many northern wetlands. Males fly widely about a foot over extensive sedge beds or with much lengthy hovering, also along vegetated lake shores at knee height and below. Patrol flights over sedges 30 feet long or shorter. Females oviposit at water level or just below in emergent grasses and sedges or matted roots, also into moss. Female either moves short distance when disturbed or flies away from water like a bullet. Many oviposit when males not present. Adult life to 70 days (prereproductive period to 35 days).

Habitat Extensive sedge marshes and mossy fens, also lakes, ponds, and ditches with emergent vegetation.

Flight Season YT Jun–Sep, BC Jun–Sep, WA Jul–Oct, OR Jul–Oct, MT Jul–Sep.

Distribution Confined to higher elevations at lower latitudes. Also across eastern Canada south to Michigan and New Hampshire and all across northern Eurasia.

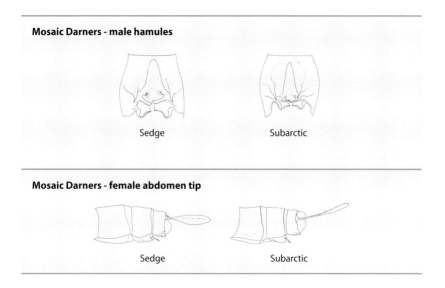

Mosaic Darners - male hamules

Sedge   Subarctic

Mosaic Darners - female abdomen tip

Sedge   Subarctic

130 Subarctic Darner   *Aeshna subarctica*   **TL 63–68, HW 40–46**

Description Lateral thoracic stripes notched and extended, less so than some other species. Short, thin streak between stripes and often narrow partial streak just in front of anterior one. Frontal stripes relatively thin, reduced. Pale spots beneath middle abdominal segments. *Male*: Eyes green, similar to Sedge Darner in being less bluish than other mosaic darners. Thoracic stripes bluish above and yellow-green below. *Female*: Polymorphic, female abdomen with blue or yellow or greenish yellow spots. Cerci longer than S9–10.

Identification Often flying with very similar **Sedge Darner**, from which it is distinguished by narrower thoracic stripes with slight notch in front and extension to rear. Male hamules

130
Subarctic Darner male—
Chenango Co., NY, September
2004, Allen Barlow (posed);
blue female—Worcester Co., MA,
August 1997, Blair Nikula (posed)

quite different, only evident in hand with magnification. Females distinguished from **Sedge Darner** by longer cerci (distinctly longer than S9–10, about 6–7 mm in length, no longer or shorter than those segments in **Sedge**, about 4–5 mm), shorter ovipositor (just reaches end of S9, surpasses it in **Sedge**). Both sexes distinguished from **Canada Darner** by black line across face, less distinct notch on first lateral thoracic stripe.

Natural History Males fly back and forth at waist height over floating beds of sphagnum and other mosses in lakes or muskeg pools, hover in one spot, then move to another. Some males remain in restricted areas for minutes at a time, others wander farther. Females oviposit in mosses and sedges at water surface, usually at the edge of open water. Both sexes cruise in darner assemblages in woodland clearings. Adult life to 70 days (prereproductive period to 35 days).

Habitat Typically fens and bogs with abundant growth of sphagnum and other mosses.

Flight Season YT Jul–Sep, BC Jun–Sep, WA Aug, OR Aug–Oct.

Distribution Also in eastern Canada south to Minnesota and New York and across northern Eurasia.

---

131 Green-striped Darner *Aeshna verticalis*　　　　TL 62–72, HW 41–47

Description Lateral thoracic stripes notched and extended, mostly green but some blue above; frontal stripes broad, green. *Male*: Eyes greenish. Simple appendages. *Female*: Polymorphic, most heteromorphs with green or yellow-green spots, rarely blue. Cerci relatively large, distinctly longer than S9–10, widest near middle; often broken off at maturity.

Identification Much like **Canada Darner,** but lateral thoracic stripes mostly green (mostly blue in **Canada**, rarely mostly blue in **Green-striped**), slightly less notched than in **Canada** (thus much less notched than **Lake**), and extension along wing base ("flag") often but not always becomes wider at end (stays narrow in **Canada**). In addition, stripes usually brighter and more contrasting. Female differs in same ways. Differs from other species in same way as **Canada**, including spots under abdomen. Fine teeth on upper surface near tip of cerci in **Canada** can be seen under magnification; lacking in **Green-striped**. Difference in female genital valves apparent in hand: grooves on ventral surface extend about to tip in **Canada**, end well before tip in **Green-striped**.

131
Green-striped Darner male—
Sussex Co., NJ, September 2005,
Allen Barlow; green female—
Sussex Co., NJ, September 2003,
Jim Bangma (posed)

Natural History Both sexes often perch on tree trunks. Males patrol over water as in other mosaic darners, often along border between open water and vegetation, sometimes over sphagnum beds. Also over dense stands of grasses and sedges. Females oviposit in bur-reed stems and pondweed. Feeds in clearings away from water, may join feeding swarms.

Habitat Forest ponds and lakes with much aquatic vegetation.

Flight Season Ontario Jun–Sep.

Distribution Also in East from Minnesota and New Brunswick south to Iowa, Kentucky, and North Carolina.

---

132 Canada Darner *Aeshna canadensis*　　　　　**TL 64–73, HW 43–47**

Description Lateral thoracic stripes notched and extended, yellow spot between them; frontal stripes well developed. Pale blue or gray spots under most abdominal segments. *Male*: Lateral thoracic stripes blue; frontal thoracic stripes well developed, greenish; abdominal spots moderate sized and blue. *Female*: Polymorphic, heteromorph with yellow-tinged brown eyes and yellow-green markings, andromorph with blue-tinged brown eyes and blue markings. May lack frontal stripes.

Identification Most similar to **Lake Darner**, but both sexes distinguished by being smaller, with no black stripe on face, and pale spots under abdomen. Posterior lateral thoracic stripe slightly indented or incurved in front in **Canada**, distinctly notched in **Lake**. Male flight style different, with much hovering in **Canada** and steady flight in **Lake**. Very similar to **Green-striped Darner**, which barely enters eastern edge of region, distinguished by mostly blue lateral thoracic stripes and minor differences in stripe shape (see that species). Female distinguished from other darners by combination of unmarked face, notched/extended thoracic stripes, and pale spots under abdomen. Others with ventral abdominal spots include **Sedge**, **Shadow**, and **Subarctic Darners**, first two with straight stripes and last with notched stripe but no extension.

Natural History Males fly slowly along shore at about waist height, usually back and forth in defined territory of no more than 30–65 feet, and very often hover over and cruise through tall beds of grasses and sedges for lengthy periods, at intervals dropping to surface for quick

132
Canada Darner male—
Snohomish Co., WA, August 2005
(posed); blue female—Kitsap Co., WA,
September 2005 (posed)

or extended female search. Copulation usually in shrubs near water. Females oviposit in plant stems, moss, algae, and mud at water level, usually in marshy area rather than at shore. Away from water, perches on tree trunks or hangs from branches; often in mixed feeding swarms. Blue, green, and yellow markings of males may be subdued gray when active at relatively low temperatures, brighten as air warms. Perhaps true of all mosaic darners, but Paddle-tailed Darners remained bright on one such occasion.

Habitat Lakes and ponds with abundant emergent vegetation in forest zone, often associated with bogs and beaver ponds.

Flight Season YT Aug, BC Jun–Sep, WA Jul–Oct, OR Jul–Oct, CA Aug–Oct, MT Jul–Sep, NE Sep–Oct.

Distribution Also across southern Canada to Newfoundland and south to Illinois and West Virginia.

---

133 Lake Darner *Aeshna eremita*   TL 66–79, HW 41–52

Description Larger than other mosaic darners, largest darner in much of its range. Lateral thoracic stripes notched and extended, pale spot or streak between them. Much blue on abdomen, none beneath it. Simple appendages. *Male*: Eyes turquoise above, black line across face. Lateral thoracic stripes blue or blue shading to green below, with spot or streak between them blue or green; frontal stripes greenish. Simple appendages. *Female*: Polymorphic, with either blue, as in male, or green stripes and spots.

Identification Larger than other mosaic darners, also recognized by continual cruising flight along lake shores. **Canada Darner** most similar species in most of range because of simple appendages and thoracic pattern. **Lake** larger than **Canada** with black line across face, no pale spots under abdomen, and notched posterior lateral thoracic stripe. Female identified by size, thoracic pattern, black face line, and relatively large cerci.

Natural History Males fly rapidly over open water or along lake shores at about knee height, often with long abdomen drooping at tip. Usually stay out of vegetation and no hovering, unlike most other mosaic darners. Sometimes move into beds of dense vegetation to search for females, however. Males remain at water late in afternoon and may continue flying until dark (under midnight sun in far north!). Perching sites for both sexes typically bare twigs, tree trunks, or ground. Females oviposit on floating logs, in stems of emergent herbaceous plants, and in tangles of rootlets on vertical banks at or just below water surface.

Habitat Typically wooded lakes and large ponds, sometimes slow streams.

Flight Season YT Jun–Sep, BC Jun–Sep, WA Jul–Oct, MT Jul–Sep.

Distribution Ranges across southern Canada to Labrador and south to Minnesota and New York.

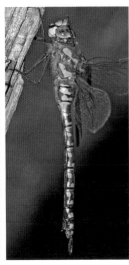

133.1
Lake Darner male—
Whatcom Co., WA,
September 2006 (posed)

133.2
Lake Darner blue female—Kootenay National
Park, BC, August 1968 (posed); green female—
San Juan Co., WA, August 2007 (posed)

134 Variable Darner *Aeshna interrupta* TL 61–72, HW 41–46

Description Lateral thoracic stripes narrow and straight, narrow and slightly irregular, or divided into upper and lower spot; frontal stripes poorly developed or absent. Fine black line across face. Spotted form mostly seen west of Cascades in Pacific Northwest. Many populations in Northwest, including coastal and montane, include both types and all intermediates between them. Some individuals in Anchorage area have stripes almost absent, represented by tiny dot at lower edge of anterior stripe. *Male*: Eyes rather dark blue, face pale greenish. Thoracic stripes mostly blue, may be yellow below; upper spots may be blue, lower ones yellow-green. Large amount of blue on abdomen, S3 usually looks all blue on sides, with only narrow black dorsal line. Isolated whitish to pale blue spots on S10 (spotted individuals may lack pale markings on that segment). Simple appendages. *Female*: Polymorphic, andromorph with blue-tinged eyes and blue body markings, heteromorph with brown eyes and yellow body markings.

Identification Both sexes distinguished from all other mosaic and neotropical darners by lateral stripes being narrow or reduced to spots. Often hard to see thoracic markings on males in flight, even with good look in passing, whereas stripes obvious on other species. Whitish spots on S10 visible on perched individual good mark to distinguish males, as those spots on other species blue. However, those few without spots on S10 could be mistaken for **Black-tipped** or **Shadow Darners**, both of which have smaller blue abdominal spots. Female much like female **Shadow** but lacks pale spots below. As much or more like female **Paddle-tailed** and may have to be distinguished in hand by smaller, flat genital valves (enclose ovipositor), those of **Paddle-tailed** a bit more robust and more obviously projecting downward at posterior end. Larger ovipositor makes S9 larger in **Paddle-tailed**, that segment looking longer than wide (about as long as wide in **Variable**). Genital valves

134.1
Variable Darner striped male—
Kitsap Co., WA, September 2005

134.2
Variable Darner spotted male—
Siskiyou Co., CA, July 2004,
Ray Bruun; yellow female—
Kittitas Co., WA, September 2007

usually dark below in **Variable**, pale below in **Paddle-tailed**, but not always distinctive. **Variable** usually has tiny tuft of hairs at end of each genital valve (typical of most mosaic darners) lacking in **Paddle-tailed** (as in all species with "paddle" appendages). One field character involves the most anterior spot on the side of each abdominal segment (look at S6–7), separated from the larger spot behind it by the lateral carina. This spot is more or less round in **Paddle-tailed**, but in **Variable** it swoops up to form a vertical line at its front end. Female **Variable** also much like female **Blue-eyed,** which has bump under S1.

Natural History Males fly rapidly or slowly along shorelines or stop to hover and inspect vegetation beds, looking for laying females. Often at edge between open water and cattails, also over dense sedge beds. Tends to fly back and forth over apparently defined territory. Both sexes feeding and at water until dark. Where common, copulating pairs regularly seen in flight, also tandem pairs, some of latter could be mismatched (males are indiscriminate). Females oviposit on floating sedge and grass stems, upright herb and shrub stems, and wet logs. Regularly in mixed feeding swarms of several species of mosaic darners. Often perch on tree trunks, also on ground at higher latitudes. Commonly encountered far from water in mountain clearings, probably "hilltops" like butterflies. Most abundant darner by far on Canadian prairies and locally common over much of range. Huge aggregations, with smaller numbers of other mosaic darners, seen in midsummer moving up east slope of Rockies in Alberta.

Habitat Lakes and ponds of all sorts, even small ones, usually with dense shore vegetation but at least some open water. From lowlands to well up in mountains.

Flight Season YT Jun–Sep, BC Jun–Oct, WA Jul–Oct, OR May–Sep, CA May–Oct, MT Jun–Oct, AZ Jun–Sep, NE Jul–Sep.

Distribution Also across southern Canada to Newfoundland and south to Minnesota and New Jersey.

Comments Several named subspecies differ in shape of thoracic stripes or structure of male cerci. *A. i. interrupta* occurs in eastern forests to east of region covered by this book. In that subspecies, lateral thoracic stripes broken into two pairs of above-and-below spots (thus *interrupta*). *A. i. lineata*, which has narrow, linear thoracic stripes, is subspecies of prairies and west to Pacific coast. In parts of Great Basin, from southern Idaho to Arizona, male cerci a bit wider, with more sinuous inner edge and dorsal ridge and small bump on inside of base, and these have been called *A. i. interna*. Extent of distribution of this form not entirely clear, but may intergrade broadly with *lineata* to north. That the color pattern may be caused by climatic factors rather than genetic differences is indicated by the populations in the wet Pacific Northwest, where thoracic pattern of spots is much like those of eastern North America.

135 Zigzag Darner *Aeshna sitchensis* TL 57–60, HW 37–40

Description Small darner with lateral thoracic stripes narrow and so strongly notched and extended that they form a zigzag shape. Stripes often broken and may have spot or streak in between. Frontal stripes reduced or lacking. *Male*: Eyes brown mixed with blue around edges. All pale markings light blue, no green or yellow. *Female*: Polymorphic, heteromorph with eyes brown with blue highlights, thoracic stripes yellow, abdominal spots yellow to white. Andromorph with eyes and pale markings blue.

Identification Small size and very blue abdomen of males obvious. From above, blue spots on midabdomen much like those of

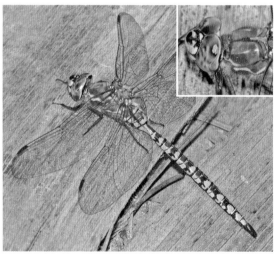

135.1
Zigzag Darner male—Heckman Pass, BC, July 2006, Netta Smith; male inset—Heckman Pass, BC, July 2006

135.2
Zigzag Darner pair—Pend Oreille Co., WA, August 2004

other darners, but spots get conspicuously larger toward rear until they cover more than half of segment, unlike all others but similar **Azure**. Only these two northern darners are so small and blue. Compared with **Azure Darner**, abdomen has alternating brown and blue along side and more brown than blue from above. See **Azure** for diagnostic difference in thoracic pattern. Pattern on top of frons also diagnostic, black transverse stripe in front of eyes bulging forward at either end like crescent in **Zigzag**, not so in **Azure**. Also check length of eye seam (see under **Azure**). Female abdomen more dark than pale from above.

Natural History Both sexes perch on ground, gravel roads, logs, tree trunks, and other usually light-colored substrates, males more commonly out in breeding habitat. Rather tame, and often land on light-colored clothing or net. Males perch on patches of light-colored moss raised above surrounding marsh, alternating low flights (knee height) over open meadows with no more hovering but more perching than other mosaic darners. Difference perhaps more apparent than real because they are easy to see when perched and other species are hidden in trees and shrubs. Pairs easier to find than in other mosaic darners, probably for same reason; tend to perch in low shrubs rather than on ground but also fly into woodland as do other species. Males at times fly above tall sedges much like Sedge Darner but usually at small, shallow pools. Females oviposit at water level in dense grass or sedges or at edge of open water, often in moss beds, into algal mats, or on mud. Retire to open woodland and clearings when away from breeding habitat, but feed both at and away from bogs, males while in sexual patrol.

Habitat Fens and cold-water pools with low sedges and mosses, often bordered by shrubby or wooded uplands and usually shallow with little open water. May dry up in midsummer.

Flight Season YT Jun–Aug, BC Jun–Sep, WA Jul–Sep, OR Aug–Sep, MT Jul–Sep.

Distribution Ranges across eastern Canada to Labrador and south to Michigan and New Hampshire.

136 Azure Darner *Aeshna septentrionalis* TL 54–63, HW 35–40

Description Small darner with lateral thoracic stripes narrow and so strongly notched and extended that they form a zigzag shape. Stripes often broken. Frontal stripes reduced or lacking. *Male*: Eyes blue. All pale markings light blue, no green or yellow. *Female*: Polymorphic, blue just as male or with yellow markings, pale markings of abdomen less extensive than in male.

Identification Should only be confused with similar-sized and also northerly occurring **Zigzag Darner**. Both sexes differ by having considerably more pale color on abdomen, blue or yellow spots sufficiently close that from side abdomen looks just about all blue or yellow, from above more pale than dark. Pale spots distinct and separate in **Zigzag**. However, both are variable, and color patterns on some individuals may be less than diagnostic. In side view, look at anterior lateral thoracic stripe. In **Azure**, thin stripe parallel and just before it extends to where lateral stripe zigzags at about mid height; in **Zigzag**, that front stripe shorter, more triangular or curved, and not extended to point of zigzagging. Markings on top of frons also distinctive, in this species black bar in front of eyes narrow at ends, not bulging forward. Perhaps best close-range characteristic is eye seam, line where eyes meet: short in **Azure**, no longer than length of occiput, distinctly longer than occiput in **Zigzag**.

Natural History Both sexes perch commonly on ground, rocks, tree trunks, and moss, usually light-colored. Males fly low over marshy areas or floating moss mats much like male Zigzags. Females oviposit in mosses and soupy mud ("muskeg slime") at edge of pools. Along with Treeline Emerald, the northernmost dragonfly.

136.1
Azure Darner male—
Old Crow, YT, July 1983,
Rob Cannings

136.2
Azure Darner yellow female—Talkeetna, AK,
July 2007, Martin Reid; blue female—Haines
Junction, YT, June 2004, Jim Bangma (posed)

Habitat Fens and shallow ponds, similar to Zigzag Darner habitat, but sphagnum and other mosses characteristically present.

Flight Season YT Jun–Sep, BC Jun–Aug.

Distribution Also around south end of Hudson Bay and east to Labrador and Newfoundland.

137 Lance-tipped Darner   *Aeshna constricta*   TL 65–73, HW 42–45

Description Lateral thoracic stripes notched and extended; frontal stripes well developed. *Male*: Thoracic stripes blue to green, extensive blue on abdomen. Paddle appendages. *Female*: Abdomen shape about as male's. Polymorphic, with duller yellow-marked heteromorph and brighter blue-marked andromorph, andromorph very bright and male-like. Heteromorphs may have largely yellowish wings. Cerci large and lance-shaped (thus common name), persistent and conspicuous. Cerci only as long as S9–10 because S9 very large, almost twice as long as S10, distinctly larger than same segment in females of other species. Also, pale spots on S9 large and square, distinctly larger than those on S8 and extending lower on sides, unique among female darners.

Identification Only species in paddletail group with anterior lateral thoracic stripe notched in front and drawn out to rear at top. Lateral stripe shape distinguishes from **Paddle-tailed** and **Shadow Darners**, only other paddletails with which it occurs. Most similar to **Paddle-tailed**, but lacks fine dark line across face and pale mark on S1 of that species as well as having different thoracic stripes. Females identified by thoracic stripe pattern in combination with large, pointed cerci and large S9 with large square spot; only other darner with similarly large cerci is very differently colored **Black-tipped**, with smaller and smaller-spotted S9.

137
Lance-tipped Darner male—
Thurston Co., WA, September
2005; blue female—Worcester
Co., MA, July 2004, Tom
Murray

Natural History Males fly beats along shores or through semiopen cattails. Females brightly colored as males and fly over water like them, perhaps mimicking them to avoid harassment by other species. Oviposit on upright stems and leaves of cattails, sweetflag, and other robust plants up to waist height above moist or dry ground; also commonly at surface, including on peat soil. Egg laying slow and methodical, may lay only few eggs in given stem. To avoid male harassment may fly with abdomen bent down sharply. Often in mixed feeding swarms. Typically hang in low shrubs or herbaceous vegetation.

Habitat Shallow marshy ponds and similar edges of larger lakes, commonly in open country. Breeding occurs in ponds that dry up every year, unusual for darners. Many of the habitats occupied by this species in the West have dried up in recent years, and it has declined in those regions.

Flight Season BC Jun–Sep, WA Jul–Sep, NE Jul–Oct.

Distribution May expand during wet years, accounting for isolated records in plains, then contract during droughts. Also ranges east across southern Canada to Nova Scotia and south to Missouri, Kentucky, and New Jersey, occasionally farther.

---

138 Paddle-tailed Darner *Aeshna palmata* TL 65–75, HW 41–46

Description Lateral thoracic stripes straight, occasionally extended; pale spot between them. Frontal stripes present, usually quite narrow. Fine black line across face. *Male*: Eyes slightly toward the turquoise side of bright blue. Thoracic stripes blue to greenish-yellow. Blue spots on abdomen large. Paddle appendages. *Female*: Polymorphic, heteromorph with brown eyes and yellow stripes and spots, andromorph with blue eyes and much blue in stripes and spots. Cerci obtusely angled at tip, about as long as S9–10.

Identification Very common western darner with paddle-shaped appendages, can be used as basic *Aeshna* from which other species must be distinguished over large part of West. First, consider other species with paddle-shaped appendages. Male most similar to **Lance-tipped**, which has anterior thoracic stripe clearly notched and more extended, also lacks black line on face and blue vertical slash on S1. (Note that aged individuals of mosaic darners often have dark smudges on face.) Also similar to **Shadow** but differs in much more extensive blue on abdomen, including S10; wider thoracic stripes; black line on face; and blue on S1. Differs from **Walker's Darner** by more colorful lateral thoracic stripes, greenish-yellow to blue (whitish in **Walker's**), and again black facial line and blue on S10. In hand, difference in appendages in these two species evident. Female with thoracic stripes rather straight and moderately wide, differs from most other mosaic darners in this. Differs from **Shadow** by lack of spots under abdomen and persistent cerci (**Shadow** females break theirs off soon after maturity). Many female **Variable Darners** have spots or quite narrow stripes, but those with widest stripes very much like female **Paddle-tailed**, differ by smaller ovipositor valves (not extending beyond end of S9, unlike **Paddle-tailed**) with tiny pencil of hairs on end of each (not present in **Paddle-tailed**). See **Variable** also. **Paddle-tailed** differs from rather similar **Blue-eyed** by lacking pale borders to T-spot and having tubercle under S1.

Natural History Males fly beats back and forth along shore and hover frequently while facing shore, over open water or in openings among emergent plants. Often single males localize in embayments where water penetrates woodland, as small as 8 feet in diameter. Typically fly at knee to waist height, lower when searching intently for ovipositing females. Even move slowly into shrubbery well away from water, perhaps looking for roosting females. Males hover for long periods, then fly out and around in short patrol flight, then back to hovering in same spot. Pairs couple over water and immediately fly into woods, hang in trees at or near waterside. Females oviposit on wet floating logs at water level or up to a foot above it on stems of cattails, bulrushes, irises, and other emergent plants. Often in mixed feeding swarms. Most common mosaic darner over large parts of the Northwest.

Habitat Lakes, ponds, and slow streams, even small ones, usually with dense shore vegetation. For the most part in forested landscapes, but some populations in open country. Colonizes small suburban ponds readily.

138.1
Paddle-tailed Darner
male—Kittitas Co., WA,
September 2007

138.2
Paddle-tailed Darner yellow female—
Kittitas Co., WA, September 2007;
blue female—Kitsap Co., WA,
September 2005 (posed)

Flight Season YT Jun–Sep, BC May–Oct, WA Jun–Nov, OR Jun–Nov, CA May–Nov, MT Jul–Oct, AZ Jun–Oct, NM Jun–Aug, NE May–Oct.

Distribution Only in mountains in southern part of range.

139 Shadow Darner *Aeshna umbrosa* TL 64–73, HW 41–47

Description Dark darner with narrow, straight lateral thoracic stripes, some with posterior "flags" at upper end; frontal stripes usually narrow but present. Abdominal spots small, but also pale spots under abdomen. *Male*: Eyes turquoise. Lateral thoracic stripes blue-green above, yellow-green below. Paddle append-ages. *Female*: Polymorphic, heteromorph with brown eyes and yellow markings, andromorph with blue-tinged eyes, greenish to yellow thoracic stripes, and blue abdominal spots. Cerci rounded at tip, longer than S9–10, usually broken off at maturity.

Identification Male distinguished from **Lance-tipped** and **Paddle-tailed Darners**, with which it often occurs, by smaller abdominal spots, black S10, and pale spots under abdomen. Blue anteroventral spot on S7 distinctly larger than posterodorsal spot, unlike most other mosaic darners; easily seen in side view. Also from **Lance-tipped** by narrower thoracic stripes. Both sexes distinguished from **Paddle-tailed** by lack of black line on face and pale marking (blue or yellow) on S10. The two occur together very commonly and constantly need distinction. May occur sparingly

139.1
Shadow Darner male—
Grays Harbor Co., WA,
October 2007

139.2
Shadow Darner yellow
female—Whatcom Co., WA,
September 2006 (posed);
blue female—King Co., WA,
August 2005 (posed)

with **Walker's Darner**, differs by yellow and green thoracic stripes and slightly differently shaped appendages. Female as other darners with fairly narrow, straight thoracic stripes but with conspicuous pale spots under abdomen. Dorsal abdominal spots generally small, makes abdomen look darker than in most other species. Lack of apparent cerci may be indicative mark, as much more frequently broken off in **Shadow** than others. Rear of head pale in **Shadow**, dark in most other species; must look carefully for this even in hand.

Natural History Males fly beats up and down streams and along lake shores, with much hovering while facing shore, even as long as 30 sec in one spot. Often fly low, closer to water than other species, but much variation. Females oviposit on logs and twigs in water or on moist tree trunks or earth banks, sometimes well above water and even in rather dry situations. Perhaps because of woody oviposition substrates, females much more likely than other mosaic darners to break off cerci as they mature. Both sexes much more common at water later in day, and morning surveys may not reveal their presence. Feeding flights in clearings and at woodland edges, from ground level to considerably higher. As name implies, often pursues all activities in shade. Regularly flies until too dark to see it clearly, much like tropical dusk-flying darners. Usually one of last species of autumn, flying along with Autumn Meadowhawks.

Habitat Lakes, ponds, even small ones, and slow streams. More common on streams than other mosaic darners in its range (only Persephone's and Walker's more characteristic of streams). Colonizes small suburban ponds readily.

Flight Season YT Jul–Aug, BC May–Oct, WA Jun–Nov, OR May–Nov, CA Feb–Nov, MT Jul–Sep, NM May–Oct, NE Jul–Oct.

Distribution In mountains in southern Rockies, lowlands in adjacent Great Plains. Also east across Canada to Newfoundland and south to Arkansas, Alabama, and Georgia.

Comments Two distinctive subspecies, eastern and western. The eastern *A. u. umbrosa*, occurring all across plains, has quite small abdominal spots that are greenish on posterior segments. The western *A. u. occidentalis*, occurring from Rocky Mountains west, has these spots blue and averaging larger. Oddly, a few individuals matching the eastern subspecies have been found in Pacific coast lowlands.

Mosaic Darners - female abdomen tip

Paddle-tailed   Walker's

140 Walker's Darner *Aeshna walkeri* TL 63–70, HW 43–48

Description Lateral thoracic stripes narrow and straight. *Male*: Eyes dull brownish-gray with blue tinge and pale blue posterior edges, face pale blue to white. Thoracic stripes pale blue above, whitish below, or entirely white. Paddle appendages. *Female*: Eyes dark brown, paler below, with hint of blue. Color pattern much as male, abdominal spots either blue or olive-green. Presumably polymorphic. Cerci narrow at base, expanded toward tip; often broken off.

Identification In parts of range, only paddletail species. Where overlapping with other paddletails (**Paddle-tailed**, **Shadow**), distinguished by whitish face, dull eye color (pale blue edges distinctive), whitish to pale blue thoracic stripes (with green and/or yellow in other species). Male appendages somewhat broader in middle than in other paddletails. Lacks black face line and blue spots on S10 of **Paddle-tailed**, spots under abdomen of **Shadow**. Overlaps extensively with **Blue-eyed Darner**, female lacks bump under S1 of that species. Whitish thoracic stripes should also distinguish female from female **Paddle-tailed**, **Shadow**, and **Variable** where they overlap. Female cerci distinctively shaped when present, broadening toward tip, whereas cerci of **Paddle-tailed** slightly larger, broadest before middle. Cerci often broken off, as in **Shadow** (relatively rarely in **Paddle-tailed**).

Natural History Males fly slowly up and down narrow streams at knee to waist height, searching for females at pools. Spend more time poking into vegetation than other darners that may be seen with them. Females oviposit in afternoon on moss and hanging rootlets on vertical earth banks or rocks. Both sexes feed above hillsides near breeding streams; rarely seen perched.

Habitat Small streams with riffles and pools bordered by riparian shrub- or woodland in lowlands and foothills. The mosaic darner most typical of streams in California, although Shadow Darner also occurs on them, mostly outside range of Walker's.

Flight Season OR Jul–Sep, CA May–Nov.

Distribution Ranges south in Mexico to Baja California.

140
Walker's Darner male—San Luis Obispo Co., CA, September 2006, Paul Johnson

Description Large brightly patterned species of southwestern mountain streams. Lateral thoracic stripes straight and broad (narrower in northern Arizona), anterior one very slightly notched and extended; frontal stripes narrow, inconspicuous in comparison with laterals. Abdominal spots arranged in rings. *Male*: Eyes bright blue. Lateral thoracic stripes yellowish to chartreuse. Abdominal

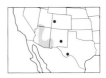

spots (rings) blue, no blue on S10. Paddle appendages. *Female*: Polymorphic. Eyes gray, otherwise colored as male, but some females with yellow markings on abdomen instead of blue. Cerci rather short, shorter than S9–10, narrow but widest toward tip.

Identification Slightly larger than most other darners in its range. Broad, straight thoracic stripes, ringed look of abdomen, and strong contrast between yellowish markings on thorax and blue on abdomen distinguish both sexes from all others in range. Population in Oak Creek Canyon in northern Arizona has much narrower thoracic stripes and could be mistaken for other species but for ringed appearance. No other paddletail-type darner reported from south of Grand Canyon, however. Ringed appearance of abdomen and yellowish thoracic stripes of **Persephone's** distinguish it from NEOTROPICAL DARNERS (**Arroyo**, **Blue-eyed**) that occur with it. Distinctly larger than **Riffle Darner**, with straight thoracic stripes. Blue and yellow coloration distinguish it from **Apache Spiketail** and **Western River Cruiser** that fly along some of same streams.

Natural History Males fly slowly along streams, especially over large pools, and search for females at shore. Also hawk for insects, even high in air around tree canopies. Both sexes hang up in dense vegetation. Generally uncommon throughout range.

Habitat Forested or shrub-bordered slow to fairly swift rocky streams in mountains.

Flight Season AZ Jul–Nov, NM Aug–Sep.

Distribution Ranges south in highlands of Mexico to Nayarit.

141
Persephone's Darner male—
Cochise Co., AZ, September
2004 (posed); blue female—
Sonora, Mexico, September
2004

Riffle, Mosaic, and Neotropical Darners - male appendages

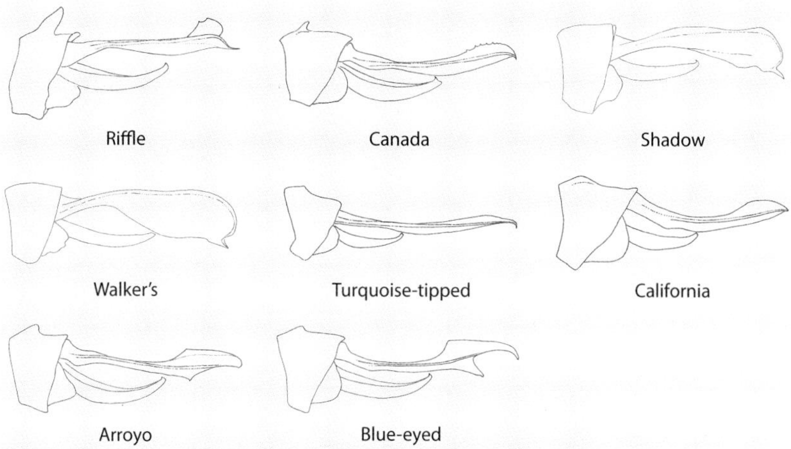

Riffle

Canada

Shadow

Walker's

Turquoise-tipped

California

Arroyo

Blue-eyed

Neotropical Darners   *Rhionaeschna*

Resembling mosaic darners, both sexes are distinguished in hand or at close range from the side by a low but fairly evident bump (tubercle) under the first abdominal segment. In addition, the black T-spot on the frons is bordered by a conspicuously paler area in neotropical darners; not so in mosaic darners. In male neotropical, blue marking on the rear of S2 is flat in front, not connected to middorsal blue line (connected in mosaic). Females of both genera are patterned the same. Forked appendages in males of some species are distinct from those of mosaic darners; eyes are bright blue in most North American species. As in mosaic darners, females of most species are polymorphic, with heteromorph more common. Males tend to fly continuously rather than hovering. The group is characteristic of South and Central America, with a few species extending north all the way to Canada. World 41, NA 5, West 4.

142 Turquoise-tipped Darner *Rhionaeschna psilus* TL 58–60, HW 36–43

Description Lateral thoracic stripes chartreuse, slightly notched and extended. Frontal stripes prominent, similarly colored. *Male*: Eyes turquoise, face blue. Abdominal markings very narrow greenish bands, look like rings from above. Posterior half of S2 and underside of S9–10 bright blue; blue on top of S10 varies from extensive to absent. *Female*: Apparently not polymorphic (polymorphism  less common in tropical species). Patterned as male but eyes and face brown, all markings greenish-yellow; no turquoise tip. Cerci very long, longer than S8–10.

142.1
Turquoise-tipped
Darner male—
Baja California Sur,
Mexico, January
2006, Steve
Mlodinow; female—
Hidalgo Co., TX,
June 2007,
Martin Reid

142.2
Turquoise-tipped
Darner male—
Lubbock Co., TX,
October 2007,
Jerry Hatfield

Identification Bright blue underside of abdomen tip diagnostic of male if visible. Otherwise, nothing else in its range like this small darner with broad greenish thoracic stripes and dull greenish abdominal rings. Might be mistaken for **Springtime Darner**, but flight season barely if at all overlaps. **Springtime** has brown eyes, blue spots on abdomen. Female distinguished by smallish size, rings instead of spots on abdomen, long cerci when present (may be broken off, but then shows broken tip of remnant).

Natural History Males fly back and forth fairly low over pools in stream beds, then suddenly hang up at head height or lower. Seem to perch more frequently than most other darners, with less time in incessant flight. This is a darner that photographers love! While moving along bank, males fly slowly, poke into shaded crevices, very thorough search for females. Females oviposit on floating wood, in vegetation in water, or on mud. Both sexes may roost in fields as well as woodland away from water.

Habitat Usually on pools of slow-flowing wooded streams, also shaded forest ponds.

Flight Season CA Oct, AZ Apr–Nov, TX Mar–Nov.

Distribution Appears to be increasing rapidly in Texas. Ranges south to Argentina.

Comments Spot-fronted Darner (*Rhionaeschna manni*) of southern Baja California could reach southwestern border. It has underside of S10 blue as in Turquoise-tipped but has narrow blue instead of broad green lateral thoracic stripes. Males look superficially more like California Darner, which lacks blue under S10 and has shorter cerci.

143 California Darner   *Rhionaeschna californica*   TL 57–64, HW 37–49

Description Small darner, unusual among neotropical and mosaic darners for its spring flight season. Lateral thoracic stripes straight and with conspicuous black borders behind, slightly more slanted toward rear than those of other darners (posterior stripe just behind base of hindwing at upper end). Frontal stripes absent. *Male*: Eyes bright blue, face whitish above, bluer in front. Appendages simple, as those of many mosaic darners, but cerci with obvious ventral tubercle in side view at about one-third length. *Female*: Polymorphic, heteromorph with eyes light brown, stripes yellow, spots yellow to white. Andromorph with eyes strongly tinged with blue, thoracic stripes and pale markings on abdomen blue. Cerci as long as S9–10 but distinctly narrower than in most similar darners.

Identification Only small darner in its range, also only one flying in early spring. Superficially most like **Blue-eyed Darner**, which is larger, flies later in summer (although much overlap), and flies higher and farther out over water. **Blue-eyed** has even more brilliant blue eyes, entirely blue face and thoracic stripes (both whitish in California). **Blue-eyed** also more likely to have frontal thoracic stripes, less likely to show fine black line across face. Equally small **Zigzag Darner** overlaps **California** slightly in range but not in flight season, habitat, or behavior. **Zigzag** is later in summer, usually over meadows in mountains, and lands on ground or tree trunks, **California** over water and perches on vegetation. Somewhat similarly patterned **Variable Darner** considerably larger and often with yellow on thoracic stripes. Female best distinguished from female **Blue-eyed** and all MOSAIC DARNERS with which it might fly by smaller size and hint of blackish lines bordering pale thoracic stripes. Neither **Blue-eyed** nor most MOSAIC DARNERS show such dark lines, although they can be evident in female **Paddle-tailed** and **Shadow Darners**, larger with later flight season. See also similar-looking but extralimital **Spot-fronted Darner** (under **Turquoise-tipped Darner**).

Natural History Males fly along shore, often right at water surface and very close to shore, moving into emergent vegetation in their search for females and hovering at that time. Pairs mate in flight, fly quickly to dense marsh vegetation or nearby trees and shrubs and copulate at length (5–20 min). Females oviposit in vertical stems or floating leaves of marsh

143.1
California Darner male—
Siskiyou Co., CA,
June 2004

143.2
California Darner yellow female—
Marin Co., CA, March 2003,
Robert A. Behrstock

plants at water level. Perching usually in shrubs or herbs, lower than many other darners, but some up in trees, even on trunks. Feed by flying around clearings, up to considerable height. One of first odonate species, almost always first dragonfly to appear in spring throughout at least northern part of range. Uncommon by midsummer when most darners just emerging.

Habitat Wide variety of ponds and lakes with fringing vegetation, also slow streams; in wooded or open country.

Flight Season BC Apr–Jul, WA Apr–Aug, OR Apr–Sep, CA Feb–Aug, MT Jun–Jul, NE Apr–Aug.

Distribution Ranges south in Mexico to Baja California.

144 Arroyo Darner *Rhionaeschna dugesi* TL 70–74, HW 48–53

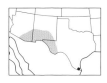

Description Lateral thoracic stripes straight and narrow, anterior one with slight extension. Frontal stripes well developed but narrow. *Male*: Eyes and face bright sky-blue. Appendages almost forked, with long upper fork and hint of lower fork. *Female*: Probably polymorphic, heteromorph with yellow-brown eyes and yellow-green body markings, rare andromorph with blue. Cerci slightly longer than S9–10, widest in middle.

Identification Only this and **Blue-eyed Darner** share such bright sky-blue eyes and entirely blue thoracic stripes in Southwest. **Arroyo** of both sexes distinguished from **Blue-eyed** by anterior thoracic stripe extending backward slightly at top. No extension in **Blue-eyed**, and this clearly visible with a good look. Good look at male appendages also distinguishes them, clearly forked in **Blue-eyed** and merely notched in **Arroyo**. Blue spot on either side of S3 may prove characteristic of yellow female **Arroyo**. In hand, ventromedial carina on middle abdominal segments of female **Arroyo** turns inward toward front end of segment, in **Blue-eyed** stays nearly parallel (best to compare). Other southwestern stream darners (**Malachite, Persephone's, Riffle, Turquoise-tipped**) all differ in color.

144.1
Arroyo Darner male—
Santa Cruz Co., AZ,
September 2004

144.2
Arroyo Darner yellow female—
Cochise Co., AZ, May 2002,
Doug Danforth

Natural History Males fly much like Blue-eyed Darners, over open water and along shore-lines at waist height or below. Not distinguishable by behavior. Appears less likely than Blue-eyed to wander far from water, but perhaps only poorly known. Females oviposit in plant stems or rootlets or on wet logs or mud of stream banks.

Habitat Slow streams and drainage ditches in open and wooded country, also isolated pools, even garden ponds. Usually higher up into mountain woodland than Blue-eyed Darner.

Flight Season AZ May–Oct, NM May–Oct, TX Jun–Sep.

Distribution Ranges south in uplands of Mexico to Oaxaca.

145 Blue-eyed Darner *Rhionaeschna multicolor* TL 65–69, HW 42–45

Description Lateral thoracic stripes straight and narrow. Frontal stripes vary from narrow to absent. *Male*: Eyes and face bright sky-blue, thoracic stripes pale blue. *Female*: Polymorphic, hetero-morph with brown eyes and yellow body markings. Andromorph with blue-tinged eyes and face, yellow to blue-green stripes on thorax, and blue spots on abdomen.

Identification Bright sky-blue eyes of male allow easy distinction from all other darners over much of range. Smaller **California Darner** has blue eyes as well, not quite as bright, and face paler blue or whitish; also looks distinctly smaller in field and typically flies lower. See **Arroyo Darner**, exactly like **Blue-eyed** at a dis-tance and overlapping with it in Southwest. Most similar to female **Arroyo**, which has slight extension on anterior lateral stripe; must be well seen to detect this difference. Female distinguished in hand from all MOSAIC DARNERS by tubercle under S1, in field with more difficulty. Those in its range with straight and relatively narrow thoracic stripes for which it could be mistaken include **Paddle-tailed**, **Shadow**, and **Variable Darners**. Differs from **Shadow** in larger blue spots on abdomen, none beneath it, but too similar to other two species to be identifiable unless tubercle or pale areas next to T-spot seen.

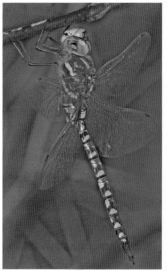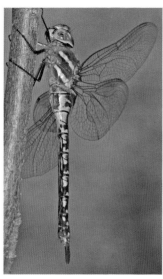

145.1
Blue-eyed Darner male—
Thurston Co., WA,
September 2005

145.2
Blue-eyed Darner yellow female—King Co., WA,
July 2007 (posed); blue female—San Benito Co.,
CA, August 2006, Paul Johnson

Natural History Males fly over open water and along shores, usually continuously but occasionally hovering briefly, typically at waist height. Usually less likely to hover than mosaic darners but may have well-defined beats along short length of shoreline. Often alternate open-water flight with slow and low patrol through dense tall emergent vegetation. Copulating pairs perch from near ground up into trees, often flying for some time back and forth over water before finding good perch. Females oviposit in dense emergent vegetation and on floating plant stems and leaves and woody branches in open water, laying eggs above or below waterline. Both sexes wander far from water when not breeding, liable to be seen anywhere away from water, including cruising over city yards and parking lots throughout West. Called "most domestic" of western odonates by C. H. Kennedy, and most common darner, along with Common Green, in much of lowland West. May occur in mixed feeding swarms. On very hot days, roost well within cover of riparian trees, sometimes in groups, and may fly until dusk. Thought to be migratory in California, where large numbers appear in fall, often with migrating Common Green Darners.

Habitat All kinds of lakes, ponds, slow streams, and canals, especially common in highly productive, open marshy lakes and in open rather than wooded areas.

Flight Season BC May–Sep, WA May–Oct, OR May–Oct, CA Mar–Dec, AZ Feb–Nov, NM Mar–Nov, NE Jun–Sep, TX Apr–Nov.

Distribution Ranges south in uplands of Mexico to Michoacan, also in Iowa and western Missouri.

Green Darners *Anax*

These large, robust darners are among the largest of dragonflies in most regions. North American species display a bright green thorax, but some Old World species are quite different, for example black and yellow banded or entirely red. The thorax is bulky, abdomen swollen at the base and a bit thicker and relatively shorter than in many other large dragonflies but excessively long in Giant Darner, the largest dragonfly in North America. Male hindwings are without the sharp angle on inner corner characteristic of other darners; also lacking are projecting auricles on S2. Adults fly swiftly and directly along shorelines or over open water and are less likely to hover than most other darners. World 28, NA 5, West 5.

146 Amazon Darner *Anax amazili* · TL 70–75, HW 49–52

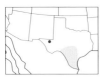

Description Robust darner with green thorax and white-spotted abdomen. *Male*: Eyes brownish gray to blue-gray, face green. "Bull's eye" on top of frons with blue border interrupted at front. Thorax bright green. Abdomen with S1–2 bright green; sides of S1 white, forming what looks like ring from side, small blue median spots on S2. Rest of abdomen black to dark brown with whitish irregular spots on sides of S3–9. With age, abdomen darkens, and spots eventually become obscure; white ring becomes green. *Female*: Colored exactly as male.

Identification Most like **Common Green Darner**, but white-spotted or entirely dark abdomen distinctive, as is white or pale green basal ring conspicuous from sides. Also somewhat like **Blue-spotted Comet Darner**, which has long legs and no marking on top of frons, as well as blue abdomen spots. **Blue-faced Darner**, also with green thorax, is smaller and much more slender as well as having fine brown lines on thorax and mostly brown abdomen.

Natural History Both sexes, or perhaps only females, spend dry season as adults, then females oviposit by themselves in herbaceous vegetation both above and below water level or in completely dry basins. Breed in temporary wetlands just as rains come, larvae developing quickly as top predators in fishless ecosystems. Males less often seen than females, rarely at water as other North American species; probably do not mate at water. Often seen in feeding flights that may be low and fast during day and at dusk; occasionally comes to lights during dusk flight. Other individuals encountered roosting at head height in woodland.

146.1
Amazon Darner male—
Travis Co., TX, August 2006,
Bob Thomas

146.2
Amazon Darner female—
Guanacaste Prov., Costa Rica,
April 1967 (posed)

Habitat Open ponds and marshes; temporary waters preferred in tropics.

Flight Season TX Apr–Nov.

Distribution Not certainly established as breeding species in United States but perhaps does so, from number of recent records in south Texas and Florida. Alternatively, perhaps regular summer movement of individuals north from tropics, with small enough numbers that males and females unlikely to find one another. Recorded also in Louisiana and Florida, ranges south in lowlands to Argentina; also Galapagos and West Indies.

147 Blue-spotted Comet Darner *Anax concolor* TL 65–74, HW 43–46

Description Long-legged green darner with blue-spotted abdomen and no markings on frons. Legs especially long. *Male:* Eyes blue to blue-green. Thorax and abdomen S1–2 green; rest of abdomen brown, reddish-brown at tip, with large blue spots on S3–9, two pairs of spots above and one below on middle segments. *Female:* Colored as male, blue abdomen spots slightly more extensive.

Identification Much like **Comet Darner**, with long legs and no marking on top of frons. Both sexes differ from that species in blue-spotted brown abdomen, looks a bit more orange at tip. Females more similar, but abdomen of **Comet** still looks more reddish, with pale spots less distinct. Male has strong yellow wash in middle of each hindwing, lacking in **Comet Darner**. **Amazon** and **Common Green Darners** also show this wash but have all blue, all violet, or white-spotted abdomens.

Natural History Feeding individuals cruise over open areas or low woodland like Common Green Darners. Known to feed on other dragonflies and butterflies, including distasteful ones. Males fly rapidly at about knee to waist height around edges and across middle of ponds with or without much shore vegetation. May also fly back-and-forth beats no more than 30 feet long. Search for females in dense grass beds where they oviposit. Also encountered roosting in woodland. Females seldom seen, even where males common in tropics.

Habitat Shallow, open ponds in or out of woodland.

Flight Season TX Jun–Nov.

Distribution Only recently discovered in United States but may be a resident, so far known only from few males seen at Santa Ana National Wildlife Refuge in south Texas. Ranges south in lowlands to Argentina, also throughout West Indies.

147
Blue-spotted
Comet Darner male—
Hidalgo Co., TX,
November 2005

Description Large darner with green thorax, red abdomen. Legs especially long. *Male*: Eyes green to blue-green. Thorax and S1–2 green, S3–10 Intensely red-orange, only faintly patterned. *Female*: Colored as male but abdomen duller, more reddish brown, more obviously patterned with paler spots, and slightly thicker and shorter. Immatures of both sexes have more prominent spots, sometimes with bluish tinge. One of few darners lacking any marking on top of frons.

Identification Unmistakable, but see closely related **Blue-spotted Comet Darner**, with which it overlaps in south Texas; females more similar than males. Confusion also possible with immature and female **Common Green Darners,** which may have reddish-purple abdomen, but that species at all ages has dark stripe along top of abdomen and "bull's eye" on top of frons.

Natural History Males fly along or off shore, cruising long distances in larger bodies of water but easily watched at length in small ones. Single male may cruise around entire shore of midsized ponds. Flight at waist to chest height. Females oviposit in emergent vegetation at water surface, may come to water in some numbers when males not present. Much less often seen away from water than Common Green Darner.

Habitat Shallow lakes and ponds, typically with extensive beds of grasses and normally lacking fish.

Flight Season TX Apr–Oct.

Distribution In East north to Missouri, Michigan, and New England, stragglers farther north.

148
Comet Darner
male—Centre Co.,
PA, June 2005, Bryan
Pfeiffer; female—
Leon Co., FL, April
2005 (posed)

Male: TL 100–116, HW 56–67; *Female:* TL 88–98, HW 56–60

Description Very large green darner (because of especially long abdomen) with mostly blue or brownish abdomen. Bull's eye somewhat like that of Common Green Darner on top of frons. *Male*: Eyes dull blue above, greenish below; face pale greenish. Thorax green, sides quite blue in some males (older ones?). Abdomen dark brown with large, irregular patches of bright blue all along its length. *Female*: Polymorphic. Abdomen considerably shorter than that of male, colored mostly brown but with large irregular spots either greenish or blue.

Identification No other dragonfly in North America has such long abdomen, and size alone distinguishes it. Coloration similar to **Common Green Darner**, with green thorax and blue abdomen, but blue of abdomen broken by black bands along most of length so looks patchier than in smaller relative. Females considerably shorter than males, more easily mistaken for female **Common Green**, but still obviously larger, with abdomen just longer than wings (shorter in **Common Green**). Middle abdominal segments almost as long as thorax in **Giant**, much shorter in **Common Green**. **Giant** also differs by having series of pale spots down abdomen, rather than continuous dark center and light sides as **Common Green**.

Natural History Males patrol long stretches of stream, 30–300 feet or more, and defend them vigorously. Usually over pools at waist to chest height, often lower over riffles. When few present, may be long time between sightings, so beats may be very long. Slow but steady flight, often alternating fluttering and gliding, but easily attains high speed and adept at avoiding capture. Abdomen of male arched in flight, perhaps necessary because of its length. Occasionally examine shore vegetation to search for females. Females oviposit in emergent streamside vegetation or on floating algal mats or submerged plants. Seldom seen away from water but sometimes few together in feeding flight as in other darners and both sexes seen in numbers feeding over river late in afternoon.

Habitat Slow streams with some gradient, often in canyons, bordered by riparian shrubbery, cattails, or bulrushes.

Flight Season CA Apr–Sep, AZ Apr–Oct, NM May–Sep, TX May–Oct.

Distribution Ranges south in uplands to Honduras.

149.1
Giant Darner
male—Travis Co., TX,
June 2005 (posed)

149.2
Giant Darner
female—Colusa Co., CA,
July 2000, Rod Miller

Description One of most common and characteristic North American dragonflies. Thorax green, abdomen with black line on upper surface, becoming wider toward tip. *Male*: Eyes dull greenish, paler below. Face greenish, frons blue above with "bull's eye" pattern. Thorax bright to dull grass-green. Abdomen with S1 green, S2–6 bright blue, S7–8 duller, often bluish-green, S9–10 mostly dark. *Female*: Polymorphic, most females with S1–2 green, S3–10 brown above and gray-green on sides. Minority colored much as male, at brightest with S2–10 black above and pale blue on sides. Immature of both sexes with S3–10 reddish violet. Wings uncolored in immatures, with age become increasingly suffused with amber, especially in females.

Similar species Only widespread western darners with bright green thorax in this genus. **Giant Darner** with much longer abdomen with blue color all across, no black stripe down middle. Female **Giant** has relatively shorter abdomen but still considerably larger than **Common Green**, with abdomen clearly longer than wings. **Comet Darner** with entirely

150.1
Common Green Darner
male—Winneshiek Co., IA,
July 2004 (posed)

150.2
Common Green Darner
female—Hidalgo Co., TX,
November 2005

150.3
Common Green Darner
immature male—Columbia
Co., FL, April 2005

Sanddragons are medium-sized clubtails with irregularly patterned thorax and slightly developed club in males. With few exceptions, they are the only clubtails with any color in the wings, a spot of brown at base of all wings. Males characteristically have flat, pale cerci that show up prominently at tip of the abdomen. Typically, they are seen along sandy shores, as their larvae are highly modified as sand burrowers. World 68, NA 4, West 2.

151 Gray Sanddragon *Progomphus borealis* TL 57–61, HW 32–35

Description Pale clubtail of desert streams with little club and distinctive white appendages. *Male*: Eyes blue-gray, face yellow-orange above, gray below, with dark crossbar. Thorax yellow in front, gray on sides, with broad T1 and T2 partially fused, T1 contacting frontal stripe; T3 represented by spot at lower end, T4 well developed. Abdomen largely black with yellow-orange on top of S1–2, long pale tan triangles on top of S3–6 and ring on base of S7–8, orange apical spots on sides of S8–10. Cerci shining white to pale yellow. Small but intense brown spot at wing bases. *Female*: Colored as male, but pale markings on abdomen whitish to gray rather than yellow to orange.

151.1
Gray Sanddragon male—Butte Co., CA, June 2004, Netta Smith

151.2
Gray Sanddragon female—Navajo Co., AZ, June 2004, Doug Danforth

Identification Easily distinguished from other clubtails in range if rather flat pale cerci of male seen. Color also distinctive, yellow to orange above on thorax and abdomen and gray on sides of thorax. Abdominal pattern rearward-pointing triangles rather than the rings typical of RINGTAILS that occur with them. See **Common Sanddragon** in very limited area of overlap.

Natural History Males perch on gravel, rocks, and sandy shores, often with abdomen well elevated at the intense midday temperatures of its habitat. Females relatively rarely seen, usually when in pairs or ovipositing. Both sexes forage from twigs as well as ground when away from water.

Habitat Sandy and rocky streams and rivers, in open or bordered by riparian woodland. Occurs over great elevation range.

Flight Season CA Mar–Oct, AZ Apr–Nov, NM May–Aug, TX May–Sep.

Distribution Ranges south in Mexico to Michoacan and Chihuahua.

152 Common Sanddragon *Progomphus obscurus* TL 47–53, HW 28–33

Description Rich brown and yellow clubtail of sandy habitats. *Male*: Eyes blue-gray, becoming olive-green; eyes of mature males can become golden above, perhaps only clubtail so colored. Face dull yellowish with two faint brown stripes crossing it. Thorax yellow with brown T1–4, T1 and T2 thick and partially to entirely fused. Abdomen dark brown with yellow rear-pointing triangles (much like burning candles) on S2–6; cerci cream-colored. Dark brown markings at base of all wings, unusual in family. *Female*: Colored as male, eyes blue-gray.

Identification Pale appendages on all dark abdomen tip distinctive of both sexes of this heavily marked species, richer yellow and brown than other clubtails of its habitat. Basal wing markings also distinctive in comparison with most species. Barely overlaps with **Gray Sanddragon**, distinguished by overall browner and more heavily marked thorax (T1 and T2 more likely fused, two instead of one lateral thoracic stripe).

Natural History Males perch on sand beaches of pools, right at edge facing water, or fly low over water in what can be lengthy patrol flights, including hovering over riffles. Also land on rocks on rocky/sandy rivers. Not aggressive in territory defense. Prefer open rather than shrubby banks. Abdomen elevated when perched, on ground or on twigs, and obelisks dramatically at midday. Pairs copulate for up to 15 min on ground or in shrubs near water.

152.1
Common
Sanddragon
male—Caldwell Co.,
TX, May 2005

152.2
Common
Sanddragon
female—Washington
Co., FL, June 2004

Female oviposits on erratic path, taps water once on yard-long approach run; also may drop eggs from above surface. Males have been seen guarding ovipositing females, unusual among clubtails.

Habitat Sandy woodland streams; sandy lakes in northern part of range. Sometimes on rocky rivers, presumably where sand available among rocks.

Flight Season NM Apr-Jul, NE Jun-Sep, TX Apr-Sep

Distribution Also throughout much of eastern USA north to Wisconsin and Maine, south to northern Florida.

Leaftails *Phyllogomphoides*

These are large gomphids with conspicuously ringed abdomen in side view and, in males, a moderate club made wider by substantial flanges on edges of S8–9. Oddly, size of flanges varies somewhat in both sexes of each species, so "tails" are more or less "leafy" in different individuals. Presence of small epiproct in male and absence of extended corners on S10 distinguishes it from greater forceptails at close range. Cerci are long, pale, and forceps-like in males, with small tooth on inside; long and white in females. Leaftails are a very diverse group of stream dwellers from Mexico to southern South America. World 46, NA 2, West 2.

153 Five-striped Leaftail *Phyllogomphoides albrighti*   TL 60–63, HW 37–40

Description Large brightly marked clubtail centered in Texas with heavily striped thorax, ringed abdomen, and wide, leaflike flanges on club. *Male*: Eyes light blue, face cream with three rather faint light brown stripes across it. Thorax almost evenly striped with cream and dark brown. Abdomen with S1–2 mostly cream with brown and black markings, S3–6 mostly black with cream ring at  base, pointed to rear on top of segment; S7 white at base, brown at tip; S8–10 dull orange with blackish markings on top of S8–9, black flanges on conspicuously clubbed S8 and less so S9. *Female*: Colored as male; abdominal club (or leaf) narrower, may be scarcely evident; end segments duller, tan rather than orangey.

153.1
Five-striped Leaftail
male—Bexar Co.,
TX, July 2004

153.2
Five-striped Leaftail
female—Bexar Co.,
TX, July 2004

Identification Most like **Four-striped Leaftail**, but stripes on side of thorax broader, more prominent, presence of fifth stripe (T5) definitive. Club plus flanges usually wider and more orange, more impressive in male than in **Four-striped**, best seen from above. **Ringed Forceptail** also quite similar, with heavily striped thorax but with overall darker abdomen (smaller pale markings); tip may be almost entirely dark (always with pale markings in leaftails). See that species for more details. Other large clubtails in range include GREATER FORCEPTAILS that have browner abdomen without obvious ringed look, although **Broadstriped Forceptail** has pale spots looking something like rings. **Narrow-striped Forceptail** has narrower thoracic stripes than any leaftail. **Flag-tailed Spinyleg** with less-striped thorax, mostly yellow club, and very long hind legs.

Natural History Males fly slowly and steadily up and down river or perch at knee to waist height on vegetation over water; more likely to perch when hot. Abdomen elevated slightly with tip curved down while perching or flying. Also may hover for many seconds over river bank, in and out of vegetation. Both sexes perch in shrubs and trees away from water.

Habitat Large clear streams and rivers with mud- or sand-bottomed pools and riffles, with or without rocks, in open with grassy banks or lined with shrubs and trees.

Flight Season TX May–Oct.

Distribution Ranges south in eastern Mexico to San Luis Potosí.

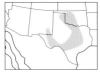

Description Large brightly marked clubtail centered in Texas with ringed abdomen, variable flanges on club, and thoracic stripes much as those of other clubtails. *Male*: Eyes light blue, face cream with two faint light brown stripes across it. Thorax cream with T1–4 all well developed. Abdomen mostly black, cream on sides of S1–2; S3–7 with cream rings at base, extending back as point on at least some segments; S8–10 cream to orange, usually paler on sides, typically with distal two-thirds of S8 and often base of S9 black above. Poorly to moderately developed flange on S8 orange to black, rarely developed on S9. *Female*: Colored as male, but less brown at abdomen tip, mostly cream except for tip of S9 and S10. Club essentially absent, but narrow leaf-like projections present on S8, less on S9.

Identification Quite similar to **Five-striped Leaftail** except for lack of fifth thoracic stripe. Both sexes, female especially, look more strikingly white-ringed than **Five-striped Leaftail** or **Ringed Forceptail**, including S8 (base of S8 whitish in **Four-striped**, orange in **Five-striped**, dark in **Ringed**). Ringed look usually definitive. See **Five-striped** for differences from GREATER FORCEPTAILS.

Natural History Males rest on elevated perches, knee to chest height, at waterside or over water, more at pools than riffles. May spend much time flying over water. Abdomen elevated slightly with tip curved down while perching or flying. Female flies close to shore among dense grass to oviposit; may drop eggs onto water while hovering. Often found in woodland or clearings at some distance from water, perched prominently and horizontally on twigs of shrubs and trees.

Habitat Slow-flowing streams and rivers and large open ponds.

Flight Season NM May–Sep, TX May–Sep.

Distribution Ranges south in eastern Mexico to Coahuila and Nuevo León.

154.1
Four-striped Leaftail male—Travis Co., TX, June 2007, Eric Isley

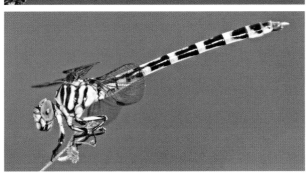

154.2
Four-striped Leaftail female—Travis Co., TX, July 2004, Greg Lasley

These are moderately large clubtails with moderately expanded abdomen tip. Curved forceps-shaped appendages are characteristic of the male, shared with greater forceptails. Extremely like genus *Aphylla*, but it differs in S10 not having projecting lower rear corners in males, females with no expansion of abdomen (female Broad-striped Forceptail somewhat similar), and hind femur with fewer and longer teeth. As with greater forceptails, lesser forceptails are diverse from Mexico south through much of South America. World 31, NA 1, West 1.

155 Ringed Forceptail *Phyllocycla breviphylla* TL 56–60, HW 30–35

Description Large clubtail of Lower Rio Grande Valley with prominently ringed abdomen. *Male:* Eyes green to turquoise (blue-gray
in younger individuals), face cross-barred brown and white. Thorax
evenly striped dark brown and white, with faint yellow tint; four
equal-sized pale stripes visible from side, otherwise entirely brown.
Abdomen with S1–2 brown, S3–7 black with well-defined whitish
ring at base, successively wider to rear and occupying from one-third to two-thirds of S7;
S8–10 reddish-brown, S8 darker above and with black flange. In darkest individuals, S8–10
much darker, mostly blackish. *Female:* Colored much as male but no club or flanges, end of
abdomen mostly dark.

Identification Although structurally like a GREATER FORCEPTAIL, colored more like a LEAF-
TAIL because of dark, pale-ringed abdomen. Differs from LEAFTAILS in thoracic pattern,
four pale stripes of about equal size rather than the unequal striping of our two LEAFTAILS.
Some **Five-striped Leaftails**, however, have similar thoracic pattern, but abdomen pattern different, **Ringed** with dorsal triangles extending down on sides while LEAFTAILS with
pale dorsal and lateral markings separated by dark line. **Ringed** has brown abdominal

155.1
Ringed Forceptail
male—Hidalgo Co.,
TX, August 2005, Sid
Dunkle

155.2
Ringed Forceptail
female—Hidalgo Co.,
TX, July 2006,
Martin Reid

Clubtail Family **243**

appendages, LEAFTAILS white. Differs from GREATER FORCEPTAILS by more strongly ringed look of darker abdomen.

Natural History Males perch on tree leaves and branches at waterside, abdomen usually down at 45° angle or even hanging vertically. Both sexes in trees in woodland near breeding sites.

Habitat Sandy/muddy rivers, including Rio Grande, and streams.

Flight Season TX May–Oct.

Distribution Occurs south through eastern lowlands of Mexico to Nicaragua. Very close to western Mexican *Phyllocycla elongata*, but *breviphylla* kept separate pending more research.

Greater Forceptails *Aphylla*

A widespread Neotropical group, these large, short-legged clubtails have a long, slender abdomen with not much of a club, but a prominent leaflike projection on either side of S8 enhances the clubtail look. Male cerci are brown, large, and forceps-like, epiproct almost lacking, but lower edge of S10 is pointed on either side, visible in side view; female cerci are yellow to orange. Thorax is dark with vivid pale stripes and the abdomen largely reddish from side and with only weakly ringed look. They are related to lesser forceptails and leaftails of similar size and are often found occurring with them, but they are more likely to occur on open lakes and ponds than other large clubtails. World 24, NA 3, West 3.

156 Broad-striped Forceptail *Aphylla angustifolia* TL 62–68, HW 36–42

Description Large Texas clubtail with broad pale stripes on thorax and slender, patterned abdomen. *Male*: Eyes blue-green, face with narrow brown bars crossing it. Thorax looks brown with yellow stripes, as all five dark stripes quite wide, T1–2 partially fused. Basal segments of abdomen brown with yellow on sides and base, forming ringed pattern on S2–4 or S5. End segments more uniformly  colored, reddish; become darker brown with age. Narrow orange flange on S8–9. *Female*: Colored much as male, but abdomen more uniform orange with yellow markings, apparently not becoming as dark with age; flange on S8–9 virtually nonexistent.

Identification Weakly ringed look distinguishes from strongly ringed **Ringed Forceptail** and LEAFTAILS. May occur with either of the other two greater forceptails. Easily distinguished from **Two-striped Forceptail** by many more stripes on thorax. Differs from **Narrow-striped** in broader thoracic stripes, in side view thorax showing about as much pale as dark (more dark than pale in **Narrow-striped**). Stripes T2 and T4 wider than T1 and T3 in **Broad-striped**, all about equal in **Narrow-striped**. **Broad-striped** has distinctly narrower flanges on S8–9, most easily seen when abdomen way up in air in obelisk position. In both sexes of **Broad-striped**, basal rings on basal abdominal segments more contrasty and bright yellow than in **Narrow-striped**, in which all pale markings tend to be orange rather than yellow. Also, end of abdomen, including clubbed area, looks almost entirely orange in **Broad-striped** with S7 usually distinctly paler, whereas in **Narrow-striped** same area with distinct black markings on top, including most of S7, that contrast with entirely pale S10. Immature male **Broad-striped** look rather orange all over, and black apical rings on middle segments show up early in immature **Narrow-striped**.

Natural History Males perch on twigs and leaves over water, from near ground to waist height or above, rarely on ground. Occasionally fly along shore to new perch, even when not disturbed, but have definite preferred perches. Both sexes feed in open areas nearby.

Habitat Lakes and ponds, pools on rivers and streams.

Flight Season TX May–Oct.

Distribution Ranges south to Guatemala and Belize, also east to southwestern Mississippi.

156.1
Broad-striped
Forceptail male—
Bexar Co., TX, July
2004

156.2
Broad-striped
Forceptail
female—Bexar Co.,
TX, July 2004

157 Narrow-striped Forceptail   *Aphylla protracta*   **TL 54–66, HW 35–39**

Description Large Texas clubtail with narrow pale stripes on thorax and reddish abdomen. *Male*: Eyes blue-green, face with narrow brown bars crossing it. Thorax brown with five yellowish-white stripes, all of similar width. Abdomen mostly pale yellow to orange with dark brown to black markings along tops of all segments except S10, extending onto sides of segments as apical rings on S3–6. *Female*: Colored as male but abdomen slightly thicker, club narrower (most of breadth from leaflike flanges).

Identification In this species, entire abdomen looks reddish from side, except for dark apical rings, whereas contrast provided in other forceptails by conspicuous basal yellow markings on at least some segments. Abdomen never with vividly ringed look of some **Broad-striped** and **Ringed Forceptails** and LEAFTAILS. See **Broad-striped Forceptail** for distinction from that species.

Natural History Males perch near or, very commonly, a little away from breeding habitat on branches of shrubs and trees or on open beaches. Fly out over water and return to former perch or move some distance along shore, probably not defending specific territory. Females perch similarly away from water, likely to be encountered in open woodland.

Clubtail Family   **245**

157.1
Narrow-striped
Forceptail
male—Sonora,
Mexico, August 2006,
Netta Smith

157.2
Narrow-striped
Forceptail
female—Sonora,
Mexico, September
2004

Habitat Variety of lakes and ponds with mud bottom, much less likely than Broad-striped at running water.
Flight Season TX Apr–Nov.
Distribution Ranges south to Costa Rica.

---

158 Two-striped Forceptail *Aphylla williamsoni*　　　　　**TL 71–76, HW 37–43**

Description Large southeastern clubtail with vivid pale thoracic stripes. *Male*: Eyes blue, face with dark brown bars crossing it. Thorax dark brown with three widely spaced yellow stripes, second and fourth light stripes of other forceptails lacking. Abdomen dark brown with yellow markings on sides of basal segments, terminal segments mostly orange. Flanges on S8 bright yellow, contrasting with red-orange sides of segment. *Female*: Colored as male but more likely to show narrow and incomplete second pale stripe on side of thorax. Abdomen mostly dark, including very narrow flanges on S8.

158.1
Two-striped
Forceptail
male—Colleton Co.,
SC, August 2006,
Sharon L. Brown

158.2
Two-striped
Forceptail
female—Sumter Co.,
GA, July 1998, Robert
A. Behrstock

Identification Unmistakably a forceptail from its overall look, distinguished from other forceptails (**Broad-striped**, **Narrow-striped**, and **Ringed**) and other clubtails that occur with it by having only two wide pale stripes visible on each side of thorax and only a single stripe visible on either side of front. Otherwise, looks much like other two greater forceptails, large and dark and with strong reddish cast to posterior abdomen. Bright yellow flanges contrasting with darker sides of S8 in male distinctive. Female abdomen darker (almost black), with smaller pale markings, than in other greater forceptails, going along with overall darker-looking thorax.

Natural History Males perch on twigs and tall reeds at the shore, often at waist height or above, and fly out at other large dragonflies, persistently at their own species. Most perching well above ground, but at least females seen on ground. Patrol flights may cover 60 feet or more. Abdomen slightly curved in flight, sometimes hovers. May remain active at water until dark, unlike most other clubtails. Oviposition may be limited to late in day and is varied, from hovering to rapid flight and from tapping water to dropping eggs from above.

Habitat Slow streams and rivers, canals, and a great variety of sand-bottomed lakes and ponds, among the most ecologically broad-based of American clubtails.

Flight Season TX Jul–Aug.

Distribution Ranges across South from Arkansas to southeastern Virginia.

Pond Clubtails *Arigomphus*

This small group is primarily distributed in central and eastern North America. Males have rather narrow clubs; females entirely narrow abdomens. Most species are relatively plain and pale, gray, or dull greenish combined with black or dark brown, but a few are brighter, chartreuse and reddish-brown. Front of the thorax is mostly pale with the frontal stripe very narrow or absent in western species. Sides of the thorax are only lightly marked behind humeral stripes. Other distinctions are structural, including long subgenital plates that project downward in females of some species. They are more likely to be found at lakes and ponds than species of related genera. World 7, NA 7, West 4.

159 Horned Clubtail *Arigomphus cornutus* TL 55–57, HW 32–37

Description Northern clubtail with lightly striped thorax and spectacularly forked appendages. *Male*: Eyes blue, face pale greenish-tan to yellow. Thorax pale gray-brown or yellow with T1 incomplete at upper end, well-developed T2, very narrow lower end of T3, and barely indicated T4. Abdomen black with pale grayish to yellow dorsal stripe, becoming spearpoints to rear and ending at base of S8; also pale spot on top of S10. S8–9 with rusty edges, contrasting with yellow everywhere else. Cerci and epiproct very wide, like branched horns. *Female*: Color pattern as male; pale color on sides of abdomen comes up higher.

Identification Male distinguished from co-occurring clubtails by absence of club, very wide branched appendages, and minimal striping on sides of thorax. Most similar of pond clubtails is **Jade Clubtail**, with less-marked thorax, mostly pale abdomen, and green eyes. Similarly

159.1
Horned Clubtail
male—Columbia Co.,
WI, June 1997, David
Westover

159.2
Horned Clubtail
female—Carver Co.,
MN, June 2006,
David Reed

colored **Plains** and **Pronghorn Clubtails** have prominent clubs, less-clubbed females distinguished by different thoracic patterns. Other clubtails in its range with minimally expanded abdomen are differently colored. **Common Sanddragon** brown and yellow, **Pale Snaketail** quite green, **Brimstone Clubtail** smaller and bright yellow.

Natural History Males perch on bare ground or rocks or flat on leaves at waterside or fly back and forth along shore or over water. Females often perch on leaves higher in woodland, where males go when disturbed.

Habitat Lakes, ponds, and slow streams, may be quite muddy.

Flight Season MT Jun–Jul, NE May–Aug.

Distribution Also east through Great Lakes region to southern Quebec.

160 Jade Clubtail *Arigomphus submedianus* TL 51–55, HW 34–36

Description Eastern prairie clubtail with green eyes and ruddy abdomen tip. *Male*: Eyes grass-green, face pale greenish-tan. Thorax pale greenish, T1 wide and distinct, T2 narrower, others indistinct or lacking. Abdomen dull greenish-gray to yellowish with dark brown dorsolateral apical triangular markings on S2–6, narrow dark rings between them; S7–9 rich brown, S10 paler orangey, appendages yellow. *Female*: Colored as male, eyes a bit duller.

Identification Eyes greener than in other pond clubtails, although not much more so than in **Bayou Clubtail**; differently colored abdomen tip easily distinguishes these two. Most

160.1
Jade Clubtail
male—Bastrop Co.,
TX, May 2003, Greg
Lasley

160.2
Jade Clubtail
female—Holmes Co.,
MS, May 2005

like **Stillwater Clubtail** except for eye color but appearing a bit paler overall, T1 distinctly wider than T2, reduced marking on sides of thorax. Of female pond clubtails in West, S9 obviously longer than S8 only in **Jade**. Other superficially similar clubtails in range include **Flag-tailed Spinyleg**, with thicker T2; middle abdominal segments white but with dark markings extensive, almost continuous; and much longer hind legs.

Natural History Perches on or near ground, males at water's edge. Forages in open and semiopen areas near woods.

Habitat Large mud-bottomed lakes, sloughs, and canals.

Flight Season NE May–Sep, TX Mar–Aug.

Distribution Ranges east to Michigan, Kentucky, and Alabama.

Description Eastern prairie clubtail with blue eyes and ruddy abdomen tip. *Male*: Eyes blue to turquoise, face pale greenish. Thorax pale greenish with T1 and T2 prominent, sometimes partially fused; T3 and T4 faintly indicated. Abdomen yellowish-green at base, duller yellow-gray on S3–6, S2–6 with apical dorsolateral blackish blotches forming somewhat ringed pattern. S7–10 rusty,
in some individuals darker above and some yellowish on sides; S8 often darker than S9. Appendages pale yellow. *Female*: Colored as male but dark markings on abdomen less extensive.

161.1
Stillwater Clubtail
male—Gonzales Co.,
TX, June 2005

161.2
Stillwater Clubtail
female—Harris Co.,
TX, May 2007,
Martin Reid

Identification Most similar to other pond clubtails with which it occurs. Both sexes differ from quite similar **Jade Clubtail** by more heavily striped thorax; latter also has quite green eyes. Differs from **Bayou Clubtail** by more reddish abdomen tip (brown in **Bayou**) and wider T1–2. In female **Jade**, S9 slightly longer than S8, about same length in **Stillwater** and **Bayou**. Both sexes colored somewhat like **Russet-tipped Clubtail,** but that species has sides of thorax prominently striped.

Natural History Males rest on shore or grass stems or twigs facing water.

Habitat Open muddy ponds and other still waters.

Flight Season TX Mar–Jun.

Distribution Also east through Arkansas to Indiana and western Kentucky.

162 Bayou Clubtail   *Arigomphus maxwelli*   TL 50–54, HW 29–32

Description East Texas clubtail with blackish abdomen tip. *Male*: Eyes green to turquoise, face pale yellow to whitish. Thorax gray-green with narrow but well-defined T1–4, T3 and T4 sometimes incomplete. Abdomen dull pale yellow with dark brown markings covering much of sides of S2 and forming irregular apical rings on S3–6; S7–9 very dark brown, S10 light brown, appendages yellow-
ish. Club distinct but not very wide. *Female*: Colored as male but paler, dark markings on abdomen smaller, tip lighter brown.

Identification Much like **Stillwater Clubtail** but with darker abdomen tip in both sexes, greener eyes. Thoracic stripes T1 and T2 narrow and well separated in **Bayou**, T1 wider and closer to T2, sometimes partially fused, in **Stillwater**. More prominent thoracic stripes and darker abdomen tip easily distinguish from somewhat similarly colored **Jade Clubtail**. Slightly more contrast between greener thorax and yellower abdomen than in other pond

162.1
Bayou Clubtail
male—Holmes Co.,
MS, May 2005

162.2
Bayou Clubtail
female—Hardin Co.,
TX, May 2006,
John C. Abbott

clubtails. Female somewhat like **Russet-tipped Clubtail** but less heavily marked thorax and more heavily marked abdomen.

Natural History Males perch on ground or rocks at edge of water bodies, move into woodland when not at water.

Habitat Shallow ponds such as borrow pits and slow-flowing ditches in open or woodland edge, more likely at streams than other pond clubtails.

Flight Season TX May–Jun.

Distribution Also east to Alabama and north to southern Illinois.

Pond Clubtails - male appendages

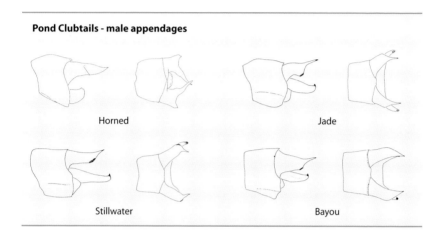

Horned   Jade

Stillwater   Bayou

Common Clubtails   *Gomphus*

This constitutes the largest genus of odonates in North America, although much less diverse in the West than in the East. It is widespread in temperate North America and Eurasia. Many species are locally distributed and often quite uncommon, notwithstanding the name of the group, but most may be common or even abundant in the right habitat during the peak of an often brief flight season. In some species, much of spring emergence at one place occurs over at most a few days. Most species are on streams and rivers but with quite varied substrates; a few are found on sandy lakes and ponds. Some members of this genus are known for the striking "roller-coaster flights" they perform, as if rapidly drawing a series of shallow U figures in the air. I have seen this in common clubtails and hanging clubtails, and it may be more widespread in the family. Species vary greatly in size (some half the bulk of others) and pattern, although not so much in color. Most change color with age with pale areas of thorax bright yellow or yellow-green when young and gray in old age, less contrasty with dark areas. Yellow color on the abdomen remains through maturity, however. Dark colors range from brown to black, even within the same species, perhaps also a function of age (black = young, brown = old). Because of this, descriptions of color patterns may use "dark" and "pale" instead of trying to pin down the variable colors.

This genus has been divided into subgenera, although authors differ in which ones they recognize. The species below are grouped in these subgenera, with an attempt for adjacency of related species. Subgenera I recognize in the West include *Gomphurus*, *Hylogomphus*, and *Gomphus*; there is still doubt of the validity of another subgenus, *Phanogomphus*, that has been used in recent publications. *Gomphurus* are typically large species with well-developed clubs and long legs. *Hylogomphus* are the smallest members of the genus, with well-developed clubs like miniature *Gomphurus*. *Gomphus* species are more varied in size and shape, most

Table 6 Common/Hanging Clubtail (*Gomphus, Stylurus*) Identification

	A	B	C	D	E
Lancet	1	2	4	23	3
Oklahoma	1	2	2	2	3
Ashy	1	2	4	2	123
Dusky	1	2	4	21	123
Sulphur-tipped	1	2	2	23	3
Pacific	· 1	2	3	2	2
Pronghorn	1	1	24	2	3
Banner	1	1	2	1	1
Midland	1	1	3	2	1
Ozark	1	1	2	2	1
Cocoa	1	1	2	2	1
Plains	1	1	2	2	3
Tamaulipas	1	1	1	2	3
Columbia	1	2	1	2	23
Cobra	2	2	2	1	1
Gulf Coast	2	2	2	21	1
Riverine	2	3	2	2	12
Laura's	2	3	2	12	1
Elusive	2	3	2	2	1
Olive	1	2	3	2	2
Russet-tipped	1	2	2	21	1
Brimstone	1	2	3	3	3

A, face: 1, unmarked; 2, cross-striped.

B, median stripes on thorax: 1, parallel; 2, wider in front; 3, complex pattern, with disconnected yellow stripes.

C, thoracic stripes from side: 1, five distinct stripes (T5 evident at rear with close look); 2, four distinct stripes (T1–2 fused in some species); 3, relatively unmarked sides (only one complete stripe, may be another partial one); 4, T3–4 essentially fused, appears to have two broad pale stripes.

D, male S8: 1, all dark above, 2, with basal yellow spot; 3, most or all of segment yellow above or all yellow.

E, top of male S9: 1, all dark; 2, with basal yellow spot; 3, complete or almost complete yellow stripe or all yellow.

without well-developed clubs and with legs shorter than those of *Gomphurus*. Other characters are in wing venation, structure of penis and hamules, and larval morphology. Individual species of one group can look much like species of another. World 51, NA 38, West 16.

Subgenus *Gomphus*

This may be a catch-all subgenus, as the species vary from moderately clubbed to scarcely clubbed, from very brightly marked to very dull. They are united by features of venation, genitalic structure, and larval morphology but may yet prove to be not so closely related. Some members of this group are the only common clubtails that inhabit lakes and ponds. None is as big and showy as a *Gomphurus*, although Pronghorn and Pacific Clubtails could easily be mistaken for one. Mating pairs fly away from water, sometimes not far, for extended copulation at rest. The duller species of the group—Lancet, Ashy, and Dusky—have all been seen in the peculiar roller-coaster flight, a series of shallow concave parabolas that may go on for several seconds. The group occurs all across North America and Eurasia, with 29 species.

163 Lancet Clubtail *Gomphus exilis*   TL 39–48, HW 23–27

Description Small drab clubtail, abundant in East and barely reaching West. *Male*: Eyes blue, face yellow. Thorax with T1–2 wide, almost touching; T3–4 also wide, minimal pale stripe between them; narrow T5 also present. Abdomen brown to black with dorsal pale spearpoints on S3–8, each mark with point almost reaching end of segment; S9 with wide yellow stripe above, S10 with bit of yellow; sides of S8–9 with wavy yellow margins. Common name from downward-projecting blade on cerci. *Female*: Colored much as male, slightly less yellow at end of abdomen.

Identification Only small common clubtail in its range in West with slender abdomen in both sexes. Both sexes differ from **Ashy** and **Dusky Clubtails** in smaller size, slightly separated T2 and T3 (fused in **Ashy** and **Dusky**), much yellow on top of S9 (rarely matched by other species). In hand, lancet-shaped cercus distinctive. Female differs from **Pronghorn Clubtail** by more slender abdomen, duller overall patterning, and presence of T5.

Natural History Males perch on ground adjacent to water, facing it, with no over-water patrols. Both sexes perch on sandy roads and other open substrates near breeding habitats, even in tiny clearings in woods; also up on leaves, but seldom far from ground. Females oviposit by flying rapidly over water, tapping every few feet, often perching between bouts.

Habitat Wide, including slow streams, ponds, sandy lakes, and even bogs. Appears to be more restricted to ponds in northern part of range.

Flight Season MO Apr–Jul.

Distribution Ranges through much of East to southern Quebec and Nova Scotia, lacking from most of Florida.

163.1
Lancet Clubtail
male—Early Co.,
GA, April 2006,
Giff Beaton

163.2
Lancet Clubtail
female—Floyd Co.,
GA, May 2006

Description Small, slender clubtail of southeastern part of region. *Male*: Eyes blue, face pale. Thorax with T1–2 separated by very narrow pale line, T3–4 wide with diffuse pale area between them. Abdomen with fairly wide dorsal stripe broken into spearpoints on S3–8, shortest on S8; S9 with pale dorsal stripe, S10 with small pale spot. Sides of S8–9 broadly pale-margined. *Female*: Colored as male, more yellow visible on sides of abdomen. Pale stripe on S8 narrow to absent, on S9–10 usually conspicuous.

Identification A bit smaller than similarly colored **Ashy Clubtail** (at its smallest in east Texas, where they overlap), its most likely associate, and distinguished by separated T3 and T4 (fused in **Ashy**). Also quite similar to **Pronghorn Clubtail**, a slightly larger species with substantially wider club in male and more contrasty pattern and more yellow at end of abdomen in female.

Natural History Both sexes perch on sandy ground or low in vegetation. Most common in spring.

Habitat Sandy ponds and lakes with mud bottoms and slow streams, in or out of woodland.

Flight Season TX Mar–Sep.

Distribution Also widely in Arkansas and Louisiana.

164.1
Oklahoma Clubtail
male—San Jacinto
Co., TX, April 2004,
Greg Lasley

164.2
Oklahoma Clubtail
female—Bastrop Co.,
TX, May 2005

Description　Slender and rather dull brown and whitish clubtail, common in East and extending into east Texas. *Male*: Eyes blue, face pale. Thorax with T1–2 wide, fused; T3–4 wide, fused; and T5 well developed; really a dark thorax with pale stripes. Abdomen dark brown to black with pale spearpoints covering much of middle segments, reduced to less than half length by S8. Poorly defined yellowish markings along sides of S8–9. *Female*: Colored as male, but spearpoints extending entire length of segments, more than half on S8; S9–10 mostly pale.

Identification　One of group of common clubtails with rather dull coloration, very wide dark thoracic stripes, and scarcely any indication of club known to overlap only in southeastern Manitoba, where difficult to distinguish. Extremely similar **Dusky Clubtail** a bit darker, with different appendages. **Dusky** with prominent lateral tooth on cerci viewed from above, **Ashy** angled but without tooth. In side view, **Dusky** also has prominent downward-projecting tooth on each cercus, **Ashy** lacks it. Females easily distinguished by subgenital plate, scarcely evident in **Ashy** and with two pointed lobes that extend almost one-third length under S8 in **Dusky**. **Ashy** distinguished from **Lancet Clubtail** by obviously larger size, different appendages. **Lancet** with downward-projecting blade on cercus in lateral view, **Ashy** without. Subgenital plate in female **Lancet** with two obvious pointed lobes.

165.1
Ashy Clubtail
male—Floyd Co.,
GA, April 2006,
Marion Dobbs

165.2
Ashy Clubtail
female—San Jacinto
Co., TX, May 1998,
Robert A. Behrstock

Natural History Males perch on sand, rocks, or logs at shore. Away from water usually on rocks, sand, or shrubs.

Habitat Slow-flowing wooded streams and small rivers, also edges of large wave-beaten lakes. Oviposits by tapping water near bank at intervals of a few inches.

Flight Season TX Mar–May.

Distribution Also east to southern Quebec and New Hampshire, south to Gulf Coast; absent from Illinois.

166 Dusky Clubtail *Gomphus spicatus* TL 46–50, HW 26–30

Description Slender clubtail, dull in maturity, that barely enters western region. *Male*: Eyes blue, face pale. Thorax with S1–2 wide, fused; S3–4 wide, mostly fused. Abdomen with narrow pale spearpoints shortening to rear, half length on S7, very short on S8; S9–10 dark. Margins of narrow S9 faintly yellow. *Female*: Colored as male but dorsal spearpoints slightly more conspicuous, as is yellow margin on S9; also pale dorsal stripe on S10.

Identification From very similar **Ashy Clubtail** (overlap in southeastern Manitoba) by different appendages. **Dusky** has prominent

166.1
Dusky Clubtail
male—Litchfield Co.,
CT, June 2004,
Giff Beaton

166.2
Dusky Clubtail
female—Price Co.,
WI, June 1997,
David Westover

257

tooth on outer surfaces of cerci, also prominent tooth beneath, visible from side; **Ashy** has neither. **Lancet Clubtail** similarly colored but much smaller. No other dull clubtail with slender abdomen in its range. Females differ by subgenital plates (see illustrations).

Natural History Males perch on ground and low vegetation, including water lilies, at edge of water, immatures and females in woodland clearings nearby. Usually perch on or near ground.

Habitat Clear-water lakes with open sandy shores or dense grass beds along shore; also large ponds, including bog ponds. Sometimes slow streams flowing into lakes.

Flight Season WI May–Jul.

Distribution Throughout Northeast from Ontario and Nova Scotia south to Indiana and New Jersey.

167 Sulphur-tipped Clubtail *Gomphus militaris* TL 47–53, HW 28–33

Description Small clubtail of Texas and surrounding states with moderate club and more yellow than similar species. *Male*: Eyes blue, face pale. Thorax with T1–2 just separated, T3–4 moderate width with well-defined pale area between them. Abdomen with usual spearpoints, shortening on S7 and half segment length on S8. S9 with broad pale stripe, S10 with narrow stripe. In arid western part of

167.1
Sulphur-tipped Clubtail male—Starr Co., TX, June 2005

167.2
Sulphur-tipped Clubtail female—Bastrop Co., TX, May 2005

range, dark markings on club much reduced, so most of S8 and all of S9 yellow. *Female*: Colored as male, more yellow on sides of abdomen, with pale areas exceeding dark lateral stripe. S9 with yellow dorsal and lateral stripes almost touching, S10 with much yellow.

Identification Pale color more extensive, especially at abdomen tip, than in most other clubtails occurring with it. Club narrower than in somewhat larger **Cocoa**, **Plains**, and **Tamaulipas Clubtails** and similar-sized **Pronghorn Clubtail** and with more yellow on club. Can look fairly much like **Pronghorn**, color patterns just about overlap. Look for pale streak on outside of femur in **Sulphur-tipped**. **Sulphur-tipped** also with much more yellow than in narrow-abdomened **Oklahoma Clubtail** found at same ponds. Could be confused with equally brightly yellow **Flag-tailed Spinyleg**, but that species larger, with narrower thoracic stripes, prominent yellow-orange club, and longer legs.

Natural History Males perch on sandy shores and in low vegetation at water and fly out and back at intervals; females on similar perches away from water.

Habitat Small to medium slow-flowing rivers and large open ponds and lakes.

Flight Season NM May–Sep, NE Jun–Jul, TX Mar–Aug.

Distribution Ranges south in Mexico to Nuevo León.

---

168 Pacific Clubtail *Gomphus kurilis* TL 48–57, HW 28–34

Description Brightly marked medium-sized clubtail of Pacific coast and mountains. *Male*: Eyes blue, face pale. Thorax with T1–2 wide, almost fused; T3 absent, T4 well developed. Abdomen black with pale dorsal spearpoints on S2–7; pale triangle on S8 just reaches half segment; remainder black above, but S9 variable, from black to fair-sized dorsal triangle. Tiny basal lateral spots on S4–7, large yellow lateral spots on S8–9 do not reach end of segment. *Female*: Colored as male, with a bit more extensive yellow and less indication of club.

Identification Only other common clubtail almost in its range is **Pronghorn Clubtail**, with side stripe on thorax (T3+4) wide and outer sides of tibiae yellow. Lack of T3 on **Pacific** quite distinctive, leaving wide pale area in its place. Colored much like **Grappletail** but larger, with thicker abdomen and more obvious club, and with pattern very different. Colored very differently from green SNAKETAILS and **Gray Sanddragon** that occur with it. When flying low over water, could be mistaken for CLUBSKIMMER, but extensive yellow markings on abdomen distinctive.

168.1
Pacific Clubtail
male—Siskiyou Co.,
CA, June 2004

168.2
Pacific Clubtail
female—Siskiyou Co.,
CA, June 2004

Natural History Males perch in shrubs or on rocks or ground at breeding sites, also fly slowly up and down stream just out from shore. Both sexes at times common in nearby upland habitats, sallying after insects from the ground or low vegetation. Copulating pairs form at water, fly into trees, and perch at head height and above. Females oviposit by tapping sporadically in pools at edge of shore vegetation. Where common, synchronized emergence of this species leads to density of individuals rarely encountered in other western odonates; thousands seen at Klamath River in northern California.

Habitat Streams and rivers with good currents, sandy to muddy bottoms. Also occurs in large ponds and lakes and may be limited to them in northern part of range and high in Sierras.

Flight Season WA Jun–Aug, OR May–Jul, CA Mar–Aug.

169 Pronghorn Clubtail *Gomphus graslinellus* TL 47–53, HW 29–34

Description Small mostly northern clubtail with bright markings and moderate club in males. *Male*: Eyes blue, face pale. Thorax with T1–2 wide, narrowly separated; T3–4 wide, partially fused. Abdomen black with wide pale dorsal stripe on S1–7, pointed on each segment and almost reaching tip; shorter triangle on S8, less than half segment; S9 with wide, S10 with narrow pale stripe. Margins of S8–9 entirely yellow. *Female*: Colored as male but more yellow visible on sides of abdomen, most of S10 yellow above. Eastern populations with thoracic stripes more thoroughly fused.

Identification Not known to overlap with other common clubtails in Northwest, but perhaps does with **Columbia Clubtail**, differing by lack of T5, smaller size, and substantially smaller club. Approaches range of **Pacific Clubtail** and similar in color, differing in wide middle stripe on side of thorax (narrow in **Pacific**) and yellow outer surfaces of tibiae. Farther east, can easily be mistaken for very similar **Plains Clubtail**; see that species for details of distinction. More brightly marked, with wider club, than other small clubtails of subgenus *Gomphus* up and down Great Plains, much darker overall than somewhat similarly shaped **Sulphur-tipped**. Also smaller and shorter-legged, with smaller club, than larger *Gomphurus* such as **Cobra**, **Cocoa**, **Midland**, and **Ozark Clubtails** with which it occurs. These four all have more dark than light on side of S8, and all but **Cocoa** have entirely dark tibiae.

169.1
Pronghorn Clubtail male—
Montgomery Co., AR,
May 2006 (posed)

169.2
Pronghorn Clubtail female—
Yakima Co., WA, June 2005

Natural History Both sexes perch typically in open on ground or rocks or in weeds or low shrubs. Assume obelisk posture at midday. May fly up and down pools and riffles on occasion, but do not hover at length over riffles as some larger clubtails do and may also search for mates away from water. Females oviposit close to bank by tapping water at irregular intervals a few feet apart, depositing 30–50 eggs with each tap.

Habitat Slow-flowing sandy or muddy streams with or without rocks and in or out of woodland, also medium to large lakes.

Flight Season BC May–Jul, WA Jun–Jul, MT Jun–Jul, NE May–Aug, TX Mar–Jun.

Distribution Ranges east to southern Ontario, Kentucky, and Arkansas.

Subgenus *Hylogomphus*

Only one species in the West is typical of this subgenus, which features brightly patterned species with prominent clubs, all of them smaller than similarly shaped *Gomphurus* and usually with larger clubs than any of subgenus *Gomphus*. The six rather similar species are restricted to eastern North America.

170 Banner Clubtail *Gomphus apomyius* TL 35–37, HW 23–25

Description Small early-spring clubtail of east Texas with prominent yellow-sided club in male. *Male*: Eyes turquoise, face cream. Thorax with T1–2 well developed, almost touching, T3–4 also well developed with pale area between them rather narrow and obscured. Abdomen black with pale dorsal stripe on S1–2, narrow basal triangles on S3–7; large basal spot on sides of S3, tiny spots on S4–6, basal spot and margin on S7, and entire sides of widely clubbed S8–9 yellow. *Female*: Color pattern about as male, eyes somewhat bluer; minimal or no club.

Identification Only quite small clubtail with wide black, yellow-sided club in West; no other species in east Texas anything like it. Most similar **Pronghorn Clubtail** may not overlap with it, recognized by much more yellow on top of abdomen, including all of S9. **Sulphur-tipped Clubtail** has even more yellow on abdomen, as well as narrow club. Very wide extent of black cerci also distinctive for male **Banner**. Female, with scarcely any club, distinctive in small size and bright markings, including bold yellow spots on sides as well as top of middle abdominal segments. Both **Pronghorn** and **Sulphur-tipped** have more extensive yellow markings along sides of abdomen. Female **Oklahoma Clubtail**, other small clubtail in range, mostly brown, with subdued pattern.

Natural History Males perch on leaves and rocks at streamside or hover over riffles with abdomen raised and club prominent. Females on leaves in woods nearby. Quite uncommon in region.

Habitat Small woodland streams.

Flight Season TX Mar–Apr.

Distribution Also Arkansas and Mississippi east to New Jersey.

170.1
Banner Clubtail
male—Richmond
Co., GA, May 2005,
Giff Beaton

170.2
Banner Clubtail
female—Cumberland
Co., NJ, May 2004,
Jim Bangma

The 13 species of this subgenus, restricted to North America, are among the favorites of club-tail aficionados. They are relatively large and strikingly marked, with a conspicuous to very conspicuous club in males (somewhat smaller in females) and long hind legs. They seem to have a predilection for butterflies, and perhaps the long legs aid in the capture of large prey. Species of other subgenera tend to be less clubbed, although—frustratingly—there are well-clubbed species in both of the other subgenera. Thus, when considering a *Gomphurus* in the West, one must also think about species in subgenera *Gomphus* (Pronghorn Clubtail) and *Hylogomphus* (Banner Clubtail) as well as species in the hanging clubtail genus *Stylurus* (Elusive and Riverine Clubtails).

171 Midland Clubtail *Gomphus fraternus* TL 48–55, HW 28–33

Description Black northern *Gomphurus* with moderate club. *Male*: Eyes greenish, face yellow. Thorax dull yellow with dark frontal stripe straight-edged; T1–2 wide, partially fused; T3 as a half-line, and T4 as a line. Abdomen yellow at base with brown dorsolateral stripes on S2; S3–10 black with yellow spearpoints on S3–9, shortening to rear and small on S8–9; S7–9 with large yellow spots on sides covering half of S7 and S8 and all of S9. *Female*: Colored as male.

Identification Plains Clubtail, another *Gomphurus* in its range, quite similar but dark areas on abdomen tip brown rather than black (also brown in some Manitoba **Midland**), two well-defined stripes on side of thorax rather than often incomplete fine lines of **Midland**. **Midland** barely overlaps on eastern edge of plains with the quite similar **Cobra Clubtail**, which has abdomen tip black, but latter has even wider club with less yellow on sides (none on S7, small spot on S8). Smaller **Pronghorn Clubtail** has more yellow on top of abdomen tip, with S9 all yellow above, and stronger stripes on side of thorax. All other clubtails in range have less-developed clubs.

Natural History Both sexes perch on roads and sandy banks. Males also fly rapidly over riffles, hovering from time to time. Often feeds on other dragonflies, including the very

171.1
Midland Clubtail
male—St. Clair Co.,
MI, June 2005,
Allen Chartier

171.2
Midland Clubtail
female—Columbia
Co., WI, June 1997,
David Westover

predatory Eastern Pondhawk. Females oviposit in rapids and breaking waves, in rapid straight flight.

Habitat Clean streams with moderate current and rock and mud substrates, also large, wave-washed lakes in northern part of range.

Flight Season WI May–Aug.

Distribution Ranges east to Maine, south to Tennessee and North Carolina.

Comments Populations on Red and Assiniboine Rivers in southern Manitoba named as separate subspecies *G. f. manitobanus*, smaller and paler than elsewhere, with yellow stripe down tibiae and prominent dorsal yellow spots on S9–10. These attributes may occur elsewhere on Great Plains.

172 Ozark Clubtail *Gomphus ozarkensis* TL 50–53, HW 29–32

Description Dark brown *Gomphurus* with moderate club and limited distribution. *Male*: Eyes turquoise, face yellow. Thorax with S1–2 wide and fused, S3–4 wide and almost fused. Abdomen with narrow yellow dorsal stripe, half-length on S7 and quarter-length on S8. Club moderate with yellow lateral spots small and basal on S8, entire length of segment on S9. *Female*: Colored as male but  slightly more yellow on sides of abdomen, including S8; dorsal stripe on S7 may be full length, and S10 with some yellow above.

Identification The few common clubtails in its range include one species with similarly fused thoracic stripes. **Pronghorn Clubtail** superficially similar but smaller and with more yellow on abdomen, including bright yellow sides of S8–9 and top of S9. Also similar are two slightly larger species with lateral thoracic stripes separate. **Cocoa Clubtail** has end of abdomen mostly brown instead of black, and **Cobra Clubtail** more vividly black and yellow and with black lines across face. **Cobra** also lacks small yellow dorsal spot on S8 present in **Ozark**.

Natural History Males perch on rocks on shore or in river much of day but also fly out over river and hover close to water facing prevailing wind. Females also perch near river, some-

172.1
Ozark Clubtail
male—Clark Co.,
AR, May 2006

172.2
Ozark Clubtail
female—Clark Co.,
AR, May 2006

times as easy to find as males, also on ground and low vegetation in clearings away from water.

Habitat Medium to large rivers with alternating pools and riffles, gravel and silt to sandy bottoms, mostly in forested areas.

Flight Season AR May–Jun.

---

173 Cocoa Clubtail *Gomphus hybridus*   **TL 50–52, HW 27–29**

Description Brown southeastern *Gomphurus* with moderate club. *Male*: Eyes blue-gray, face dull yellow. Thorax yellow with wide and almost fused T1–2, well-developed T3–4 with space between them somewhat grayish. Abdomen yellow at base with blackish dorso-lateral stripes on S1–2, S3–6 black with continuous yellow dorsal spearpoints; S7–10 becoming more brownish with yellow spear-point on S7, small basal spot on S8, yellowish markings on sides of S8, and much yellow on side of S9. *Female*: Colored as male, slightly more yellow on sides of abdomen.

Identification Overlaps with several other *Gomphurus* clubtails, some of them only slightly. Differs from **Plains** by brown versus black markings, from brown form of **Cobra** in Texas by blue rather than green eyes and smaller club, from **Gulf Coast** by smaller size, smaller club, and brown and yellow rather than black and yellow coloration. No other clubtail in range similar enough to *Gomphurus* to be confusing.

173.1
Cocoa Clubtail
male—Howard Co.,
AR, April 2006,
Mike Dillon

173.2
Cocoa Clubtail
female—Floyd Co.,
GA, April 2006,
Marion Dobbs

Natural History Both sexes perch on ground or in vegetation (generally low) at and near water. Males usually on sandy and gravelly beaches.

Habitat Sand- and silt-bottomed rivers.

Flight Season TX Apr–May.

Distribution Also across the Southeast to southern Indiana and North Carolina, absent from most of Florida.

---

174 Plains Clubtail *Gomphus externus*   TL 52–59, HW 30–33

Description Widely distributed brown *Gomphurus* with moderate club. *Male*: Eyes turquoise, face yellowish. Thorax yellowish-green with T1–2 wide and almost fused, T3–4 wide and close together. Abdomen yellow at base with wide brown dorsolateral stripes on S2, S3–7 black to brown with narrow yellow spearpoints above, widening to rectangle that covers S9; S8–9 also with entirely yellow edges; S10 mostly brown with yellow dorsal spot. Some individuals with dark markings more restricted, thoracic stripes narrower and more yellow on abdomen tip. Narrower thoracic stripes in north-western part of range. *Female*: Colored as male.

Identification Most similar to member of another subgenus, **Pronghorn Clubtail**, and much overlap in range. In fact, **Pronghorn** essentially identical in color pattern, although very slightly smaller. Best mark may be edge of S8, usually entirely yellow in **Pronghorn** but with some dark markings in **Plains**. Also, look at hind legs. When back along body, they extend to or beyond middle of S2 in **Plains**, not to middle in **Pronghorn**. Easily distinguished in hand by male appendages (epiproct wider than narrow cerci **Plains**, same width as "horned" cerci in **Pronghorn**) and female subgenital plate (each half long and pointed in

174.1
Plains Clubtail
male—Caldwell Co.,
TX, May 2005

174.2
Plains Clubtail
female—Gonzales
Co., TX, May 2005

Plains, short and wide in **Pronghorn**); see illustrations. Also, slender arms of epiproct of **Plains** rather dramatically curved upward in side view, whereas thicker epiproct of **Pronghorn** less curved; may be visible in close view. Size difference may help once familiar with species. In color pattern less similar to other *Gomphus* with which it might occur because of extensive yellow on top of S8–9 (very little or none in **Cobra** and **Cocoa**) and very close T3 and T4 (well separated in **Cobra**). See **Tamaulipan Clubtail**, which approaches edge of range of **Plains**. **Sulphur-tipped Clubtail** smaller with greater extent of yellow on abdomen tip and much yellow on femora (black in **Plains**).

Natural History Males perch on ground near water with abdomen elevated and occasionally fly long beats over river. Usually perch flat on leaves when away from water. Copulating pairs in low vegetation near water. Females oviposit by straight flight over river, tapping water at intervals. One female produced 5100 eggs when abdomen dipped in glass of water.

Habitat Sandy or muddy streams and rivers with moderate current, open grassy or wooded banks.

Flight Season MT Jun–Aug, NM Mar–Jul, NE May–Sep, TX Mar–Jul.

Distribution Ranges east to Wisconsin, Ohio, and Kentucky.

175 Tamaulipan Clubtail *Gomphus gonzalezi*   TL 46–50, HW 31

Description Brown *Gomphurus* of Lower Rio Grande Valley with moderate club. *Male*: Eyes blue-gray, face yellowish. Thorax yellow with well-developed T1–4, narrow T5. Abdomen yellow on S1–2 with wide black dorsolateral stripes; black on S3–6 with white dorsal spearpoints covering most of segment; S7–10 yellow again with smudgy dark brown dorsolateral markings, yellow extending most of length of segments on top and sides. *Female*: Colored as male, slightly less yellow on tops of terminal abdominal segments and more extensive white spotting on sides of middle segments.

Identification No other member of *Gomphurus* subgenus known within its range. Closest in appearance would be **Sulphur-tipped** in *Gomphus* subgenus, which is smaller, with narrower club (S9 longer than S8, the reverse in **Tamaulipan**), and much more yellow at abdomen tip. Most similar to **Plains Clubtail** of Texas hill country (not known to overlap in range), which is a bit larger and blacker and lacks T5 entirely. Other clubtails that might occur with it are very different FORCEPTAILS, LEAFTAILS, RINGTAILS, and SANDDRAGONS.

Natural History Males perch on rocks or overhanging vegetation at and near water.

Habitat Found at both clear, spring-fed rocky rivers and muddy channels.

Flight Season TX Apr–May.

Distribution Rarely encountered in United States, even at localities where previously found. Ranges south in coastal Mexico to San Luis Potosí.

175.1
Tamaulipan Clubtail male—Starr Co., TX, April 1998, Blair Nikula (posed)

175.2
Tamaulipan Clubtail female—Hidalgo Co., TX, May 2006, David Hanson (posed)

Description　Black *Gomphurus* of limited distribution in Northwest interior with wide club and pruinosity. *Male*: Eyes blue, face yellow.
Thorax yellow with all five stripes well developed, T2–3 very close and
almost fused. Abdomen yellow at base with wide dark dorsolateral
stripes on S1–2. S3–10 black, S3–6 with yellow spearpoints and S7–9
with wider, brighter yellow triangles above; S7–9 with much yellow
on sides. Thorax and abdomen base becoming pruinose and somewhat duller with age. *Female*: Colored as male but not pruinose.

Identification　No other common clubtail known from same locali-
ties, but both **Pacific** and **Pronghorn Clubtails** occur nearby. Both have smaller clubs, **Pacific** with only one stripe on side of thorax (lacks T3) and **Pronghorn** substantially smaller.
Both lack T5 stripe characteristic of **Columbia**.

Natural History　Both sexes perch on ground or rocks or on dead twigs of shrubs. Males
most commonly seen in sagebrush and other shrubs but also fly out over water at intervals.
Not found far from water in its arid habitat.

Habitat　Good-sized sandy to muddy rivers in open shrub steppe or bordered by riparian
woodland.

Flight Season　WA Jun–Aug, OR Jun–Aug.

Distribution　Known only from lower reaches of Yakima River in Washington and John Day
and Malheur Rivers in Oregon.

176.1
Columbia Clubtail
male—Benton Co.,
WA, July 2006,
Netta Smith

176.2
Columbia Clubtail
female—Benton Co.,
WA, July 2005

Description Eastern *Gomphurus* with wide club like cobra's hood. Variable in color, northern populations with abdomen black, Texas populations with abdomen brown. *Northern male*: Eyes green, face yellow with dark horizontal lines. Thorax with T1 very wide but interrupted above, T2 narrower; T3–4 narrow but prominent, well separated. Abdomen with narrow dorsal stripe on S3–7 extending only half length on most segments, thus quite interrupted; looks like series of pointed dashes. Lateral yellow spots on club small and basal on S8, large and full length on S9. *Northern female*: Colored as male but more yellow on sides of abdomen, dorsal stripe less interrupted. *Texas male and female*: Differ in browner coloration of dark markings, especially on abdomen,

177.1
Cobra Clubtail eastern male—Juneau Co., WI, July 2003, Mike Reese

177.2
Cobra Clubtail Texas male—Gonzales Co., TX, May 2005

177.3
Cobra Clubtail Texas female—Gonzales Co., TX, May 2005

and more extensive yellow markings. Pale dorsal line almost continuous on middle ab-dominal segments; club distinctly brown to reddish brown, with base of S7 entirely yellow.

Identification On east edge of plains, black populations occur with several other well-clubbed species. **Ozark Clubtail** very similar but with wider thoracic stripes, T3–4 fused, and tiny spot of yellow on top of S8. **Plains Clubtail** slightly smaller and browner overall, with T3–4 broader and club with more yellow on sides, S9 all yellow above. **Pronghorn Clubtail** distinctly smaller and with much yellow on top of club. Brown populations in Texas occur with latter two and still other species. **Plains Clubtail** more similarly colored in Texas but still distinguished by thoracic stripes. **Gulf Coast Clubtail** quite black, much less yellow on S7. **Cocoa Clubtail**, occurring just east of hill country, differs in having blue eyes, narrower club. **Sulphur-tipped Clubtail** considerably smaller, with more heavily striped thorax, smaller club with more yellow on it.

Natural History Males perch on shore of river with abdomen elevated, then fly beats up and down. Usually fly a few feet above water, slow when facing wind and may hover in slight breeze. Both sexes typically perch on leaves with abdomen inclined downward. In numbers in favorable habitat. Copulating pairs perch in tall weeds, shrubs, and trees at head height and above. Females oviposit by flying rapidly, sometimes far from shore, and tapping water at intervals to release eggs.

Habitat Rivers and streams with slow to moderate current and sandy or silty bottoms, with or without rocks.

Flight Season NE May–Jul, TX May–Aug.

Distribution Ranges east to Atlantic coast, from New Brunswick to Georgia.

---

178 Gulf Coast Clubtail *Gomphus modestus* **TL 55–63, HW 34–37**

Description Large black southern *Gomphurus* with prominent club. *Male*: Eyes green; face yellow, crossed by two fine black lines. Thorax prominently striped; T1–2 wide, almost touching, T3–4 narrower but still well developed. Abdomen with narrow yellow dorsal stripe widening on S7, small basal spot or not on S8, no black on S9–10; prominent club, with large yellow lateral spots on S8–9, not reaching tip of segments. *Female*: Colored as male.

178.1
Gulf Coast Clubtail
male—Carroll Co.,
TN, June 1999,
Ken Tennessen

178.2
Gulf Coast Clubtail
female—San Jacinto
Co., TX, May 2007,
Martin Reid

Identification All other *Gomphurus* in its range—**Cobra**, **Cocoa**, and **Plains Clubtails**—have much more brownish abdomens and are at least a bit smaller. T1 joins black on top of thorax in **Gulf Coast**, separated from it in **Cobra**. No other common clubtail in Texas anything like this species. Superficially similar to **Black-shouldered Spinyleg** but differs in quite different thoracic pattern, also much more yellow on top of club in **Spinyleg**.

Natural History Males perch along streams on bank or leaves. Quite uncommon.

Habitat Large, slow-flowing rivers over rock and sand substrates.

Flight Season TX May–Aug.

Distribution Also east to Tennessee and northwestern Florida.

Common Clubtails - male appendages

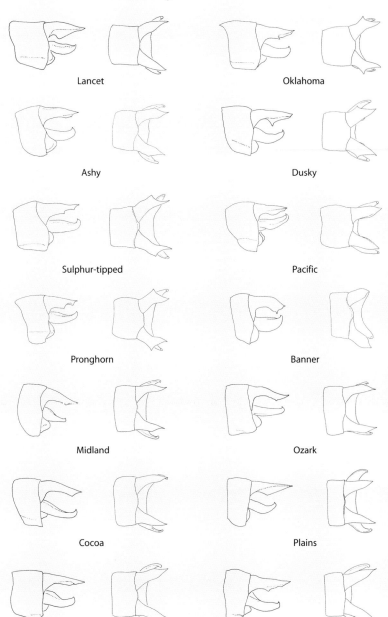

Lancet

Oklahoma

Ashy

Dusky

Sulphur-tipped

Pacific

Pronghorn

Banner

Midland

Ozark

Cocoa

Plains

Tamaulipan

Columbia

Common Clubtails - male appendages (*continued*)

Cobra

Gulf Coast

Common Clubtails - female sudgenital plates

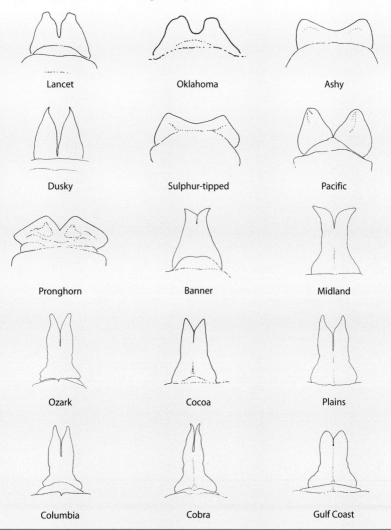

Lancet

Oklahoma

Ashy

Dusky

Sulphur-tipped

Pacific

Pronghorn

Banner

Midland

Ozark

Cocoa

Plains

Columbia

Cobra

Gulf Coast

As the name implies, members of this genus often hang down, even to the vertical, when they perch, and this distinguishes them from most other North American clubtails. Do not be confused by tenerals of other species that may hang down for a while immediately after emergence. Hanging clubtails also perch on broad leaves with rear end pointing downward, as do some common clubtails. Some also perch flat on ground or even with abdomen elevated slightly, but not on twigs or rocks with abdomen elevated as typical of many other clubtails. Otherwise, they look much like common clubtails. A thoracic pattern is distinctive of certain species: yellow stripes on front of thorax that are not connected to yellow above or below. World 30, NA 11, West 6.

179 Riverine Clubtail *Stylurus amnicola*   TL 47–49, HW 29–33

Description Boldly marked clubtail with distinctive thoracic stripes and moderate club. *Male*: Eyes dark turquoise, face pale with dark irregular markings. Thorax with complex pattern on front, like flower in vase with leaf on either side; T1–2 very wide, only faint line between them, T3–4 narrow and well separated. Abdomen with usual spearpoints on S2–7, short yellow triangle on S8; small yellow spot on side of S7, large spots almost filling sides of S8–9. *Female*: Colored as male but complete dorsal stripe on S1–7, yellow basal triangle on S8–9. Yellow stripe on sides of S1–3, interrupted stripe and spots on middle segments, more prominent spots on S8–9.

179.1
Riverine Clubtail male—Hampden Co., MA, June 1998, Blair Nikula (posed)

179.2
Riverine Clubtail female—Black Hawk Co., IA, July 2004, John C. Abbott

Identification Pattern on front of thorax diagnostic. One of few clubtails with very wide T1–2 and much less prominent side stripes. Superficially like **Black-shouldered Spinyleg** because of that, and front of thorax even somewhat similar, but smaller and shorter-legged, with narrower club. Of other hanging clubtails, most like **Elusive Clubtail**, but different thoracic patterns distinguish them. Not especially like any COMMON CLUBTAIL because of thorax, but note other black and yellow species, **Cobra Clubtail** with distinctly wider club and **Pronghorn Clubtail** with more yellow on abdomen tip.

Natural History Rarely seen except during emergence. Adults hang out in treetops except for brief visits to breeding habitat, when males tend to stay over midriver for brief periods. Most likely to find immature individuals near river after emergence.

Habitat Medium to large slow-flowing to rapid rivers with varied bottom types, in or out of woodland.

Flight Season NE Jun–Sep.

Distribution Ranges east to Maine, south to Louisiana and Georgia, poorly known in most regions.

---

180 Laura's Clubtail *Stylurus laurae*   TL 60–64, HW 36–42

Description Large clubtail with strongly striped thorax and club with yellow sides. *Male*: Eyes green, face pale with two dark cross-bars. Yellow stripes on front of thorax not connected below to yellow collar. Thorax with T1 wide, narrowly separated from T2; T3–4 narrow but distinct. Abdomen mostly black, becoming brown on terminal segments. Pale dorsal stripe complete on S1–5, then  shorter on S6–7. Sides of S3–6 with tiny pale basal spots, S7 with narrow pale edge, S8–9 entirely yellow on sides with slightly darker edges. *Female*: Colored as male, more yellow on sides of middle segments of abdomen.

180.1
Laura's Clubtail male—Rockdale Co., GA, July 2002, F. M. Stiteler

180.2
Laura's Clubtail female—Cherokee Co., GA, August 2005, Giff Beaton

Identification Greener eyes, darker abdomen tip, and much narrower pale stripe between T1 and T2 distinguishes **Laura's** from **Russet-tipped Clubtail**, only other hanging clubtail in eastern Texas; front of thorax also different. Somewhat similar to *Gomphurus* COMMON CLUBTAILS, but completely dark top and yellow sides of S8–9 of club distinctive, as is thoracic pattern. Also, flight season generally later than *Gomphurus*.

Natural History Males perch on overhanging leaves and branches from near the water to above head height, less often on rocks and logs. Fly out over riffles briefly and then back to perch. Present at midday but may stay active at water into early evening. Oviposits by tapping water at short intervals, not as wild a flight as many other clubtails. Feeds from trees at forest edge. Rarely encountered in region.

Habitat Clear shallow forest streams with rocky riffles, sand or mud bottoms.

Flight Season TX Jul.

Distribution Also in East from Michigan and Maryland south to northwestern Florida.

---

181 Elusive Clubtail *Stylurus notatus* TL 52–64, HW 30–35

Description Large, dark clubtail with moderate club, adults of which are almost never seen. *Male*: Eyes blue, face pale but with brown bar across upper part. Yellow stripes on front of thorax not in contact with yellow at either end. Thorax with T1–2 wide, joined at top; T3–4 narrow but distinct. Abdomen black, pale stripe on S1–2; small pale triangles on S3–8, shorter than typical clubtail spearpoints (perhaps longer in some individuals). Sides of S3–6 with tiny pale basal spots, S7–9 with large pale spot at base reaching half length or more. *Female*: Colored as male, more yellow on sides of middle segments of abdomen.

Identification Lack of any reddish color distinguishes **Elusive** from **Russet-tipped Clubtail**, thoracic pattern from similarly colored and often co-occurring **Riverine Clubtail**. Abdomen more slender than in **Cobra**, **Midland**, **Plains**, and **Pronghorn Clubtails**, color pattern brighter than in **Ashy** and **Dusky Clubtails**.

Natural History Justifies its name, as rarely seen except during emergence. Adults fly into forest canopy to feed and come back to breeding habitat briefly to breed. Even when at water, apparently fly far out from shore and difficult to capture, photograph, or observe!

Habitat Large, slow-flowing rivers, less often large lakes.

181.1
Elusive Clubtail
male—Wayne Co.,
MI, September 2005,
Allen Chartier

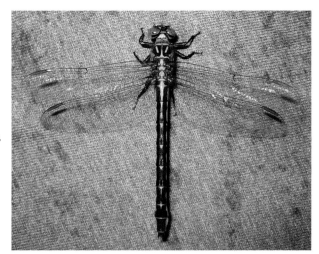

181.2
Elusive Clubtail
female—Lafayette
Co., WI, August 2006,
John and Cindy
Anderson

Flight Season NE Jun–Sep.
Distribution Also east through southern Quebec to Maine, south to northern Alabama, northern Georgia, and Maryland.

Description Large, rather dull clubtail of western rivers with moderate club. *Male*: Eyes bright blue, face pale. Thorax with brown stripes, very wide T1 and wide T2 almost fused; T3 narrow, half height from bottom, T4 narrow, half height from top, so sides of thorax little marked. Abdomen black with pale stripe on S1–2, spearpoint on S3, same reduced to triangles with narrow midline extension on S4–9. Sides of S1–3 pale, basal spots and short streak in middle on S4–6, large pale spots reaching about halfway down segment on S7–9, entire side of S10. *Female*: Colored as male, but
stripes on sides of thorax may be indistinct; much more yellow on sides of abdomen, most segments largely pale. Populations of arid interior paler all over, with thoracic stripes narrower, T3–4 absent; much less black on abdomen, with continuous dorsal stripe and sides entirely pale. Female looks ringed because black dorsolateral stripes widely interrupted.
Identification Only clubtail in its range with little or no indication of side stripes on thorax. Because of late flight season, overlaps for only short time with **Columbia** and **Pacific Clubtails**. From above, note that only this species has pale spot at base of S9 short and wide. Occurs locally with smaller, brighter **Brimstone Clubtail**, **Pale Snaketail**, and **White-belted Ringtail**.
Natural History Males alternate flying low over water in leisurely zigzag flight, sometimes far from shore on big rivers, and perching nearby; both for extended periods. Males perch near breeding habitat, typically on leaves of woody vegetation from knee height to well up in trees, and may make beeline out over river when flushed. Perch usually by hanging vertically or inclined on top of drooping leaf, less commonly on sandy or rocky ground. Females in same areas or farther from water. Pairs copulate for extended periods hanging at chest height or above in trees. Females oviposit in rapid and direct flight.
Habitat Typically large, warm, mud-bottomed rivers, usually with sand banks, in or out of woodland. Occurs down Columbia River to where it rises and falls with tidal influences.
Flight Season BC May–Aug, WA Jul–Nov, OR Aug–Sep, CA Jun–Sep.

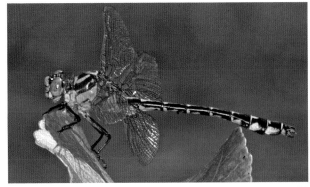

182.1
Olive Clubtail
male—Clark Co., WA,
September 2006

182.2
Olive Clubtail
female—Humboldt
Co., NV, July 2005

183 Russet-tipped Clubtail *Stylurus plagiatus*   TL 57–66, HW 30–40

Description Wide-ranging clubtail of southern regions with well-defined reddish or yellow and black club. *Male*: Eyes turquoise to blue, face unmarked but slightly clouded with age. Thorax with S1–4 all present, well separated. Abdomen blackish with pale spearpoints on S3–6, small pale triangle on S7; most of S7–10 reddish orange, yellower on sides, with narrow blackish margins on club. *Female*: Colored as male, abdomen less conspicuously marked and appearing mostly brownish with orangey tip. Little indication of club. Populations in Colorado Desert (southwestern Arizona and  southeastern California) rather different and do not bring to mind "Russet-tipped" when sighted. They are paler overall, with narrower thoracic stripes, middle abdominal segments whitish, terminal segments yellowish rather than russet, and with contrasting brown markings, and abdomen with bold black stripes that give them much more patterned look than their relatives to the east. In this population, eyes quite blue and club wider. Farther east, however, even in Texas hill country, populations seem somewhat intermediate. North of Texas, individuals on Great Plains with abdomen scarcely patterned at all, entirely reddish in side view and slightly paler in club area. Eyes also greener.

Identification Eastern population overlaps with several COMMON CLUBTAILS, club narrower than in **Cobra**, **Midland**, **Plains**, **Pronghorn**, and **Tamaulipas Clubtails** and russet-tipped abdomen distinctive. Most like **Sulphur-tipped Clubtail**, overlapping in Texas and surrounding states. **Sulphur-tipped** smaller, with much more contrasty abdominal markings, including dark stripes down abdomen onto club; **Russet-tipped** has

183.1
Russet-tipped Clubtail
male—Gonzales Co., TX,
July 2004

183.2
Russet-tipped Clubtail
female—Gonzales Co., TX,
July 2004

183.3
Russet-tipped Clubtail
southwestern male—
Imperial Co., CA, October
2006, Paul Johnson

183.4
Russet-tipped Clubtail
southwestern female—
Imperial Co., CA, September
2006, Robert Parks

duller abdomen, obscurely patterned club. FORCEPTAILS and LEAFTAILS that occur with Texas population all have greater extent of dark markings on thorax. Individuals of desert population look as much like RINGTAIL as another kind of clubtail, but abdominal pattern, with no obvious pale rings, different from that of **White-belted** and **Serpent Ringtails**, two that overlap. **Brimstone Clubtail**, common associate of this population, smaller and even brighter yellow, with dark markings on S7 but not S8–10.

Natural History Probably mostly in trees when not at water. Males often seen hanging vertically in trees or less often in more horizontal position, especially when on leaves. Sexual patrol in back-and-forth flight over breeding habitats, usually over riffles. Females rarely seen, mating pairs usually in trees well above head height. Females oviposit in rapid flight, tapping open water at long intervals.

Habitat Slow-flowing rivers and streams down to fairly narrow ones, in or out of woodland. Mostly in large drainage canals at western end of range.

Flight Season CA Jun–Oct, AZ Jul–Nov, NM Jun–Aug, NE Jun–Oct, TX Apr–Nov.

Distribution Ranges south in eastern Mexico to Nuevo León, also widely in East from Wisconsin and New York south; mostly on Coastal Plain in Northeast.

184 Brimstone Clubtail   *Stylurus intricatus*   TL 41–45, HW 26–32

Description Small, mostly yellow clubtail of the arid West with moderate club. *Male*: Eyes blue, face pale. Thorax with T1–4 narrow, all well separated; T3–4 even narrower than T1–2. Abdomen mostly yellow, black hourglass markings on sides of S3–6, scarcely any other dark markings. *Female*: Colored as male, a bit more black on S2 and S7.

Identification Only clubtail in range of this small yellow species even vaguely similar is **White-belted Ringtail**, with more vividly striped thorax and bold ringed pattern on abdomen. Co-occurring **Olive Clubtail** is larger, olive to gray rather than yellow. **Pale Snaketail**, larger with greener thorax, often occurs with it.

184.1
Brimstone Clubtail male—Imperial Co., CA, July 2006, Paul Johnson

184.2
Brimstone Clubtail female—Humboldt Co., NV, July 2005

Natural History Males fly rapidly back and forth low over water in short beats, hovering at intervals, or perch in streamside shrubs hanging at 45° angle or on logs; relatively rarely on ground. Females more often in shrubs near water, not perching as high as some other hanging clubtails. Copulation at rest in shrubs for an hour or more. Females oviposit by tapping at intervals while in rapid flight.

Habitat Typically slow-flowing, warm muddy rivers in open country but with associated riparian shrubs and/or trees; sometimes in irrigation canals.

Flight Season CA Jun–Oct, AZ Jun–Nov, NM Jun–Aug, NE Jul–Sep, TX Jun–Oct.

Distribution Sparsely distributed, primarily on major rivers, but range likely more extensive than shown.

Hanging Clubtails - male appendages

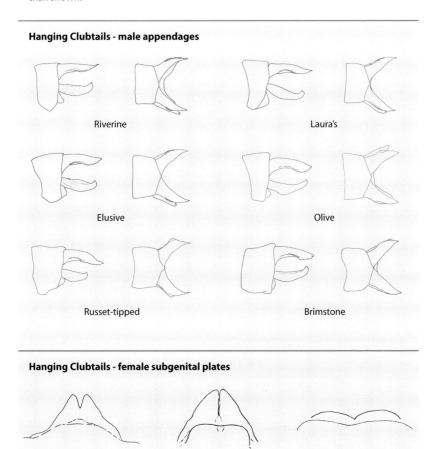

Riverine

Laura's

Elusive

Olive

Russet-tipped

Brimstone

Hanging Clubtails - female subgenital plates

Riverine

Laura's

Elusive

Olive

Russet-tipped

Brimstone

Found only in central and eastern North America, spinylegs stand out by very long hind legs with prominent large spines on femur, presumably adaptations for capturing large prey. Otherwise they resemble some *Gomphus* species, especially subgenus *Gomphurus*, which also have long hind legs, but they are larger than all but a few of them and with club not as wide in males. World 3, NA 3, West 2.

185 Black-shouldered Spinyleg *Dromogomphus spinosus* TL 53–68, HW 32–40

Description Large, long-legged rather dark clubtail with moderate club and characteristic wide dark stripes on otherwise pale thorax. *Male*: Eyes green, face yellow. Thorax yellow to gray with frontal stripe split into widely separated narrow lines joined at top and bottom to produce characteristic pattern. T1–2 very wide and fused or almost fused to make very broad dark brown stripe on either side of front; sides entirely yellow except for dark half-line for T3 and complete line for T4. Abdomen mostly black with yellow to gray line down middle of all segments, wider on basal segments  and interrupted and expanded into triangles on S8–9 and spot covering much of S10. Yellow markings low on sides of most segments, forming obvious spots on sides of S7–9. *Female*: Colored as male, abdomen slightly thicker so club not as pronounced.

Identification This distinctive species most likely mistaken for COMMON CLUBTAILS of subgenus *Gomphurus*, with their large size, long legs, and striking color patterns, but club not as prominent. Fortunately, **Black-shouldered Spinyleg** only clubtail with its broad thoracic stripes. Very different-looking than closely related **Flag-tailed Spinyleg**, and not much else even vaguely similar. Note colored somewhat like much larger and similarly long-legged **Dragonhunter**, but latter has broad, fused T3–4 and even less club.

Natural History Males perch on leaves up to waist height or above or on ground or rocks near water, facing stream. Also on rocks in water, either pools or riffles. May obelisk in mid-day sun. Alternate perching and flying up and down or hovering over stream at knee to waist height, abdomen tilted upward. Females and immatures in woodland nearby, usually

185.1
Black-shouldered
Spinyleg male—
Ottawa Co., ON, July
2005, Greg Lasley

185.2
Black-shouldered
Spinyleg female—
Hillsborough Co.,
FL, June 2004

perched up in vegetation and often in shade. Pairs fly into trees for lengthy copulation. Females oviposit in rapid flight, skimming water and tapping it every few feet.

Habitat Rocky and muddy streams and rivers from small to large, more often in woodland. Also in rocky lakes in northern part of range.

Flight Season TX May–Jul.

Distribution Also in East from southern Ontario and Nova Scotia south.

186 Flag-tailed Spinyleg *Dromogomphus spoliatus* TL 56–65, HW 32–36

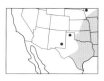

Description Large, long-legged brightly marked clubtail with largely yellow-orange abdomen tip. *Male*: Eyes light blue or turquoise, face yellowish-tan. Thorax dull yellow with rather narrow T1–4, T3 often broken. Abdomen with S1–2 mostly yellowish, dark brown dorsolateral blotches on S2; S3–6 white with blackish dorsolateral markings, smaller anterior one connected or not to larger posterior one; from side, may present appearance of two black and white rings alternating on each segment; S7–10 yellow to orange, S7–9 often with brown markings above. *Female*: Colored as male, abdomen less clubbed.

Identification Common name comes from club being laterally compressed, very conspicuous like flag from side but not very wide in top view. A few other clubtails in range share general appearance of ringed midabdomen with orange S7–10. **Jade** and **Stillwater Clubtails** both a bit smaller, with pale yellow or cream-colored abdomen and only one large dark marking on each middle segment. **Flag-tailed Spinyleg** has white midabdomen with two apparent markings per segment. **Cobra**, **Plains**, and **Sulphur-tipped Clubtails** have dark markings continuous along sides of abdomen and darker, more conspicuous markings on yellow club. FORCEPTAILS and LEAFTAILS, often in same habitats, have much less well-developed club; also, if abdomen looks ringed, one ring per segment. **Eastern Ringtail**, often found with it, conspicuously smaller with green thorax. **Black-shouldered Spinyleg** very differently colored, with smaller club.

Natural History Males perch on elevated perches at waterside, also low on rocks, or fly out over water with much hovering. Abdomen up to vertical obelisk on sunny midday but also hanging down in some individuals perched in shrubs.

Habitat Slow-flowing rivers and large, sometimes muddy, ponds.

Flight Season NM Jul–Aug, TX May–Oct.

186.1
Flag-tailed Spinyleg
male—Kimble Co.,
TX, July 2004

186.2
Flag-tailed Spinyleg
female—Caldwell
Co., TX, July 2004

Distribution Also south in Mexico to Tamaulipas, east to Michigan, West Virginia, and western Georgia.

Dragonhunter *Hagenius*

Dragonhunter is the largest of North American clubtails, with poorly developed club and wide, flat larva that resembles wood chip. It is closely related to the equally large and fierce *Sieboldius* of Asia. World 1, NA 1, West 1.

187 Dragonhunter *Hagenius brevistylus*   TL 73–90, HW 47–58

Description Long-legged, black and yellow monster dragonfly with small head that often flies with abdomen tip curled down. *Male*: Eyes green, face yellow with fine black line across it. Thorax yellow with broad S1 and S2 almost fused, broad S3 and S4 almost fused; yellow stripe between pairs narrower than dark area. Abdomen black with yellow on sides of basal segments and narrow yellow spearpoints, shortening to rear and becoming triangles by S7–8; yellow basal spots on sides of middle segments expanding into larger spots on sides of S8–9. *Female*: Colored as male but much more yellow visible on sides of abdomen, yellow on top and sides of S8 in contact.

Identification No other clubtail as large as this one, might be more likely mistaken for SPIKETAIL or RIVER CRUISER as it flies by, but notice different pattern of yellow on black on both thorax and abdomen. Perched individual not like anything else, although note somewhat similar appearance but smaller size and different thoracic pattern of **Black-shouldered Spinyleg**.

Natural History Males perch on tree and shrub branches over water, often hanging down a bit and seeming awkward, too large for perch. Perch held distinctively in crook of long hind legs, but awkwardness confirmed when one crashes into tangle of twigs, flutters, and then flies out again! Also perches flat on ground or rocks near water. Males fly leisurely up and down streams or along shorelines, looking for prey and/or females, with abdomen a bit elevated, tip

curled down. Females may fly in same manner, then look exactly like males. Preys on other odonates and butterflies and presumably other large insects. Ebony Jewelwings are common prey, and Dragonhunters may be most common at streams where jewelwings are abundant. Other clubtails also common prey, including largest species occurring with it. Typically oviposits facing vertical stream banks, hovering and dropping forward to the water to tap once, then rising (and backing up!) to drop and tap again, changing position every tap or every few taps. This may be continued for several minutes. Also by long flights over open water and reported to drop eggs from above water. Immatures of both sexes in shrubby areas near water, usually quite wary.

Habitat Wide variety of streams and rivers, less commonly lakes. As other top predators (cougars, wolves, great horned owls), seem to be able to utilize all habitats. This may be because larva not a burrower (rests among detritus), thus not tied to any particular substrate.

Flight Season TX May–Oct.

Distribution Also throughout East, north to central Ontario and Nova Scotia.

187.1
Dragonhunter
male—Washington Co.,
VT, July 2004,
Bryan Pfeiffer

187.2
Dragonhunter
female—Kerr Co.,
TX, June 2001,
Robert A. Behrstock

With diversity centered in Mexico, these are neotropical relatives of temperate snaketails, although none of them have horned females. The two groups probably do not overlap in distribution anywhere. Male ringtails are moderately clubbed, females not; legs are short. Ground color of thorax is green except in White-belted (yellow and white) and California populations of Serpent (gray). From the side, middle of the abdomen is black with conspicuous pale rings visible on each segment, often yellow as they extend across top; S7 has wide white band; and club is reddish, orange, or yellow and black. Snaketails instead show yellow spots on each segment, no white rings visible from the side, and the club is not differently colored. The species can be divided into two groups based on the shape of male cerci from the side; White-belted and Blue-faced taper smoothly, whereas the other four are angled above and quite pointed. Females are mostly similar to one another and must be distinguished by thoracic markings. Males typically perch on the ground or on rocks or twigs over water; females perch similarly but visit water only for breeding. Females oviposit by flying rapidly over water and tapping at intervals. World 22, NA 6, West 6.

Table 7   Ringtail (*Erpetogomphus*) Identification

	A	B	C	D	E
Blue-faced	1	1	1	1	1
Eastern	2	1	1	2	1
Serpent	1	14	2	2	1
Yellow-legged	3	2	2	2	2
Dashed	2	2	2	2	2
White-belted	1	3	1	1	21

A, thoracic pattern: 1, complete set of four prominent stripes; 2, T1 incomplete, side stripes reduced; 3, virtually unstriped.

B, thoracic color: 1, dark blue-green; 2, medium green; 3, yellow and white; 4, gray.

C, leg color: 1, tibia all dark; 2, outer surface of tibia with yellow stripe.

D, male cerci: 1, slightly downcurved, tip rounded; 2, angled above, tip sharply pointed.

E, pale color at abdomen tip: 1, predominantly orange; 2, predominantly yellow.

Description Small Texas ringtail with blue face and front of thorax, conspicuous pale ring in front of brown abdomen tip. *Male*: Eyes bright blue, face turquoise with faint brown bars across front. Front of thorax turquoise, sides green; T1–4 narrow but conspicuous, T1 broken at top. Abdomen green to turquoise on S1–2 with brown markings; S3–7 blue on top, whitish on sides, with each segment

188.1
Blue-faced Ringtail
male—Gonzales Co.,
TX, July 2004

188.2
Blue-faced Ringtail
female—Gonzales
Co., TX, July 2004

mostly brown at rear. S8–10 reddish brown, darker on top of S8–9. *Female*: Colored as male but eyes duller blue, blue not evident on thorax, white more prominent on sides of abdomen and across base of S7; tip of abdomen all brown.

Identification Only North American clubtail with blue face, evident in front view of both sexes. Eyes darker blue than other ringtails. Note also that white rings are on middle of segments, not at base. Only ringtail with conspicuous white ring in front of brown abdomen tip in its range, shared by **Serpent Ringtail** farther west. Smallest ringtail in the region and smallest clubtail where it occurs (still smaller LEAST and PYGMY CLUBTAILS not in Texas). Only other American clubtails with blue body coloration (many have blue eyes) are related small ringtails in Mexico and Central America.

Natural History Both sexes perch low in vegetation at edge of open areas near water. Also fly slowly in and out of vegetation, looking rather like damselflies. Very tame. Males at times common over Guadalupe River in numbers that have been called a swarm.

Habitat Clear spring-fed sandy and rocky rivers, usually in woodland.

Flight Season TX May–Oct.

Distribution So far known only from three Texas counties in the United States, on the Guadalupe and San Marcos rivers and Rio Grande; also occurs widely in eastern and southern Mexico and south to Costa Rica.

---

189 Eastern Ringtail *Erpetogomphus designatus*   TL 49–55, HW 30–35

Description Ringtail with dark stripe on front of thorax incomplete below, small patches of color at wing bases. *Male*: Eyes light blue, face pale greenish with two faintly indicated light brown stripes across front. Thorax dull light green with narrow but well-developed stripes T1–4; T1 incomplete at both ends. Abdomen with S1–6 mostly black with pale yellow to off-white basal ring and yellow dorsal stripe on each segment; S7 similar but with more orange, S8–10 entirely orange or S8–9 with much brown or black above. *Female*: Colored as male but yellow more extensive on top  of abdomen, black markings more reduced. All wings with well-defined brown spots at base, inconspicuous in minority of individuals but entirely lacking in some populations along Rio Grande in west Texas. These populations also generally paler, with duller markings and more yellowish club.

Identification With moderately well-striped thorax, most similar to **Serpent Ringtail**, with which it barely overlaps. Differs by incomplete dark stripe on front of thorax and

189.1
Eastern Ringtail
male—Kimble Co.,
TX, July 2004

189.2
Eastern Ringtail
female—Kimble Co.,
TX, July 2004

189.3
Eastern Ringtail
southwestern
male—Brewster Co.,
TX, November 2007,
Netta Smith

often all-orange club of males, conspicuous even in flight. At close range, longer points of cerci in **Eastern** should be evident. Thoracic striping much heavier than in **Dashed** and **Yellow-legged Ringtails** and gives quite different impression from moderately striped **Blue-faced** and **White-belted Ringtails**. In hand, slightly swollen occiput of both sexes distinctive.

Natural History Both sexes on ground or on twigs up to head height, can be some distance from water. Males fly up and down riffles on beats 10–20 feet long, hover for short periods facing wind, then move again. Also perch on rocks and in low vegetation. Abdomen usually elevated slightly, up to vertical at midday. Females oviposit in pools, fairly erratic with one tap on each yard-long run.

Habitat Sandy and gravelly streams and rivers, even large ones.

Flight Season MT Jul, NM May–Sep, TX May–Nov.

Distribution Ranges south in uplands of Mexico to Durango and Nuevo León, east to Maryland and northwestern Florida.

---

190 Serpent Ringtail *Erpetogomphus lampropeltis* TL 41–56, HW 27–35

Description Widespread desert ringtail, subspecies *E. l. natrix* with blue-green heavily striped thorax, vivid white-ringed abdomen with black and orange club. *Male*: Eyes blue; face pale tan, greenish above. Thorax bluish-green with well-developed dark brown stripes; T1 and T2 wide and very close, almost fused; T3 and T4 narrow but conspicuous. Abdomen white with heavy black markings on S1–7 leaving white basal rings; yellow to orange dorsal stripe. S7–10 orange, often with much black above. California coastal subspecies *E. l. lampropeltis* rather different looking, with thorax ground color gray, slightly tinged with bluish-green, T1 and T2 almost completely fused, T3 and T4 relatively wide and joined by dark area between them. *Female*: Colored as male, dorsal abdominal stripe may be darker orange.

Identification Desert subspecies could only be confused with other RINGTAILS in its range, differs from all of them by bluish-green thorax. Also differs from **Dashed** and **Yellow-legged Ringtails** in much more heavily striped thorax, from **White-belted** in lack of white lateral thoracic stripe. Overlaps with **Eastern Ringtail** in eastern part of range, differs from that species in more bluish-green thorax and mostly dark club. Higher in mountains, green-thoraxed **Arizona Snaketail** differs in having sparsely striped thorax and lacking white rings on abdomen; probably do not overlap in altitude. In range of coastal subspecies, only other clubtails are **Pacific Clubtail** and **Grappletail** with black and yellow thorax and **Desert Sanddragon** with less prominent thoracic striping, yellow on front of thorax, and white markings on abdomen prolonged as spears rather than ring-shaped.

Natural History Males of both subspecies perch on rocks and twigs at breeding habitat, usually at riffles. Occasionally fly out over stream on patrol, frequent clashes with other males. May arrive on territory early in day before sun hits stream, then vibrate wings to warm up. Females more often found in open areas nearby, typically perching on ground or rocks. Females oviposit by flying rapidly over pools, dipping to surface at intervals of a few feet.

Habitat Shallow, rocky streams, both wide and narrow, with pools and riffles in open.

Flight Season CA Jun–Oct, AZ May–Oct, NM Aug, TX Sep–Oct.

Distribution Coastal California populations are subspecies *E. l. lampropeltis*, *E. l. natrix* elsewhere and ranging south in Mexico to Baja California Sur and Durango. Substantial difference in coloration between coastal and desert populations caused past workers to consider them different species.

190.1
Serpent Ringtail
male—Yavapai Co.,
AZ, September 2005

190.2
Serpent Ringtail
female—Sonora,
Mexico, September
2005

190.3
Serpent Ringtail
California male—Los
Angeles Co., CA,
August 2007,
Don Roberson

191 Yellow-legged Ringtail *Erpetogomphus crotalinus* TL 45–59, HW 29–35

Description Mexican Plateau ringtail with mostly unmarked thorax and yellow abdomen. *Male*: Eyes dull grayish blue, face lime green. Thorax entirely bright lime green without dark stripes in most, but some show faintly indicated T1. Abdomen with S1–2 lime green with small brown dorsolateral marks on 2, rest mostly yellow; white on sides of S3–6, with black dorsolateral stripes on S3–9. Tibiae black with yellow stripes on outside. *Female*: Colored as male, but S1–2 mostly green.

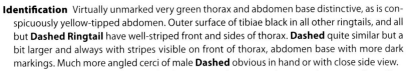

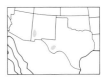

191.1
Yellow-legged Ringtail
male—Chihuahua, Mexico,
September 2005

191.2
Yellow-legged Ringtail
female—Chihuahua,
Mexico, September 2005

Identification Virtually unmarked very green thorax and abdomen base distinctive, as is conspicuously yellow-tipped abdomen. Outer surface of tibiae black in all other ringtails, and all but **Dashed Ringtail** have well-striped front and sides of thorax. **Dashed** quite similar but a bit larger and always with stripes visible on front of thorax, abdomen base with more dark markings. Much more angled cerci of male **Dashed** obvious in hand or with close side view.

Natural History Typical of genus, both sexes perching on rocks or sand at or near breeding habitat. Rarely found north of Mexican border.

Habitat Sandy, rocky hill streams and rivers, spring runs, and irrigation ditches; in desert habitats with wooded or shrubby banks.

Flight Season NM Aug–Sep.

Distribution Ranges south in uplands of Mexico to Guerrero.

192 Dashed Ringtail   *Erpetogomphus heterodon*   TL 50–55, HW 31–36

Description Bright southwestern ringtail with sparse markings on thorax and much yellow on abdomen. Thorax medium green, incomplete brown stripe on either side of front, no markings on sides. *Male*: Eyes dull blue-gray, face lime green. Thorax lime green with well-developed but usually isolated T1, narrow brown suture lines for T2 and T4. Abdomen with S1–2 and base of S2 green, rest yellow with sides of S3–6 white and large dorsolateral marks, leaving pale rings at base of segments. S7–10 yellow-orange with black dorsal markings. *Female*: Colored about as male, no hint of club.

Identification Less heavily striped than most other ringtails. Overlaps in range with **Yellow-legged Ringtail**, which has even less indication of thoracic striping and yellow outer surfaces of tibiae. Also occurs with **Serpent Ringtail**, from which it differs in a much less striped thorax.

Natural History Typical of genus, males perching on sand or rocks and flying out over riffles from time to time.

192
Dashed Ringtail
male—Chihuahua,
Mexico, September
2005

Habitat Shallow, rocky streams in mountains, in open desert country or pine-oak woodland. Typically a riparian band of shrubs.

Flight Season NM Jul–Aug, TX Jun–Sep.

Distribution Ranges south in uplands of Mexico to Chihuahua.

193 White-belted Ringtail   *Erpetogomphus compositus*   TL 46–55, HW 30–35

Description Brightly marked ringtail with white front and side stripes on yellow-green and brown thorax. May have distinct orange suffusion at wing bases. Individuals from southwestern California and Baja California smaller than those from elsewhere. *Male*: Eyes light blue, face whitish. Thorax yellow to chartreuse in front and on sides with four dark brown stripes, white between T1 and T2 and between T3 and T4. Abdomen with S2–7 black and white ringed, S8–10 yellow to orange and black. Club varies from largely orange in most populations to moderately marked with black in some

193.1
White-belted Ringtail male—
Benton Co., WA, July 2004

193.2
White-belted Ringtail female—Benton
Co., WA, August 2007, Netta Smith

northern individuals. *Female*: Colored as male, thorax may be paler in front, abdomen tip with more black. No evident club.

Identification Thorax palest of North American ringtails, yellow instead of green, and characterized by white "belt" on side. Most other ringtails that occur with it have green thorax, only **Eastern** and **Serpent Ringtails** with thorax as heavily striped. Occurs sparingly with California form of **Serpent Ringtail**, which has no trace of yellow or white on thorax.

Natural History Males perch on ground or on rocks or branches, often in very open areas and usually at riffles. At water during warmer parts of day and near it at other times. Both sexes found away from water perching in shrubs and trees, even above head height.

Habitat Open sandy streams, rivers, and irrigation ditches, typically in desert country; also at sink holes.

Flight Season WA Jul–Aug, OR Jun–Sep, CA Apr–Oct, AZ May–Nov, NM Jun–Aug, TX Apr–Oct.

Distribution Ranges south in Mexico to Baja California Sur and Sonora.

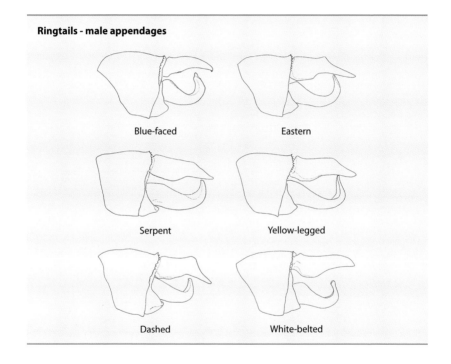

Ringtails - male appendages

Blue-faced

Eastern

Serpent

Yellow-legged

Dashed

White-belted

These beautiful green clubtails are much sought by dragonfly enthusiasts, as they usually live along pristine rapid streams, especially in the East, and some are quite uncommon. Western species may be on slower, larger rivers, and most are common. Males are slightly but distinctly clubbed, females scarcely so; legs are short. Like ringtails, these clubtails have a green thorax, but light markings on the abdomen are not in shape of distinct rings. Abdomen is black and yellow, not as strongly tinged with orange as in most ringtails. Females of most species have one or two pairs of spinelike "horns" on or behind the occiput. Males perch on sandy or gravelly shores or streamside vegetation, usually fairly low, and alternate perching with steady flight up and down beats over shallow riffles and pools. Females are only at water to breed but often are found nearby in clearings and open country. There is substantial emergence at times, with immature individuals sometimes satisfyingly common in clearings near rivers. World 29, NA 19, West 8.

Table 8   Snaketail (*Ophiogomphus*) Identification

	A	B	C	D
Arizona	1	1	1	1
Pale	1	1	1	3
Bison	3	1	1	1
Great Basin	2	1	3	13
Sinuous	2	1	1	2
Boreal	2	2	2	12
Riffle	2	2	2	13
Rusty	2	3	2	2

A, anterior thoracic stripes: 1, absent or incomplete stripe; 2, two strong but separated stripes; 3, stripes entirely fused.

B, abdomen tip: 1, black and yellow striped; 2, mostly black with white spots; 3, mostly brown.

C, male cerci: 1, longer than epiproct; 2, about same length as epiproct; 3, shorter than epiproct.

D, female horns: 1, two occipital; 2, two occipital plus two postoccipital; 3, none.

Description Snaketail of southwestern mountains with virtually unmarked thorax. *Male*: Eyes turquoise, face pale chartreuse. Thorax bright green (dull when older), with faintly indicated dash for T1 that may be almost illegible, also fine brown lines for T3 and T4. Abdomen yellow-orange with black stripes along sides of S2–9 forming pale dorsal stripe or backward-pointing triangles where  broken on some segments by black. *Female*: Colored as male. Most have pair of stout horns projecting backward from occiput.

194.1
Arizona Snaketail
male—Apache Co.,
AZ, July 2007

194.2
Arizona Snaketail
female—Apache Co.,
AZ, July 2007,
Doug Danforth

Identification Very similar to **Pale Snaketail**, which may overlap with it but is not surely known to do so. **Pale** usually but not always has larger mark for T1, may have to check male appendages for length of epiproct for certain distinction (half length of cerci in **Arizona**, about three-fourths in **Pale**). Females even more similar, **Arizona** usually with two small spines at rear edge and depression in middle of occiput; **Pale** has no spines or depression. Several species of RINGTAILS usually at elevations below **Arizona Snaketail**, but all have ringed rather than striped abdomens, most likely overlap (**Serpent Ringtail**) with heavily striped thorax.

Natural History Males perch on rocks in stream, retire to nearby grass and trees when cloudy. Females more likely to oviposit when males absent, for example when cloudy.

Habitat Small, rocky, mountain streams with moderate current and open banks; typically in open pine forest.

Flight Season AZ Jun–Oct, NM Jun–Aug.

---

195 Pale Snaketail *Ophiogomphus severus*   TL 49–52, HW 28–34

Description Wide-ranging western snaketail with lightly marked thorax. *Male:* Eyes light sky-blue to blue-gray, face light green. Thorax pale green, dull in older individuals, either virtually unmarked or with small to large dark brown oval spot for T1, fine to slightly wider dark line in suture for T2. Abdomen pale green on S1–2, otherwise yellow, with black lateral stripes on S2–10 meeting in middle on most segments, forming backward-pointing triangles on top. Variable in color pattern, populations in cooler regions with well-developed short thoracic stripes and more black on abdomen, those in hotter regions virtually lacking such stripes. *Female:* Colored as male but whitish prominent below black side stripes because of thicker abdomen. No occipital horns.

Identification All western snaketails potentially occur with this species, although coexistence with **Arizona** and **Bison Snaketails** not definitely known. Differs from most in having least-developed dark thoracic stripes, only T1 obvious but always short, or no stripes.

195.1
Pale Snaketail
male—Harney Co.,
OR, July 2005

195.2
Pale Snaketail
female—Harney Co.,
OR, July 2006,
Netta Smith

Paler overall than other species, thorax not as bright, but thoracic stripes must be seen for definitive identification. For possibility of occurrence with **Arizona Snaketail**, see that species. Recall that female and immature male **Western Pondhawk** also bright green; it has relatively short, wide green abdomen.

Natural History Males rest on sand, rocks, and twigs at stream up to head height or fly up and down, usually localized at riffles. Perch facing water with abdomen slightly, sometimes greatly, elevated. More time spent perched than flying. Obelisk during hot midday. Copulation lengthy, at rest. Females have been seen to fly out from rock to riffle, tap water once, and return to rock, then do it again. Might not be typical oviposition for species.

Habitat Sandy and rocky streams and rivers, in both forested and entirely open country. Often at smaller streams than other western snaketails. Also sometimes at sandy lakes in northern part of range.

Flight Season BC May–Aug, WA Jun–Aug, OR May–Aug, CA May–Jul, MT Jun–Aug, NM Jun–Oct, NE Jun–Aug.

Comments Most heavily marked individuals are in wetter, cooler parts of range in northern Rocky Mountains. These have been distinguished as a distinct subspecies, *O. s. montanus*, but it is not recognizable as a geographic entity.

Description Pacific Coast snaketail with wide rich dark brown stripes on thorax. *Male:* Eyes blue to dark turquoise, face chartreuse. Thorax green with T1–2 wide and fused, narrow line at T4. Abdomen yellow-orange, mostly dark brown to black on sides forming series of pale triangles down entire abdomen; white low on sides of S1–3, yellow-orange on sides of S7–10. *Female:* Colored as male. Conspicuous pair of horns project from either side of center of occiput.

Identification Darkest green and most vividly marked of western snaketails, with thick dark humeral stripes on thorax. Overlaps with **Great Basin** and **Sinuous Snaketails**, less with **Pale Snaketail**, and heavy fused thoracic stripes distinguish **Bison** from all. Other than those species, no clubtails in range with green thorax. Occipital horns of female closer together than in other western snaketails.

Natural History Males perch on rocks, ground, and twigs at water; females on similar perches away from it.

Habitat Swift rocky streams bordered by willows, mostly in forested habitats, from lowlands to well up in mountains.

Flight Season OR Jun, CA Apr–Oct.

Comments One hybrid with Great Basin Snaketail reported from northeast California.

196.1
Bison Snaketail
male—Shasta Co.,
CA, May 2002,
Ray Bruun

196.2
Bison Snaketail
female—Butte Co.,
CA, June 2003,
Chris Heaivilin

Description Western interior snaketail with strongly striped thorax. *Male*: Eyes turquoise, face light green. Thorax light green with T1 and T2 parallel, T1 disconnected above; T4 represented by narrow suture line. Abdomen yellow with black stripes along sides, enclosing rearward-pointing triangles on S2–9; S10 and appendages pale. Individuals in eastern part of range (Nevada) with narrower stripes than farther west but stripes with same configuration. Rarely anterior stripe reduced to oval spot. *Female*: Colored as male, white more conspicuous below black on sides of distinctly thicker abdomen. Some females with pair of horns on occiput adjacent to eyes.

Identification Both sexes quite similar to **Sinuous Snaketail** in general appearance, and difference in T1 and T2 will have to be seen: parallel and well separated in **Great Basin**, almost touching with narrow curved line between in **Sinuous**. Eyes more intense blue and thorax a bit yellower than in **Sinuous**. Usually easily distinguished from **Bison Snaketail**, with fused T1–2, and **Pale Snaketail**, with those stripes much reduced, but occasional **Great Basin** with reduced stripes may have to be distinguished by male appendages. **White-belted Ringtail,** also superficially similar, has ringed abdomen, white on thorax.

Natural History Males perch on twigs and rocks at breeding habitats, also fly back and forth low over water. Females in open areas away from water. Ovipositing females fly rapidly upstream and tap water at intervals of 3 feet or so, then move some distance and do it again; then fly quickly away from water.

Habitat Streams and rivers, usually with gravelly or sandy bottom and moderate current; also flowing irrigation canals with mud bottom and, less often, lakes. Tolerates extremely alkaline situations, as in Pyramid Lake, Nevada.

Flight Season OR Jun–Sep, CA May–Aug.

Comments One hybrid with Bison Snaketail reported from northeast California.

197.1
Great Basin Snaketail
male—Siskiyou Co.,
CA, June 2004

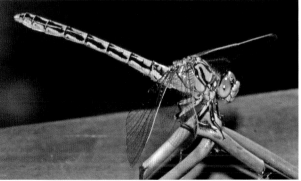

197.2
Great Basin Snaketail
female—Siskiyou Co.,
CA, June 2004,
Netta Smith

Description Western interior snaketail with distinctive anterior thoracic stripes. *Male*: Eyes blue-gray, face yellow or pale green. Thorax green with T1 and T2 partially separated by narrow sinuous pale stripe. Abdomen mostly yellow with black stripe down sides and black at rear margin of S2–9 producing series of rearward-pointing triangles; S8–9 slightly clubbed, with prominent yellow margins. S10 yellow. *Female*: Colored as male but conspicuous white along sides below black stripes; thicker, less clubbed abdomen. One pair of horns on occiput points upward, second pair behind them points backward.

Identification Western snaketails, all with green thorax, best distinguished by thoracic pattern. Wavy stripes on front of thorax distinguishes **Sinuous** from most similar, **Great Basin**, in which stripes very similar but not wavy. Thorax also clearly different from heavily marked **Bison** and sparsely marked **Pale Snaketail**. Thorax brighter green than in **Pale** and slightly brighter than in **Great Basin**.

198.1
Sinuous Snaketail
male—Gilliam Co.,
OR, June 2005,
Netta Smith

198.2
Sinuous Snaketail
female—Siskiyou Co.,
CA, June 2004

Natural History Both sexes scattered around shrubby flats near breeding habitats, perched on rocks, open ground, and especially branches of low shrubs up to chest height. Usually become common at water later in day. Copulation at rest, often high in trees. Females oviposit by tapping water in low, rapid flight, then splashing into water several times before flying away. At times synchronized emergence of this species leads to density of individuals rarely encountered in other western odonates; thousands have been seen at the Klamath River in northern California.

Habitat Slow-flowing sandy and gravelly streams and rivers, in forested or open country. Frequently found on lakes in northern, upland parts of range, not known if they breed there.

Flight Season BC May–Aug, WA May–Aug, OR Jun–Aug, CA Mar–Aug, MT Jun.

Comments Populations at southern end of range, in Central Valley of California, more sparsely marked with dark pigment than those farther north.

199 Boreal Snaketail *Ophiogomphus colubrinus* TL 41–48, HW 27–31

Description Bright green, heavily marked northern snaketail with dark facial stripes. *Male*: Eyes green, face green crossed by four narrow blackish lines. Thorax green with T1–2 fairly wide and partially fused, T4 prominent; lower end of T3 represented by dark point projecting upward into green side. Abdomen black along sides; green on top of S1–2, duller and paler green candles on top of next segments, turning to pale yellow spots by end of abdomen. Whitish on sides expanded to prominent spots on S8–10, each with variable brown smudges. *Female*: Colored as male but abdomen thicker, club less prominent, white more obvious on sides. Pair of blunt horns on back of occiput adjacent to eyes, may be a tiny pair behind them.

Identification Looks like **Riffle Snaketail** but distinguished by dark lines across face and pale basal segment of legs. Easily distinguished from **Rusty Snaketail** by more heavily striped thorax, black and yellow abdomen. In western Canada, overlaps with **Pale Snaketail,** which has scarcely any thoracic striping.

199.1
Boreal Snaketail male—
Vilas Co., WI, July 2007,
Ken Tennessen;
inset male—
Timiskaming Co.,
ON, July 2005,
Greg Lasley

199.2
Boreal Snaketail female—
Vilas Co., WI, July 2007,
Ken Tennessen

Natural History Males fly up and down low over water, sometimes hovering, and come to rest on rocks, logs, or shrubs up to head height or more. Most activity in morning and midafternoon, not at midday. Females oviposit in riffles or upstream of them by tapping water rather methodically, then perching, usually on an exposed rock in the stream, curving abdomen upward and extruding eggs for 5–15 sec, all the while wing-whirring. Then fly over water again to do the same, tapping 4–10 times for up to 30 sec, then fly up into trees.

Habitat Clear rapid streams and rivers with pools and riffles, gravelly or rocky beds.

Flight Season YT Jul–Aug, BC Jul.

Distribution Also east through Canada to Newfoundland, south to Michigan, New York, and Maine. Northernmost clubtail in North America.

200 Riffle Snaketail *Ophiogomphus carolus* TL 40–45, HW 24–28

Description Small, vividly striped, rather dark snaketail of north-  eastern forests with well-defined club. *Male*: Eyes green, face light green. Thorax bright green with dark brown stripes, T1 and T2 wide and mostly fused, T4 narrow and confined to suture; lower end of T3 represented by dark point projecting upward into green side. Abdomen black on sides, green on top of S1–2, yellow from there back, markings candle-shaped in midabdomen but smaller and smaller to merely spots at end. Whitish markings low on sides turn into prominent spots on sides of S7–9 smudged with brown markings of variable size. *Female*: Colored as male but abdomen thicker, club scarcely evident. Some females have pair of horns on occiput.

Identification Overlaps only with **Boreal** and **Rusty Snaketails** in our region, not with any western species. Differs from **Rusty** in black and yellow abdomen, from very similar **Boreal** in unmarked face (**Boreal** with fine cross-stripes) and black femora (mostly pale in **Boreal**).

Natural History Males perch on rocks out in current or on leaves of shrubs and low trees at shore, then fly short patrols up and down riffles just above surface. Difficult to follow because of low, swift flight over agitated water surface. Female oviposits in smooth flight over water, may rest beforehand extruding obvious mass of eggs. Both sexes in clearings near breeding habitat.

Habitat Clear, rocky, and sandy streams and rivers with pools and riffles bordered by riparian shrubs and trees.

200.1
Riffle Snaketail
male—Hampden Co.,
MA, June 2004,
Glenn Corbiere

200.2
Riffle Snaketail
female—Hampden
Co., MA, July 2003,
Glenn Corbiere

Flight Season WI May–Jul.
Distribution Throughout Northeast from southern Ontario and Quebec to Nova Scotia, south to Michigan, Virginia, and New Jersey.

201 Rusty Snaketail *Ophiogomphus rupinsulensis* TL 45–54, HW 27–32

Description Distinctive eastern snaketail with mostly brownish abdomen. *Male*: Eyes green, face green with tan labrum. Thorax lime green with T1–2 fairly narrow and partially fused, no other stripes. Abdomen mostly brown, typical snaketail pattern present but quite obscure except for green markings on S1–2, hint of whitish spots on sides of S8–9. Club quite obvious from above. *Female*: Colored as male, again typical snaketail pattern present but obscured, overall impression light brown with white on lower sides and dark dorsal spots on posterior segments. Pair of prominent horns on back of head adjacent to eyes, some females with second upward-pointing tiny pair on occiput.

Identification Only eastern snaketail with largely brownish rather than black and yellow abdomen in both sexes. Only barely overlaps with **Boreal** and **Riffle Snaketails** in West. Be aware that tenerals of other snaketails could look something like this species before fully colored.

Natural History Males perch on twigs over water or fly low and erratically over riffles. Present throughout day, although at many sites visit in greatest numbers late in afternoon. Copulation once observed at dusk. Female oviposits in straight fast runs upcurrent, perches between brief bouts and extrudes orange-tan egg mass. Both sexes perch above ground in weedy fields near breeding sites.

Habitat Large streams and rivers with moderate current, often larger and siltier than those used by other snaketails.

Flight Season WI May–Jul.

Distribution Also throughout Ontario and southern Quebec east to Nova Scotia, south to Tennessee and Virginia.

201.1
Rusty Snaketail
male—Renfrew Co.,
ON, July 2005,
Greg Lasley

201.2
Rusty Snaketail
female—Mohican
Co., OH, July 2007

Snaketails - male appendages

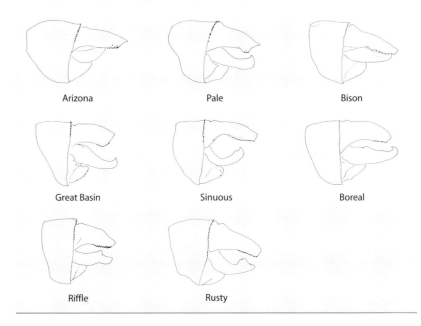

Arizona

Pale

Bison

Great Basin

Sinuous

Boreal

Riffle

Rusty

Least Clubtails *Stylogomphus*

These are the smallest of North American clubtails and among the smallest clubtails in the world. Related to pygmy clubtails (*Lanthus*) by small size and similarities in larval shape and antennae, both genera occur only in eastern Asia and eastern North America. World 11, NA 2, West 1.

202 Interior Least Clubtail *Stylogomphus sigmastylus*   TL 34–37, HW 20–22

Description Tiny clubtail of rocky streams with dark, yellow-ringed abdomen, scarcely any club, and white appendages. *Male*: Eyes turquoise, face pale yellow. Thorax yellow with stripes T1–4 well developed. Abdomen black with yellow dorsal stripe on S1–3, yellow rings on S3–7 almost broken on top, and yellow side spots at base of S8–9. Cerci white. *Female*: Colored as male with slightly

larger areas of yellow on abdomen, especially on sides of basal segments. Yellow colors change to gray with old age.

Identification Small size and slender body distinctive among clubtails in its range, as are white cerci in both sexes. Much smaller than black and yellow COMMON CLUBTAILS that occur with it.

Natural History Males perch flat on rocks in river, less often on leaves above it, and patrol in circles low over rapids, when very difficult to follow. Probably come from surrounding trees, as they often fly back up into them. Tenerals sometimes rather common in trees above river. Females relatively rarely seen, perhaps usually up in trees, but when found

202
Interior Least Clubtail
male—Metcalfe Co.,
KY, June 2006

usually perching on leaves above river. Females oviposit by tapping at intervals in circling path just above riffles.

Habitat Small, clear rivers with moderate current and sand to rock substrates, usually in woodland.

Flight Season East Apr–Aug.

Distribution Also east into Kentucky and Tennessee.

Comments Recently separated from very similar Eastern Least Clubtail, *Stylogomphus albistylus*, so some references to that species include this one.

Grappletail *Octogomphus*

This small species has no close relatives in the West but by larval morphology is related to least and pygmy clubtails (*Stylogomphus* and *Lanthus*) of the East, both of which have representatives in Asia. World 1, NA 1, West 1.

203 Grappletail *Octogomphus specularis* TL 51–53, HW 29–32

Description Rather delicate, slender, brown-eyed clubtail of Pacific streams. *Male*: Eyes dull greenish to bluish above, yellowish below, but may look mostly dark brown, unusual color among clubtails; face yellow. Thorax yellow, with T1–2 fused into very wide stripe, also narrow T4. Abdomen almost entirely black with fine pale stripe down S1–6 or S7. Cerci yellow and very wide, wider than abdomen, so end of abdomen looks expanded rather than club-shaped as in other clubtails. *Female*: Abdomen slightly thicker, straight-sided throughout. Both sexes bright yellow when young, become duller grayish-cream with age.

Identification Nothing else in West like this slender clubless species with widened abdomen tip. Occurs commonly with **Pacific Clubtail**, which is similarly colored but larger with obvious club and more yellow on sides of abdomen. SNAKETAILS that occur with it are green. **Pacific Spiketail** on same streams is much larger, with bright blue eyes and yellow-spotted abdomen, and flies rapidly up- and downstream.

Natural History Males perch flat on leaves, along twigs, or on rocks over water, usually in sun; abdomen held flat on substrate, not elevated. Aggressive to other males but with territories only 10 feet long. Also fly slowly up- or downstream for short periods with rather slow flight, less rapid and direct than other clubtails. Females nearby in wooded areas. Both

203.1
Grappletail
male—King Co.,
WA, June 2005

203.2
Grappletail
female—Klickitat Co.,
WA, July 2003,
Netta Smith

sexes perch from near ground to 30 feet up in trees. Copulating pairs hang up in trees. Females oviposit by striking water at intervals of 2–6 feet while frequently changing direction, then ascend back into trees. Very tame, with care can be nudged to crawl onto hand!

Habitat Small rocky or muddy woodland streams with moderate current, rocky and muddy bottom. In northern part of range, typically in relatively warm streams that flow out of lakes.

Flight Season BC Jul–Aug, WA May–Aug, OR May–Aug, CA Apr–Oct.

Distribution Ranges south in Mexico to northern Baja California.

The four genera of this family are all very much alike in appearance and habits, large black or brown dragonflies with prominent yellow stripes on the thorax and yellow-patterned abdomen. Most species occur in Eurasia, where they fly up and down small woodland streams and behave much as the North American species of our single genus do. World 51, NA 9, West 4.

Spiketails *Cordulegaster*

Spiketails are large black or dark brown dragonflies with similar bright yellow thoracic stripes and species-specific abdominal markings. Relatively small, slightly separated or just-touching eyes are either bright blue or bright green. They fly over sunny clearings, often quite low, foraging for small insects and then perch by hanging in herbaceous or shrubby vegetation, typically right out in the open and presumably watching for flying prey. Wasps and bees seem to be common prey for these large dragonflies. Some individuals hang much like darners; others extend out from a branch at a 45° angle. Males are much more often seen in this situation than females, and they may be looking for females as well as prey. They do not wander as far from breeding habitats as do the similarly large darners that feed from the air. Males are also seen flying long beats up and down streams, from moderate-sized rushing rocky ones to shallow, mossy forest trickles, depending on the species. Cruising beats are often long, so expect to wait until a given individual returns. Mating pairs are rarely seen, but copulation is lengthy, lasting an hour or more, usually in trees near or away from water. Females tend to oviposit after midday, flying along and stopping suddenly to hover over a shallow spot and jab eggs into the substrate with pointed and spikelike (thus the group name) ovipositor, really a "pseudo-ovipositor" formed from the prolonged subgenital plate. Occasional males may have enough feces projecting from end of abdomen to look like females! Other black and yellow dragonflies are easily distinguished. Superficially similar river cruisers have much larger eyes, single stripe on either side of thorax; they tend to be on larger streams and hang straight down when perched. Dragonhunter has a different pattern on thorax and abdomen. World 25, NA 9, West 4.

204 Pacific Spiketail *Cordulegaster dorsalis*　　　TL 70–85, HW 43–49

Description Large, dark, blue-eyed far western spiketail with spotted abdomen. *Male*: Eyes pale blue above, darker below; face pale tan with one or two dark bars across it. Thorax dark brown with vivid yellow stripes on front and sides. Abdomen blackish with yellow spots on S2–9, slightly divided on midline and getting wider and shorter toward rear. Western Great Basin populations, from Owens Valley of California into Nevada, have much more yellow on body, often with narrow yellow stripe between wide ones on thorax and extra small spots along abdomen. *Female*: Colored as male, recognized by long pointed ovipositor.

Identification Nothing else just like it in range. **Black Petaltail** smaller, with brown eyes and thorax as well as yellow-spotted abdomen, different behavior. **Western River Cruiser** has gray eyes, single stripe on side of thorax, less contrasty pattern. Coexisting DARNERS with thoracic stripes less contrasty, abdomens more complexly spotted. **Apache Spiketail** to south has rings across top of abdomen rather than paired spots, also green eyes and much more black on face.

Natural History Males usually seen patrolling streams in low, slow, and steady flight over long beats; usually turn around when obstruction or large open pool reached. Just as often in sunny clearings in or at the edge of forest, also brushy ravines. Away from water, alternate lengthy bouts of perching with brief foraging flights below chest height. Often return

204.1
Pacific Spiketail male—
Skamania Co., WA, August
2005

204.2
Pacific Spiketail female—
Monterey Co., CA,
September 2006, Don
Roberson; inset female—
Inyo Co., CA, August 1973

204.3
Pacific Spiketail Great Basin
male—Inyo Co., CA;
June 2003

to same perch site in sunny spot. Females much less often seen, usually when searching for
oviposition sites; perhaps do not mix with foraging males, which will harass them. Females
oviposit in shallow water of stream edge by rising and falling, driving eggs into substrate
by vertical thrusts. One female did this in one spot for 8 min, about one thrust per second;
another changed locations and made 4–10 thrusts at each spot.
Habitat Small cool streams with some current in forested areas; substrate can be sand or
mud mixed with rocks. In northern part of range, warmer streams coming out of lowland
lakes; in southern, faster streams in mountains and foothills. Great Basin populations at
spring-fed streams in desert regions.
Flight Season BC Jun–Aug, WA Jun–Aug, OR Jun–Sep, CA May–Oct, MT Jul–Sep, NM Jul–Aug.
Distribution Ranges south along Pacific coast to Baja California.

Comments Specimens from Owens Valley, California, were described as a separate species, *Cordulegaster deserticola*, now generally considered a subspecies of *C. dorsalis*. The extent of its range is poorly known, and intermediates between it and other populations have not been seen, but not all populations in other parts of Great Basin are so colored.

205 Apache Spiketail *Cordulegaster diadema* TL 74–88, HW 43–53

Description Large, green-eyed southwestern spiketail with yellow-ringed abdomen. *Male*: Eyes light green or blue-green above, darker below; face mostly black with yellow crossband. Thorax black with broad yellow stripes. Abdomen black with irregular yellow markings producing spots on sides and bands across top. *Female*: Colored as male, easily distinguished by slightly thicker abdomen, long ovipositor.

Identification Only dragonfly like it in southwestern mountains. See **Pacific Spiketail**, farther north; the two overlap slightly. Differs from DARNERS in small green eyes, much more vividly contrasting thoracic stripes, and clearly ringed abdomen. Most like **Persephone's Darner**, also large and shares its habitat although tends to fly later in season. Male and most female **Persephone's** have blue markings rather than yellow on abdomen, some female

205.1
Apache Spiketail male—
Cochise Co., AZ, July 2001,
Doug Danforth

205.2
Apache Spiketail female—
Greenlee Co., AZ, August
2007, Kim Davis and Mike
Strangeland

Persephone's with yellow distinguished from **Apache** by larger, not green, eyes and different-looking abdomen tip. Spiketails also have much more conspicuous stripes on front of thorax than darners. **Riffle Darner** smaller than spiketail and with mostly blue markings. **Western River Cruiser** duller, with large, gray eyes and single stripe on side of thorax.

Natural History Males fly fairly rapidly up and down small streams, even deep in woodland, looking for females. Numerous males may visit same segment of stream. Females rarely seen at water. Pairs copulate for about 2 hr up in trees, then female oviposits unguarded. Both sexes cruise around clearings near home streams, from near ground to floating up at treetops. Typically perches and flies higher than Pacific Spiketail.

Habitat Small, forested streams in mountains.

Flight Season AZ Apr–Oct, NM Aug–Sep.

Distribution Ranges south in uplands of Mexico to Michoacan and Veracruz, replaced by less heavily marked subspecies *C. d. godmani* from southern Mexico south to Costa Rica.

---

206 Twin-spotted Spiketail *Cordulegaster maculata* TL 64–76, HW 38–49

Description Large, blue-eyed eastern spiketail with yellow-spotted abdomen. *Male*: Eyes pale blue above, duller and may be brownish below; face pale with dark bar across it, sometimes quite dark above. Thorax dark brown with two yellow stripes on front and pair of yellow stripes on each side. Abdomen blackish-brown with paired yellow spots or triangles on S2–9, two pairs each on S2–4, where each pair pointed toward the other. *Female*: Colored as male, but spots/triangles distinctly smaller and only on S2–7 or S2–8. Recognized by long pointed ovipositor extending well beyond S10.

206.1
Twin-spotted Spiketail male—Fannin Co., GA, May 2005, Giff Beaton

206.2
Twin-spotted Spiketail female—Cape May Co., NJ, May 1995, Patricia Sutton

Identification Differs from only other spiketail in range in West, **Arrowhead Spiketail**, by paired rather than central markings on abdomen; also slightly smaller. Other large black and yellow dragonflies (**Dragonhunter**, RIVER CRUISERS) patterned differently.

Natural History Males fly rapidly up and down breeding streams a few inches above water. On wider streams stay near one bank but often cross over on narrow streams. Not averse to wending way through dense branches and flying through tunnels. Stand in water and they will fly around your legs. May become more common later in afternoon. Both sexes feed low in clearings in woodland, where they can be found when away from water, and hang up not far above ground. Copulating pairs into treetops. Females oviposit by hovering a few inches above water with abdomen hanging down obliquely, then dropping quickly with it vertical to push eggs into mud, sand, or fine gravel, quickly ascending and doing it again, as many as 100 times/min. May then move a few feet and repeat same actions.

Habitat Small to midsized rocky streams with good current and muddy pools, typically in forest.

Flight Season TX Jan–Apr.

Distribution Ranges east in southern Canada to Nova Scotia, south to Louisiana and northern Florida.

207 Arrowhead Spiketail   *Cordulegaster obliqua*   TL 72–81, HW 41–50

Description Large eastern spiketail with arrowheads down abdomen. *Male*: Eyes pale green above, brownish below; face pale with one or two black bars across it. Thorax dark brown, almost black, with two broad vivid yellow stripes on each side and two narrower ones on front. Abdomen black with series of yellow arrowhead-shaped marks on S2–7, large blotch on S8, and smaller spot on S9. *Female*: Colored as male, usually lacks spot on S9. Recognized by long pointed ovipositor extending well beyond S10.

Identification Distinguished from **Twin-spotted Spiketail** by central rather than paired markings on abdomen; also slightly larger. All other large black and yellow dragonflies (**Dragonhunter**, RIVER CRUISERS) patterned very differently.

Natural History Males perch in open on twigs or herbaceous stems or fly back and forth over breeding habitat, cruising slowly and steadily at knee to waist height and often having

207.1
Arrowhead Spiketail
male—Sussex Co., NJ,
May 2004,
Allen Barlow

207.2
Arrowhead Spiketail
female—Sussex Co.,
NJ, June 2004,
Allen Barlow

to thread way among tall herbaceous plants. May patrol entire length of small rivulets. Both sexes hang in shrubs at waist to chest height at or near water.

Habitat Small swift streams and muddy seeps in forested habitats.

Flight Season TX Apr–Jun.

Distribution Throughout East from Minnesota and southern Quebec south to Louisiana and northern Florida.

Cruiser Family   *Macromiidae*

Cruisers are easily recognizable large dragonflies with large eyes and long legs that fly up and down streams and rivers or along lake shores and then hang up vertically in trees and shrubs. They are all brown to black with pale markings, a single stripe on each side of the thorax, and a spotted or ringed abdomen. The abdomen is often slightly clubbed in males, a good distinction from darners as they cruise overhead in feeding flight. Some species of cruisers have brilliant emerald green eyes, and some authors have considered this a subfamily of the emeralds, Corduliidae. Other authors combine both of those families with the skimmers, Libellulidae, but the long-legged sprawling larvae of cruisers are quite distinct from larvae of either emeralds or skimmers, and no other North American dragonflies are very similar to cruisers. Wing venation of the three groups is different, only emeralds and skimmers showing a distinct anal loop, and the triangles of cruisers are arranged more like those of clubtails and darners. World 121, NA 9, West 6.

Brown Cruisers   *Didymops*

These are smaller, duller editions of the large and showy river cruisers, body color brown instead of black. Eyes are greenish but much duller than those of river cruisers and meet only over a short distance. The flight season is earlier, and the two genera probably seldom overlap in time. They are North American endemics, the second species of *Didymops* is restricted to the extreme Southeast, where it is a lake-dweller. World 2, NA 2, West 1.

208 Stream Cruiser   *Didymops transversa*   TL 56–60, HW 34–38

Description   Small brown cruiser of eastern streams. *Male*: Eyes brown with green highlight above. Thorax brown with narrow whitish to pale yellow stripe on each side. Abdomen brown with dull pale yellow to whitish spots on S3–8, becoming progressively shorter and wider toward rear. Abdomen distinctly clubbed. Tiny brown spot at base of each wing. *Female*: Eyes brown. Colored as male, but spots on basal abdominal segments may be larger, creating a more mottled effect. Abdomen thick, no hint of club.

Identification   Considerably smaller and paler than RIVER CRUISERS, generally flies earlier in season. Most like **Bronzed River Cruiser**, palest of that group, but note size and eye color

208.1
Stream Cruiser
male—Travis Co., TX,
April 2005,
Greg Lasley

and sometimes visible brown basal wing spots in **Stream Cruiser**. Also, cerci of **Stream Cruiser** whitish in both sexes, dark in RIVER CRUISERS. Abdominal spots often but not always conspicuous in flight. Might be mistaken for CLUBTAIL except for single pale stripe on side of otherwise dark thorax, also large eyes and flying and perching behavior. No EMERALDS have dark abdomens with light spots.

Natural History Males fly rapidly and usually below waist height up and down streams in beats up to 100 yards long, also along lake shores; sometimes hover, unlike river cruisers. Otherwise hangs up from almost horizontal to almost vertical fairly low in trees and shrubs. Ovipositing female flies along shore or around tree trunks and taps water rapidly at varying intervals. Feeds on insects as large as other dragonflies, during short foraging beats in or out of woodland, from near ground to considerably higher; most foraging at lower heights than river cruisers.

208.2
Stream Cruiser female—Travis Co., TX, April 2004, Greg Lasley

Habitat Sandy forest streams and rivers, less commonly large lakes.

Flight Season TX Mar–Apr.

Distribution Also east through southern Canada to Nova Scotia, south to Gulf Coast.

River Cruisers *Macromia*

These large to very large dragonflies with big eyes and somewhat metallic bodies cruise long distances along river and stream banks. They are easily distinguished from other large fliers (darners, spiketails) by a single pale stripe on each side of the thorax and slightly clubbed abdomen in males of most species. They are very diverse in tropical Asia, with only a few species extending into the temperate zone of Asia and one in Europe. A secondary but small radiation in North America is represented by five showy eastern species with emerald-green eyes and two duller central and western species. Some eastern species apparently hybridize. Eyes of green-eyed species are brown when immature. World 77, NA 7, West 5.

209 Swift River Cruiser *Macromia illinoiensis* TL 65–76, HW 44–49

Description Vividly banded club-tailed cruiser with big green eyes. Description refers to subspecies *M. i. georgina*, occurring from Kansas south. *Male*: Eyes brilliant green, face black with yellow spots on top of frons, and crossband on front. Thorax metallic green-black with upwardly pointed yellow stripe on front and wide yellow stripe on side. Abdomen shiny black with yellow ring on S2; paired triangles on S3–6, smallest on S6; and median spots on S7–8. See below for geographic variation. *Female*: Colored as male, but thorax browner, yellow abdominal markings averaging larger and not much smaller on S6 than S7; abdomen not clubbed. Wings, especially tips, get increasingly orange-brown with age, not the case in males.

Identification No other species of group on northern Great Plains, where river cruisers are scarce, but most common species in most parts of range. Distinguished from larger **Royal River Cruiser** by abdominal club, complete stripe around S2 (broken in **Royal**), and pale spots on top of frons. Also, in both sexes, **Royal** usually with small paired spots or no markings on S8, **Swift** with single median triangle or crossbar. **Swift** differs from **Gilded** by much less pale color everywhere: small instead of large spots on top of frons, shorter and narrower stripes on front of thorax, and smaller abdominal markings. In any view, abdomen of **Swift** looks black with pale spots, that of **Gilded** banded or ringed black and yellow. Brilliant green eyes, blacker coloration, and less pale color on face distinguish **Swift** from **Bronzed**. Flight usually more rapid, higher, and farther out from shore than vaguely similar **Twin-spotted Spiketail**.

Natural History Both sexes cruise over clearings, from just above ground to well up in tree-tops, and hang up on bare limbs of shrubs and trees, even weeds, from knee height to high in canopy. Fly regular beats in small clearings, sometimes only at certain times, probably hanging up in intense heat. Males at water fly long beats not far above water, often sticking close to shore, often with abdomen slightly arched. Accelerate like rocket in lengthy chases of one another, especially when males after females. When ovipositing, females fly low and fast from near shore to midriver, much as males, tapping water at intervals of up to 30 feet in straight or curved path. At least in northern areas, both sexes cruise over water from dawn until just about dark.

Habitat Large streams and rivers, clear to muddy with slow to fast currents, rocks or not. At open lakes in northern part of range.

Flight Season TX May–Oct.

Distribution Also throughout East, north to southern Ontario and Quebec and Nova Scotia.

209.1
Swift River Cruiser southern
male—Gadsden Co., FL, June
2005, Giff Beaton

209.2
Swift River Cruiser southern
female—Travis Co., TX, June
2003, Greg Lasley

209.3
Swift River Cruiser northern
male—Cleburne Co., AL, June
2005, Giff Beaton

209.4
Swift River Cruiser northern
female—Eau Claire Co., WI,
June 2007

Comments Most individuals in range of this book are subspecies *M. i. georgina*, often called Georgia River Cruiser. Very sparse populations on northern plains are *M. i. illinoiensis*, often called Illinois River Cruiser (also former name for species). They are much less marked in both sexes. Typically thorax with no anterior stripe in *illinoiensis*, stripe usually prominent in *georgina*. In male of northern subspecies, abdomen with narrow, interrupted ring on S2, very small spots on S3–4, and no yellow on S5–6. In northern females, abdominal spots vary from virtually absent to rather large, but yellow spot much larger on S7 than on S6. In southern females, spots all about same size.

Description Most brightly marked green-eyed river cruiser, showing as much yellow as black. Smaller in southern part of range. *Male*: Eyes brilliant green, face white with wide black bar across front. Thorax metallic green-black with wide cream stripe on front and side. Abdomen shiny black with wide cream to yellow markings, ring on S2 and large basal spots on S3–8. Spots on S3–6 partially split by black middorsal line. *Female*: Colored as male, but pale spots on abdomen more separated; abdomen not clubbed.

Identification Largest expanse of yellow/cream of any river cruiser, with wide, complete stripes on front of thorax and large spots on abdomen, occupying almost half of each segment S3–8. **Bronzed River Cruiser** most similar in color pattern but much duller in life, with gray to greenish eyes and overall brown rather than black coloration. **Swift River Cruiser** also similar in size and shape, with pale color on top of frons, but **Swift** usually with incomplete antehumeral stripe and smaller abdominal spots, encompassing much less than half of each segment and minuscule on S6, at least in males. Larger

Royal River Cruiser differs in same way, also has no evident club at abdomen tip. **Royal** also has top of frons almost entirely black, **Gilded** with much pale color.

Natural History Males fly long beats over river, often near shore, even when this involves deviating around beds of water plants and overhanging shrubs. Females only occasionally encountered hanging in woods.

Habitat Clear rivers with long pools and moderate current, in or out of woodland.

Flight Season TX Apr–Sep.

Distribution Ranges northeast to Wisconsin and Ohio.

Comments Some river cruiser specimens taken in Oklahoma and Texas are not typical of any species and may be hybrids between two of them, involving Bronzed, Gilded, and/or Royal River Cruisers.

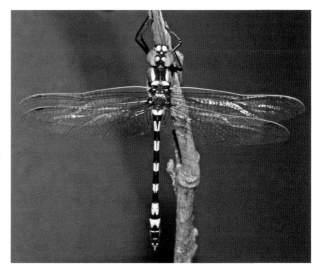

210.1
Gilded River Cruiser male—Kimble Co., TX, July 2001 (posed)

210.2
Gilded River Cruiser female—Howard Co., AR, August 2005, Herschel Raney

Description Largest river cruiser, with no abdominal club, yellow markings relatively restricted. *Male*: Eyes brilliant green, face mostly metallic green-black with yellow crossband. Thorax metallic green-black and brown with wide yellow side stripe and more restricted yellow antehumeral stripe pointed above. Abdomen shiny black with yellow markings: interrupted ring on S2, small paired spots on midsegment S3–8, those on S7 larger and fused into one. *Female*: Colored as male, but abdominal spots slightly larger, usually double on S7 and may be lacking from S8. Wings become orange with maturity.

Identification Often occurs with **Swift**, barely overlaps with **Bronzed** and **Gilded River Cruisers**. Distinguished from all three by larger size (hard to determine in field), very little indication of abdominal club in male, generally less yellow all over (much less than **Gilded**). Other three have large central pale spots on S7–8, **Royal** smaller and/or divided spots, easily visible on perched individuals and perhaps discernible on passing ones.

Natural History Males cruise up and down rivers at knee height or a bit above, often near shore but at times farther out, going long distances in one direction before returning on same path. More common at water in morning. Both sexes hang up in trees near breeding

211.1
Royal River Cruiser
male—Long Co., GA,
June 2005, Giff Beaton

211.2
Royal River Cruiser
female—Travis Co., TX,
June 2004, Greg Lasley

areas and feed well above ground in open areas or among treetops, sometimes in feeding swarms with other large dragonflies. Copulating pairs also hang up in trees. Oviposition may be leisurely at one spot or in rapid flight with tapping at long intervals.

Habitat Wooded streams and rivers.

Flight Season TX May–Jan.

Distribution Also throughout Southeast, north to Wisconsin, Michigan, and Delaware.

Comments Probably hybridizes ("*Macromia wabashensis*") with both Bronzed and Gilded River Cruisers, situation still poorly understood.

---

212 Bronzed River Cruiser *Macromia annulata* TL 68–73, HW 44–50

Description Pale-banded, gray-brown cruiser of the southern plains. *Male*: Eyes blue-gray, some with green highlights; face cream with brown crossbands. Thorax brown, overlaid with gray, with wide cream-colored stripes on front and side. Abdomen dark brown with white to pale yellow band around S2 and large whitish basal spots on S3–8. *Female*: Colored as male, eyes blue-gray; abdomen not clubbed.

Identification Four species of river cruisers cruise Texas rivers. Compared with others (**Gilded**, **Royal**, **Swift**), **Bronzed** with eyes dull green to gray and body overall grayer to

212.1
Bronzed River Cruiser male—
Gonzales Co., TX,
August 2003, Greg Lasley

212.2
Bronzed River Cruiser
female—Frio Co., TX,
June 2005, Martin Reid

browner. Stripe on front of thorax more prominent than in last two species, extending closer to top, less prominent than in **Gilded**, which also has larger abdominal spots. At close range, look at color of vertex (just in front of eye seam), pale in **Bronzed** and dark in other species. Also check male cerci, rounded at ends in **Bronzed**, pointed in others.

Natural History Males cruise rapidly up and down streams like other river cruisers. Both sexes hang up in shade in riparian woodland, often at head height or higher. Females oviposit by rapid flight over water, tapping at intervals of 6–12 feet.

Habitat Large streams and small rivers with some current, often rocky, in open country or with bordering woodland.

Flight Season NM Jun–Aug, TX Apr–Oct.

Distribution Ranges south in eastern Mexico to San Luis Potosí.

Comments May hybridize with Royal River Cruiser at some locations; poorly understood.

---

213 **Western River Cruiser** *Macromia magnifica* TL 69–74, HW 40–46

Description Rather dull, gray-eyed, but still impressive river cruiser of the West. *Male:* Eyes blue-gray, face white with black band across upper part. Thorax brown, heavily overlaid with gray, with pale yellow stripe on front and on side. Abdomen black with cream markings: complete or interrupted ring on S2, basal spots on S3–10, largest on S8, and narrowed to bars on S9–10. See below for geographic variation. *Female:* colored as male, but cream spots on abdomen considerably larger, prominent on S9 but narrow bar on S10; abdomen not clubbed.

Identification No other river cruiser in its range, and not very similar to any other large dragonfly. Easily distinguished from large DARNERS by overall gray/brown/yellow appearance, with no trace of blue. Differs even from dull female darners by single pale stripe on side of thorax and single row of spots down abdomen. When cruising high above in feeding flight, slightly clubbed abdomen of males distinguishes them from DARNERS.

213.1
Western River
Cruiser male—
Benton Co., WA,
July 2006 (posed)

213.2
Western River Cruiser female—
Colusa Co., CA, August 2006,
Paul Johnson

213.3
Western River Cruiser northwestern
male—Cultus Lake, BC, July 1971
(posed)

Natural History Males fly lengthy but regular beats (50–100 yards long) up and down streams and rivers or around edge of lakes. Typically at knee height but higher at times and usually quite rapid. Most consistently present at water before midday. Both sexes forage by cruising over clearings and at woodland edge from waist to treetop height. Often fly with tip of abdomen curved downward. Pairs in trees at some distance from water in lengthy copulation. Females oviposit in open water, tapping water every 3–6 feet or so in full rapid flight or hovering in one spot, facing shore, and tapping water vigorously, then backing up and dropping down to tap again. Larvae may travel spectacularly long distances to emerge, sometimes high on tree trunks away from water.

Habitat Moderate streams to large rivers with slow to rapid flow and mud or sand bottoms, with or without aquatic vegetation; also flowing irrigation canals. Rocky streams less preferred but inhabited in some areas. Large lakes with open shoreline used in northern part of range.

Flight Season BC May–Sep, WA Jun–Aug, OR Jun–Aug, CA Apr–Sep, AZ Jun–Aug.

Distribution Ranges south on Mexican Plateau to Hidalgo.

Comments Populations in southwestern British Columbia represent distinct subspecies *M. m. rickeri*; distribution poorly known but common at Cultus Lake. Overall much darker than populations to south and east, with much reduced antehumeral stripe and abdominal markings. In males, small spots on S3–5, almost none on S6, slightly larger ones on S7–8, none thereafter.

This family is named for brilliant green eyes (red-brown when immature) characteristic of most species and metallic green bodies of some. The body is generally dark, with pale markings at abdomen base, the same color as pale areas on thorax, and typically a pale ring at junction of S2 and S3, continued on other segments in some "ringed" species. No other North American odonates have such coloration, but not all emeralds exhibit it. Most are of northern affinities and seem well protected against low temperatures by hairiness of thorax and, often, abdomen. Hairs are usually golden, producing a golden haze around an otherwise very dark thorax. Few structural characters distinguish emeralds from skimmers, but note short anal loop in hindwings, not very foot-shaped, and soft keel variously developed on tibiae of at least the third pair of legs. Male emeralds also have a small auricle on each side of S2 and slightly angulate hindwing bases as do other dragonfly families—but unlike skimmers. Emeralds are all fliers except for the rather aberrant boghaunters that perch like skimmers. When they perch, the larger, long-bodied emeralds hang up like darners, but the smaller ones often perch flat on leaves. Many emeralds are characteristic of northern habitats—bogs and fens. World 240, NA 50, West 38.

Boghaunters   *Williamsonia*

The two species in this genus are quite aberrant, the smallest emeralds on the continent, and the only ones that typically perch on ground and tree trunks. Eyes are darker than in most other emeralds and bodies dark and nonmetallic with yellow markings. Both species are restricted to acid bogs and thus local over much of their ranges, and only one of them barely reaches this region. They are among the earliest species to emerge in spring. World 2, NA 2, West 1.

214 Ebony Boghaunter   *Williamsonia fletcheri*   TL 29–35, HW 21–23

Description Very small dark emerald that behaves like a skimmer. *Male*: Eyes pale greenish-gray, becoming bright green with age; face metallic greenish-black. Thorax brown with hints of metallic green and fine whitish transverse lines at upper edge of front. Abdomen glossy black with white rings at base of S3–4, rarely further. *Female*: Eyes duller and darker than those of male, abdomen thicker.

Identification Size, shape, color, and perching habits make it much more likely to be mistaken for **Black Meadowhawk** or WHITEFACE.

214.1
Ebony Boghaunter
male—Worcester Co.,
MA, June 2005,
Glenn Corbiere

214.2
Ebony Boghaunter
female—Worcester
Co., MA, June 2005,
Glenn Corbiere

All WHITEFACES have white faces (dark in **Boghaunter**), brown eyes (gray to greenish in **Boghaunter**), and red or yellow markings on thorax and/or red or yellow spots on abdomen. **Black Meadowhawk** has brown eyes, more slender abdomen, and, except for mature males, yellow spots on thorax and abdomen. Also less likely on ground or tree trunk than **Boghaunter**. Among other emeralds, **Racket-tailed Emerald** small enough to be confusing, but note different abdomen shape and perching and flight habits.

Natural History Both sexes rest flat on ground, logs, or tree trunks, unique among emeralds of our region. Although other emeralds from time to time land on flat surfaces, most hang from leaves or twigs. Male boghaunters perch on twigs and bog mat at small bog pools and make short flights over the pools, but copulating pairs often seen away from water. Oviposits in small pools with or without floating sphagnum. Early-season species.

Habitat Acid bogs, usually surrounded by woodland and full of sphagnum and other mosses but always with at least some open water, if only small pools.

Flight Season WI May–Jun.

Distribution Ranges east through southern Canada to Nova Scotia, south to Michigan and New York.

Common Emeralds *Cordulia*

This genus, most similar to striped emeralds (*Somatochlora*) and little emeralds (*Dorocordulia*), is distinguished by the forked epiproct of males and minor differences in venation. Single European and Asian species closely related to American one. World 3, NA 1, West 1.

215 American Emerald *Cordulia shurtleffii*　　　　　TL 43–50, HW 29–32

Description Medium-sized, unmarked dark emerald with slight abdominal club. *Male*: Eyes brilliant green, face dull brownish, iridescent green-black above. Thorax brown with metallic green stripes in sutures between sections. Abdomen shiny black, S1–2 mostly orange-brown on sides. *Female*: Colored as male, abdomen obviously thicker. Wings become brownish with age.

Identification Smaller and shorter-bodied than most STRIPED EMERALDS with which it might occur, no trace of yellow markings on thorax. Superficially most like smallest species **Brush-tipped** and **Ocellated Emeralds**, but those with prominent thoracic markings. In hand, thorax bronzy brown, not metallic green-black of most STRIPED EMERALDS. Darker and with brighter green eyes than BASKETTAILS of similar size and flight habits. **Racket-tailed Emerald** most similar but smaller, with more pronounced club as well as bright yellow spots at base of S3.

Natural History Males fly over open water or along shore vegetation beds, typically at waist to knee height, alternating rapid flight and brief hovering. Shoreline patrols 30–80 feet long but not territorial; males change patrolling areas frequently. Can be very common. Activity greater on cool mornings and afternoons than at warmer middays, when most leave water. Perch by hanging up from twigs or landing horizontally on leaves, sometimes ground vegetation. Copulation at least partly in flight, fairly lengthy. Females oviposit by flying rapidly in straight line, touching water every few feet, typically along shores overhung by vegetation. Oviposition continues late into afternoon in shade. Sometimes in mixed feeding swarms in open areas with other emeralds, especially baskettails in spring; flight from just above ground to head height.

Habitat Great variety of lakes and ponds, mostly but not always in forested country. Beaver ponds and bog lakes typical.

Flight Season YT Jun–Jul, BC Apr–Aug, WA May–Sep, OR May–Aug, CA May–Sep, MT May–Aug.

Distribution Also east in Canada to Newfoundland, south to Wisconsin and North Carolina. Restricted to mountains in southern part of range.

215.1
American Emerald male—
Okanogan Co., WA,
June 2005

215.2
American Emerald female—
Siskiyou Co., CA,
June 2005, Ray Bruun

Little Emeralds *Dorocordulia*

This small genus of northern North American species includes some of the smallest of our emeralds. With bright green eyes and dark, metallic body, they could be mistaken for the smallest striped emeralds (*Somatochlora*), and in fact, they are very closely related. They commonly perch on top of leaves. World 2, NA 2, West 1.

Description Small, dark emerald with conspicuous club. *Male:* Eyes brilliant green, face shiny black. Thorax metallic green or bronze, mixed with brown. Abdomen shiny black, yellow markings low on sides of S1–2 and small paired spots at base of S3. End of abdomen dramatically widened (like tennis racket). *Female:* Colored as male but pale spots on S3 larger, more obvious from above; "racket" not so wide.

Identification Smaller than most other green-eyed emeralds, with more prominently expanded abdomen. With metallic thorax and

216.1
Racket-tailed Emerald
male—Timiskaming
Co., ON, July 2005,
Greg Lasley

216.2
Racket-tailed Emerald
female—Penobscot
Co., ME, June 2006,
Netta Smith

black abdomen, most similar to **American Emerald**, which is larger, with more pale color on face, less prominent club, and no pale spots on S3. Smallest STRIPED EMERALDS, **Brush-tipped** and **Ocellated**, with yellow markings on sides of thorax and expanded part of abdomen farther forward and not as wide. No overlap in range with SUNDRAGONS.

Natural History Males fly back and forth somewhat irregularly over marsh vegetation or farther out over open water at knee to waist height, a bit higher than the larger American Emeralds often found in the same spots. Females oviposit by tapping surface in open water or in vegetation beds. When feeding, both sexes cruise around over short beats in open areas, usually below tops of tall shrubs, and land much more frequently than most other emeralds, even occasionally on ground. Because of this, easiest emerald to photograph. Also, as other emeralds, cruise around people to take attendant black flies!

Habitat Lakes and large ponds, commonly associated with bogs.

Flight Season WI Jun–Jul.

Distribution East through Canada to Nova Scotia, south to Indiana and New Jersey.

Sundragons *Helocordulia*

These small, dark, green-eyed emeralds of eastern forests barely make it into the West. They have bright yellow spots on a black abdomen and usually visible dark spots to mark cross-veins at the front of both wings. They fly up and down streams or cruise lightly over sunny clearings, then hang up on stems at oblique angles. Perhaps closely related to baskettails, they differ in color and wing venation as well as details of the genitalia. Dark dots on wings are evident at close range, especially in hand. World 2, NA 2, West 2.

217 Selys's Sundragon *Helocordulia selysii* TL 38–41, HW 26–28

Description Small dark southeastern emerald with small but obvious wing spots. *Male*: Eyes green, face yellow-orange. Thorax brown with metallic green highlights. Abdomen black, S3 orange at base and tiny orange basal spots along sides of S4–8. *Female*: Colored as male, spots on S4–8 larger. Both sexes with series of dark brown markings at crossveins along front edge of wings, more prominent toward base; larger spots at base.

217.1
Selys's Sundragon
male—San Jacinto
Co., TX, March 1999,
Robert A. Behrstock

217.2
Selys's Sundragon
female—Taylor Co.,
GA, March 2003,
Giff Beaton

Identification Only **Uhler's Sundragon** very similar, not presently known to overlap in West. **Uhler's** has orange marking mixed with black at hindwing base, **Selys's** not. Other small emeralds in range of sundragons include BASKETTAILS and SHADOWDRAGONS, both with brown or mostly brown abdomen. Female **Double-ringed Pennant** somewhat similar, with black, yellow-ringed abdomen, but that species has black and yellow thorax, reddish eyes, and different behavior.

Natural History Males fly up and down streams, not far above water. Pairs copulate in flight, then perched. Both sexes cruise for food in clearings and perch low in vegetation.

Habitat Woodland streams.

Flight Season TX Mar–Apr.

Distribution Also from Arkansas east to Delaware, south to northwestern Florida.

---

218 Uhler's Sundragon *Helocordulia uhleri* TL 41–46, HW 25–29

Description Small dark northeastern emerald with small but obvious wing spots, orange at wing bases. *Male*: Eyes green, face yellow-orange. Thorax brown with metallic green highlights, some obscure yellow markings in front of wings. Abdomen black, S3 orange at base and orange basal spots along sides of S4–8. *Female*: Colored as male, spots on S4–8 larger. Both sexes with series of dark  brown markings at crossveins along front edge of wings, more prominent toward base; larger spots at base with orange spots just behind them.

Identification See **Selys's Sundragon** for differences from that and other species. No other emeralds fly low and fast up and down streams during day. Looks small and dark. Green eyes visible if light is right, perhaps not as bright as in some other emeralds.

Natural History Males fly rapidly along streams below knee height, often just offshore, and may land on rocks, although most perching is by hanging in trees and shrubs. Sexual patrol flight often lasts until dark. Oviposition by tapping water among floating debris. Both sexes feed in flight in sunny clearings, including over paths through woodland.

Habitat Small rocky woodland streams to fair-sized open rivers, all with good current.

Flight Season MO May–Jun.

218.1
Uhler's Sundragon
male—Bartow Co.,
GA, April 2004,
Giff Beaton

218.2
Uhler's Sundragon
female—Sussex Co.,
NJ, May 2003,
Allen Barlow

Distribution Occurs in East from Ontario, Quebec, and Nova Scotia south to Arkansas, Alabama, and North Carolina.

Striped Emeralds *Somatochlora*

This represents the most diverse group of emeralds, and they are among the most sought by dragonfly enthusiasts, as many are limited to very specific habitats and thus are often relatively rare and local. Several eastern species barely make it into this region; others are common and widespread in the North. Appearing dark from a distance, they are surpassingly beautiful up close, almost all with brilliant green eyes (not included in species descriptions) and dark metallic bodies, some of them conspicuously marked with yellow. Yellow thoracic markings are brightest in immatures but may become obscured in older individuals. The largest are about twice the size of the smallest. They are typical of bogs and fens and forest streams, with a few species at lakes and none at the marshy ponds favored by skimmers.

Because the group is diverse, identification is a real challenge. Some species can be identified in the field, especially once the observer is familiar with local fauna and habitats in which each species can be expected. However, as many are uncommon and local, familiarity with common species may not provide the knowledge to identify another one entering the picture. If there is any doubt, they should be captured for identification. In hand, species are distinguished by thoracic markings and appendages, including cerci in males and subgenital plates in females, as well as size and abdomen shape. Female subgenital plates vary from short and skimmer-like to large and projecting, a "pseudo-ovipositor" for inserting eggs into substrates denser than water. The latter may be short or long, parallel or obliquely slanted, or dramatically perpendicular to the abdomen. Some field marks allow quick assignment to a small group, for example, brown spots at hindwing bases (Delicate, Whitehouse's, Muskeg) or narrow but obvious white rings around the abdomen (Ringed, Lake, Hudsonian). In some others (Clamp-tipped, Brush-tipped, Whitehouse's), male appendages are immediately distinctive. World 42, NA 26, West 21.

Table 9 Striped Emerald (*Somatochlora*) Identification

	A	B	C	D
Ringed	3	2	1	2
Hudsonian	3	2	1	1
Treeline	1	2	1	2
Quebec	3	12	1	3
Lake	1	2	1	1
Whitehouse's	31	1	1	53
Muskeg	31	1	1	2
Forcipate	4	1	3	3
Mountain	42	1	3	3
Delicate	2	1	3	4
Kennedy's	1	1	3	3
Williamson's	1	1	2	5
Ocellated	4	1	2	5
Brush-tipped	5	1	27	4
Plains	5	1	4	5
Mocha	1	1	4	5
Coppery	6	3	2	5
Fine-lined	6	1	6	4
Ozark	6	1	5	4
Texas	6	1	5	4
Clamp-tipped	5	1	5	5

A, sides of thorax: 1, no pale markings (may be brown and metallic green); 2, one spot; 3, one stripe; 4, two spots; 5, stripe followed by spot; 6, two stripes.

B, abdomen: 1, unmarked black; 2, faint to distinct white rings; 3, brown.

C, male cerci: 1, parallel, then sharply bent inward toward tip; 2, more or less parallel, approaching at tip or not; 3, forceps-shaped, curved inward toward tip; 4, forked or truncate at tip in side view; 5, angled downward toward tip; 6, diverging at tip; 7, brush of conspicuous long hairs at tip.

D, female subgenital plate: 1, not projecting, tip rounded or slightly indented; 2, not projecting, tip clearly notched; 3, visible in side view, not reaching end of S9; 4, visible in side view, extending to or past end of S9; 5, extending downward at or almost at right angle.

219 Ringed Emerald *Somatochlora albicincta* TL 45–52, HW 28–33

Description Medium-sized, white-ringed northern emerald. *Male*: Face black, yellow on sides. Thorax metallic green-black with small vertically elongate pale yellow spot on side of mesothorax. Abdomen black with faint brown markings at base, narrow but conspicuous white ring formed by joint at ends of S2–9. Cerci typical of ringed group, straight in top view and then sharply angled inward; in side view, thin upcurved tips. *Female*: Colored much like male, thorax brown mixed with green. Subgenital plate projecting but short, not half length of S9, and notched or bilobed. Individuals from near Pacific coast somewhat larger than those from interior. Populations in Queen Charlotte Islands considerably larger (male TL 58 mm) than those on mainland.

Identification Distinguished from **Lake Emerald** by smaller size, less than 2 inches in length (**Lake** more than 2 inches), as well as marked thorax. Populations in western British Columbia and Washington a bit larger, and those from Queen Charlotte Islands as large as **Lake** (but **Lake** not on Queen Charlottes). **Ringed** distinguished from **Quebec** and **Treeline Emeralds** by more prominent rings on abdomen, although Pacific coast **Ringed** (where **Quebec** and **Treeline** lacking) less obviously ringed. Very similar to **Hudsonian Emerald**, very slight color differences. Look for pale spots on each side of S10 at base of cerci, often in **Ringed** but not in **Hudsonian**. Males of
Hudsonian with prominent basal tooth on edge of cerci that makes appendages look wider at base (narrow throughout in **Ringed**, without prominent tooth). Subgenital plate of female **Ringed** shorter than in **Hudsonian** and strongly notched.

Natural History Males fly over open water and typically just off shorelines, usually quite low. Rapid flight over water, sometimes hover over vegetation. May be common in morning, then decrease when darners dominate flight lanes. Females oviposit in open water or near shore or floating objects; fly along slowly and tap once, then rise and move farther before dropping to tap again. May reverse direction constantly to remain in one small favorable area.

Habitat Lakes of all sizes down to rather small bog ponds, in open or forest, also slowly flowing wide streams. Typically no aquatic vegetation except shore emergents.

Flight Season YT Jun–Aug, BC Jun–Aug, WA Jul–Oct, OR Jun–Sep, CA Jun–Aug, MT Jul–Sep.

Distribution Also in eastern Canada, south to New York and Maine.

Comments Population of large individuals in Queen Charlottes named as subspecies (*S. a. massettensis*) but validity questionable. Hybridizes with Hudsonian and Treeline Emeralds in northern Yukon.

219.1
Ringed Emerald
male—Kittitas Co.,
WA, September 2007

219.2
Ringed Emerald female—
Baker Co., OR, August 2004
(posed)

Description White-ringed northern emerald, very similar to some others. *Male*: Face black, yellow on sides. Thorax metallic green with barely evident dull yellowish elongate spot on each side. From above, cerci well separated at base, straight and then sharply angled inward and then curved gently outward to almost meet at tip; extreme tip pointed and curled upward. From side with a tooth on both exterior and interior surface, exterior one near base. Abdomen black with whitish rings at rear of S2–9. *Female*: Colored as male. Subgenital plate projecting, rounded, less than height of S9.

Identification Member of small group with prominently ringed abdomen, identical to **Ringed** in field and distinguished by appendages in hand. **Hudsonian** male cerci with prominent exterior tooth near base lacking in **Ringed**. Female subgenital plate distinctly longer than that of **Ringed** and not notched. Only other conspicuously ringed species is **Lake**, which is much larger. Rings of these species often not visible in flight, when **Hudsonian** looks like many others. Capture always advised for striped emeralds until local fauna well understood, and even then, additional species are always possible in this diverse group.

Natural History Males fly low over water along shore much like **Ringed Emerald**. Females oviposit by tapping open water.

Habitat Lakes, ponds, and slow streams, especially large muskeg pools with abundant sedge growth. Often with Ringed Emerald.

Flight Season YT Jun–Aug, BC May–Aug, MT Jul–Aug.

Distribution In Canada east to northeastern Ontario.

Comments Hybridizes with Ringed and Treeline Emeralds in northern Yukon.

220
Hudsonian Emerald
male—Deadman
Lake, AK, June 2003,
Robert Armstrong
(posed)

Description Stocky far-northwestern emerald of ringed group but all dark. *Male*: Face black, yellow on sides. Thorax entirely metallic green. Abdomen shiny black with faint yellow apical ring on S2. Narrow tips of cerci strongly angled inward in top view, appear to cross and then swept up in flattened curve. *Female*: Colored as male. Subgenital plate less than half length of S9, not projecting; notched at end.

Identification In far northern range, distinguished from **Hudsonian** and **Ringed Emeralds** by lack of white abdominal rings. However, hybrids in this group problematic to identify. Distinguished from **Delicate** and **Kennedy's Emeralds** by much more robust abdomen, shorter than wing (longer than wing in other two). Most like **Whitehouse's Emerald** but slightly larger, slightly thicker abdomen (less pronounced waist), lacks brown in hindwings.

Natural History Males fly patrol flights over pools and along margins of larger ones, staying a bit offshore. Tends to fly over open water rather than at edge of vegetation. Both sexes perch in shrubbery away from water. Copulating pairs quickly fly into nearby woods.

221.1
Treeline Emerald
male—Russia, July
2003, Oleg Kosterin

221.2
Treeline Emerald
female—Utsjoki,
Finland, August 1998,
Sami Karjalainen
(posed)

Females oviposit in open water underlaid with aquatic mosses, also away from shoreline vegetation; usually in afternoon.

Habitat Small pools, open ponds, and small lakes surrounded by cold fens and bogs, mostly in shrub tundra habitats within 60 miles of treeline. Aquatic mosses almost invariably present and water bodies deep and cold.

Flight Season YT Jun–Aug.

Distribution Also occurs across far northern Eurasia, from Finland to Siberia.

Comments Hybridizes with Hudsonian and Ringed Emeralds in northern Yukon.

---

222 Quebec Emerald   *Somatochlora brevicincta*   TL 43–50, HW 27–31

Description Dark northern emerald of ringed group but with no conspicuous field marks. *Male*: Face black, yellow on sides. Thorax shiny green-black with vertically elongate faint yellow spot on side of mesothorax. Abdomen shiny black with faint indication of whitish rings between terminal segments. Cerci with tiny lateral tooth at base, tips angled inward and then curved upward in long, slender point. *Female*: Colored much like male, thorax brown and green. Subgenital plate not projecting but almost as long as S9. Distinctly smaller in British Columbia than in East.

Identification Not easily distinguished from similar species except in hand. No brown at wing base as is present in **Delicate**, **Muskeg**, and **Whitehouse's Emeralds** (first more slender). Pale abdominal rings recall **Hudsonian** and **Ringed Emeralds** but much less distinct, even when present interrupted in center. In group of species with cerci slightly angled in top view and curled up at tip in side view, most similar to those of similar-sized **Hudsonian** and **Ringed** (**Lake** is much larger), but show only one projecting tooth near base in side view (others show two, much more prominent in **Hudsonian**). Epiproct of **Quebec** long, over half length of cerci, as in **Hudsonian** (in **Ringed** less than half). Female subgenital

222.1
Quebec Emerald male—
McBride, BC, August 2000,
Sid Dunkle (posed)

222.2
Quebec Emerald female—
McBride, BC, August 2000,
Sid Dunkle (posed)

plate as long as S9 and rounded in **Quebec,** half length of S9 and notched in **Ringed,** projecting ventrally in **Hudsonian.**

Natural History Males fly slowly at knee height over fens. Females oviposit among water-logged vegetation in small pools. Has been found in feeding swarms of several striped emerald species in East.

Habitat Shallow patterned fens with sedges and mosses, those also inhabited by Muskeg and Whitehouse's Emeralds.

Flight Season BC Jun–Jul.

Distribution Also from northern Minnesota and Ontario east to Newfoundland and Maine. Very likely occurs between British Columbia and Ontario.

Description Large, white-ringed northern emerald of large lakes. *Male*: Face black, yellow on sides. Thorax metallic green and brown, otherwise unmarked. Abdomen black, with whitish rings at ends of all segments 2–9. Cerci with prominent lateral tooth at base, tips sharply angled inward and then curved upward in slender point. Epiproct broad at tip, unusual for this genus. *Female*: Colored as male. Subgenital plate short, not projecting.

Identification Very large size for an emerald distinctive, combined with rather conspicuous white rings (more so than in smaller species). Both **Hudsonian** and **Ringed Emeralds** somewhat similar, and they also fly over open water, but **Lake** considerably larger, by 10–15 percent. Green eyes and black body distinguish it from MOSAIC DARNERS, the other large dragonflies that fly over northern lakes, but in silhouette much like a darner.

Natural History Males fly low and swiftly over lakes with no hovering, usually 6–15 feet offshore but sometimes approaching shore, easy to observe but difficult to capture. Males curve abdomen down at end, reminiscent of cruiser in flight rather than one of the smaller, slower-flying striped emeralds. Pairs form at water and immediately fly off into woods, but also seen far from water, perhaps also pairing away from it. Females oviposit by flying rapidly back and forth or in broad circles, tapping water every few feet.

Habitat Lakes, with or without much floating and/or emergent vegetation; also large, slow-flowing rivers.

Flight Season BC Jul–Aug.

Distribution Ranges throughout eastern Canada and south to Wisconsin and New York.

223
Lake Emerald male—Solco
Lake, BC, August 2000 (posed);
female—Cook Co., MN, July
2005, June Tveekrem

Description Small northern emerald with brown patch at base of hindwing. *Male*: Face black with yellow on sides. Thorax metallic green and brown; narrow, elongate dull yellow-orange spots below each wing that become obscure with age. Abdomen black with orange-brown dorsolateral spots on S2, white ring at base of S3. Dark brown triangle at base of each hindwing. Cerci in top view narrowing from base, then suddenly expanded and finally angled together; in side view curled up at tip. *Female*: Colored as male.

Subgenital plate projecting at right angle and scoop-shaped but very short, half length of S9.

Identification Only two other striped emeralds have moderately visible brown spot at hindwing base, **Delicate** and **Muskeg**. **Delicate** with longer and much more slender abdomen than **Whitehouse's**. **Muskeg** looks identical, must be distinguished in hand by differences in male appendages and female subgenital plate (see that species). **Whitehouse's** usually more common than **Muskeg** at preferred habitats. If brown spot can not be seen, **Whitehouse's** could be mistaken for rarer **Quebec Emerald**, which has hint of white rings

224.1 Whitehouse's Emerald male— Juneau, AK, July 2004, Robert Armstrong (posed); female— Pend Oreille Co., WA, August 2004 (posed)

224.2 Whitehouse's Emerald immature male—Heckman Pass, BC, July 2006, Netta Smith

on abdomen and different appendages. Other striped emeralds have longer abdomen and/or brighter yellow spots on thorax.

Natural History Males fly back and forth at knee height and below in relatively short beats over water with open stands of low vegetation, then wander across ridge to next depression or into woods to perch before returning to water. They drop to water surface from time to time, presumably looking for females. Females oviposit by flying low and slowly through vegetation in shallow ponds, tapping water or wet moss once and moving short distance before doing it again; usually near shore or emergent plants where slow flight and leisurely tapping renders them quite inconspicuous.

Habitat Classical muskeg, small to moderate-sized open ponds with abundant sedges, buckbean, and algae, clean water but soft mud bottom and quaking substrate.

Flight Season YT Jul, BC Jun–Jul, WA Jul–Aug.

Distribution Also in northern Ontario, Quebec, and Labrador.

225 Muskeg Emerald *Somatochlora septentrionalis* TL 39–48, HW 26–30

Description Far-northern emerald with brown in wing bases. *Male*: Face blackish with yellow on sides. Thorax metallic green and brown, faintly indicated yellow spot under forewing. Abdomen glossy black with brown markings on S1–2, white ring and brown spots at base of S3. Dark brown triangle at base of hindwing. Cerci sharply angled inward in top view, strongly swept up at tip, with basal tooth visible in side view. *Female*: Colored as male, pale spots on S3 slightly larger. Subgenital plate short and not projecting, with two lobes.

Identification In size, abdomen shape, and brown at base of hindwing, both sexes look exactly like comparable sex of **Whitehouse's Emerald**. Check appendages to distinguish males (in top view, **Whitehouse's** cerci converge and cross, **Muskeg** cerci remain parallel than turn sharply inward), side view of subgenital plate (projecting downward in **Whitehouse's**, flat against abdomen and bilobed at tip in **Muskeg**) to distinguish females. Differs from all other species in same ways as **Whitehouse's**. Male appendages somewhat like

225
Muskeg Emerald
male—Heckman
Pass, BC, July 2006

those of ringed group, and could be confused with **Quebec** or **Treeline Emeralds** (not known to occur with latter), which also lack rings, but prominent downward-directed spine under base of cerci should distinguish **Muskeg**.

Natural History Males fly low over water in continued patrol flights, only hovering occasionally. Apparently roost near breeding sites, in shrubs and at edge of tiny pools with overhanging sedges that effectively hide them. Copulating pairs fly into woods, and one observer reported female returning in 15 min to lay eggs. Females oviposit in slow flight by tapping steadily both in open water and on mat of decayed vegetation floating on it.

Habitat Open fens with shallow pools of open water, typical muskeg in boreal forest habitats.

Flight Season YT Jun–Jul, BC Jun–Jul.

Distribution Also eastern Canada east to Newfoundland.

226 Forcipate Emerald *Somatochlora forcipata* TL 43–51, HW 29–33

Description Slender northern emerald with spotted thorax and faintly dotted abdomen. *Male*: Face black, some yellow on sides. Thorax mostly metallic green except brown just behind prothorax and with two oval pale yellow spots on each side. Abdomen black with light brown markings on sides of S1–2, whitish ring at end of S2, and obscure orange lateral dots at front end of S5–7. Cerci with long, slightly incurved and sharply pointed tips. *Female*: Colored as male. Subgenital plate broad, projecting downward obliquely, about as long as S9.

Identification Most like **Mountain Emerald** of numerous species with which it occurs because of forcipate (forceps-like) cerci and two more or less round yellow spots on each side of thorax. Distinguished from **Mountain** by shape of cerci, not almost meeting in top view (just about meeting in **Mountain**) and more arched in side view (more or less straight in **Mountain**). **Mountain** (except rarely in females) also lacks tiny pale spots on midabdomen of **Forcipate**. Only other species with two well-defined round spots on thorax is **Ocellated**, much smaller, with abdomen clubbed in middle and quite different appendages. At a distance could be easily mistaken for other species of similar size but without yellow spots, for example **Delicate** and **Kennedy's**, or the slightly smaller and stockier **Muskeg** and **Whitehouse's**.

226
Forcipate Emerald male—Scotch Ridge, NB, June 1996, Blair Nikula (posed); female—Worcester Co., MA, June/July 2003, Michael Veit

Natural History Males fly up and down above tiny streams, from just above surface to waist height. Females oviposit in open water, even tiny pockets, by tapping water leisurely; also in mats of moss and stonewort. Foraging flight in clearings or over roads, usually at not much more than head height, and ascending up to 20 feet in trees to perch.

Habitat Small spring-fed boggy streams, in or out of woodland.

Flight Season BC Jul.

Distribution Also east to Newfoundland, south to Wisconsin and West Virginia. In mountains in southern part of range.

<table>
<tr><td>227 Mountain Emerald   Somatochlora semicircularis</td><td>TL 47–52, HW 27–32</td></tr>
</table>

Description Common emerald of western mountains with faint spots on thorax. *Male*: Face mostly blackish, yellow on sides, and stripe at frontoclypeal suture. Thorax metallic green with pair of yellow spots on each side that gradually become obscured with age. Abdomen black with faint yellowish markings on S1–2. Cerci straight in side view, narrowing toward tip; from above like pair of curved forceps. *Female*: Colored as male but yellowish paired dorsolateral spots on S3; may show tiny basal spots on S6–7. Subgenital plate scoop-shaped but short, not reaching end of S9 and projecting obliquely.

Identification This species occurs with many others in its extensive range. Differs from **Hudsonian**, **Lake**, and **Ringed** by lacking white rings, also much smaller than **Lake** and typically in different habitat (flies over sedge meadows, not along lake shores). Probably can be distinguished from **Brush-tipped**, **Muskeg**, **Ocellated**, **Quebec**, and **Whitehouse's** by longer and more slender abdomen. Long, slender abdomen recalls **Delicate**, **Forcipate**, and **Kennedy's**, and these species present biggest identification problem. Dull yellow spots on side distinguish **Mountain** from **Kennedy's** (no spots) and perhaps **Forcipate** (brighter spots). **Delicate** has even more slender abdomen and basal brown hindwing spots. Capture will often be necessary to be certain.

227.1
Mountain Emerald male—
Lassen Co., CA, August 2006,
Ray Bruun

227.2
Mountain Emerald female—
Skamania Co., WA, July 2005
(posed)

Natural History Probably the most likely striped emerald to be seen over much of the West. Forages in clearings from near ground to up in trees, alternating flight and hanging up on twigs. Males fly incessantly a foot or two above tall sedges and grasses of wet meadows, back and forth over relatively small territory (up to 30 feet in length) with regular hovering, or ranging farther and changing cruising beat to another location. Also fly over open water of small ponds, streams, and ditches but always near vegetation. Pairs in copulation fly back and forth over meadows for periods up to several minutes, then land in herbaceous vegetation or low shrubs but just as often suddenly ascend into surrounding forest trees to hang up, where copulation continues for as long as 25 min. Females oviposit by flying low and slowly over water, stopping to tap in one spot, often changing direction, then moving short distance to another spot; very inconspicuous. "Dragonfly graveyards" of this species have been encountered, with numerous dead individuals floating on water of small pond, possibly at end of life and getting stuck on water surface.

Habitat Wet meadows of tall sedges and grasses, often with small ponds or streams associated; shuns lakes except with extensive tall-sedge margins. Many breeding habitats dry up in late summer, and larvae can withstand lengthy dry periods.

Flight Season YT Jun–Jul, BC May–Aug, WA Jun–Oct, OR Jun–Sep, CA Jun–Aug, MT Jun–Sep.

Distribution Restricted to mountains in southern part of range.

---

228 Delicate Emerald *Somatochlora franklini* TL 44–54, HW 25–30

Description Very slender far-northern emerald with brown at wing bases. *Male*: Face black, yellow on sides. Thorax metallic green mixed with some brown, when young with dull yellowish spot evident under forewing base that disappears with maturity. Abdomen black with dull brown side spots on S1–2, dull yellow ring at end of S2. Dark brown triangle at each hindwing base. Cerci from above like forceps, from side a simple slightly droopy and pointed finger. *Female*: Colored as male but thorax brown mixed with

228.1
Delicate Emerald male—Parsnip Valley, BC, August 2000, Sid Dunkle (posed)

228.2
Delicate Emerald female—Somerset Co., ME, July 2003, Blair Nikula

green, abdomen with dull orange dorsolateral spots on S3. Eyes stay reddish for much of life, finally turn green. May be indistinct brown stripes extending out front of all wings. Subgenital plate length of S9, not projecting.

Identification Very slender abdomen good clue, even more slender than others similar enough for confusion, including **Forcipate**, **Kennedy's**, **Mountain**, and **Quebec**. Comparison of appendages in hand may be necessary. One of few species with obvious brown at base of hindwing, not that easy to see in field. **Muskeg** and **Whitehouse's**, also with brown in wing and often in same habitat, have shorter abdomen and very different appendages in both sexes.

Natural History Males fly back and forth and hover frequently at waist height over beds of tall grasses and sedges and among shrubs, very much like Mountain Emerald. Females oviposit by tapping onto wet moss or in tiny pools.

Habitat Sedge- and moss-filled fens, at foot of hillsides below seepage or in wide-open meadows; usually not near open water.

Flight Season YT Jun–Aug, BC Jun–Aug, WA Jul–Aug.

Distribution Widespread in eastern Canada, south to northern tier of states.

229 Kennedy's Emerald *Somatochlora kennedyi* TL 47–55, HW 29–34

Description Slender northern emerald with no distinctive markings. *Male*: Face brown with paler sides. Thorax metallic green and brown, in younger individuals with dull yellowish spot below forewing. Abdomen black with paler brownish areas low on sides of posterior segments. Cerci somewhat forcipate, long and fingerlike, narrowly pointed at tip. Epiproct relatively short, less than half length of cerci. *Female*: Colored as male but dull yellowish dorsolateral spots on S2–3. Subgenital plate projecting slightly, scooplike, as long as S9.

229.1
Kennedy's Emerald male—
Sussex Co., NJ, May 2004,
Allen Barlow (posed)

229.2
Kennedy's Emerald female—
Sussex Co., NJ, June 2005, Allen
Barlow (posed)

Identification If well seen, **Kennedy's** should only be mistaken for other slender-bodied species without strong thoracic markings: **Delicate**, **Mountain**, and **Williamson's**. Differs from **Delicate** in slightly larger size and in having abdomen slightly more expanded and lacking brown at hindwing base; from **Mountain** in having abdomen widest past middle (about at middle in **Mountain**, in which narrow "wasp waist" not so long) and in lacking even dull pair of thoracic spots of that species; and from **Williamson's** in smaller size. Male cerci slender and almost touching at tips in **Kennedy's**, well separated at tips in **Delicate** and **Mountain**, and upcurved and tipped with hairs in **Williamson's**. Note also probability of confusion at a distance with other slender species of northern latitudes such as **Forcipate Emerald** with brightly spotted instead of plain thorax.

Natural History Males patrol over water at knee to waist height, hovering regularly and changing orientation while doing so. Fly into nearby shrubbery and hang up in shade for short period, then resume patrolling. Females oviposit by flying leisurely among clustered plant stems or over floating moss, tapping surface frequently in one spot of open water for few seconds, then moving to another nearby to repeat.

Habitat Open bogs or fens with sedge and mosses, often at small ponds.

Flight Season YT Jul, BC Jun–Jul.

Distribution Ranges east to Newfoundland, south to northernmost states.

---

230 Williamson's Emerald *Somatochlora williamsoni* TL 53–59, HW 35–40

Description Large, slender northeastern emerald without pale thoracic markings and with abdomen widest at midlength. *Male*: Face blackish with dull yellow-orange sides. Thorax metallic green and brown. Abdomen black, with faint pale ring at base of S3. Cerci straight in side view, curling up at tips, and quite hairy above; exterior basal tooth visible from side or above. *Female*: Colored as male but orange-brown anterior ventrolateral spots on S4–8. Subgenital plate projecting at right angle, longer than height of S9 and very sharply pointed. Cerci longer than S9–10.

Identification Rather similar to other slender striped emeralds, distinguished from **Delicate**, **Forcipate**, and **Kennedy's Emeralds** by larger size. **Kennedy's** also distinguished by

230
Williamson's Emerald male, female—Sussex Co., NJ, July 2004, Allen Barlow (posed)

usually evident pale lateral thoracic stripe, straight rather than upcurled tips of male cerci, and short, rounded subgenital plate. **Forcipate** distinguished by two pale spots on each side of thorax, **Delicate** by brown spots at hindwing bases. Much larger than **Brush-tipped Emerald**, only other species with visibly hairy cerci, and tip of cercus long and pointed in side view rather than blunt. Long, pointed, perpendicular subgenital plate of female different from all other emeralds in range in West, but see **Plains Emerald**.

Natural History Males patrol streams and lake shores, alternating rapid flight and brief hovering, typically at knee to waist height. Foraging flight high in air along woodland borders. Females usually oviposit on wet muddy banks above water, alternating series of thrusts into mud with tap in water, presumably to clear ovipositor.

Habitat Slow forest streams and clear sand- or rock-margined lakes with wave-washed shores.

Flight Season WI Jun–Aug.

Distribution Also across southern Canada to Nova Scotia, south to Minnesota, Michigan, and New Jersey.

Description Very small northern emerald with small but conspicuous thoracic spots. *Male*: Face metallic green, dull yellow-brown at sides. Thorax brown and metallic green with two pale yellow oval spots on each side. Abdomen black with conspicuous yellow spots on side of S2 and base of S3. Cerci slender, straight, but slightly angled inward in top view with two minute teeth along outside; meet at end where pointed and slightly upswept. *Female*: Colored as male. Subgenital plate projecting at right angle, pointed, not as long as height of S9.

Identification Only one other striped emerald is this small, **Brush-tipped Emerald**, and both sexes have quite different terminal

231.1
Ocellated Emerald male—
Coos Co., NH, June 2006,
Blair Nikula

231.2
Ocellated Emerald female—
Riding Mountain, MB, July 2005,
Jim Bangma (posed)

appendages. Color patterns similar, but anterior spot on side of thorax of **Brush-tipped** elongate, whereas both spots oval and same size in **Ocellated**; also, **Brush-tipped** has tiny pale spots on sides at midabdomen. From above, **Ocellated** abdomen looks spindle-shaped (widest at middle), **Brush-tipped** closer to club-shaped (widest slightly behind middle) in both sexes. Also like **American Emerald**, latter in different habitat and with no yellow on thorax. In Manitoba, distinguish from still smaller **Racket-tailed Emerald**, with its wide club nearer end of abdomen.

Natural History Males fly back and forth along small stream or hover above it for lengthy periods, usually close to water, as close as a few inches above it; patrol length averages around 30 feet. Often perch on grasses, sedges, or pale rocks at stream after lengthy flying bout, also on tree trunks nearby. Foraging flight relatively low, usually below head height in woodland clearings, sometimes a bit higher with high-flying larger emeralds. Some females oviposit by alternating taps on water surface near shore and wet mossy bank above, then short move to new location to repeat process. Others just tap surface of shallow flowing water.

Habitat Small to medium, slow to moderately fast clear streams with pools, in both woodland and open. Good habitat is where stream drops out of forest onto meadow.

Flight Season YT Jun–Aug, BC May–Aug, WA Jul–Aug, OR Jun–Jul, MT Jul–Sep.

Distribution Ranges across Canada to Newfoundland, south to Wisconsin, Ohio, and New Jersey.

232 Brush-tipped Emerald   *Somatochlora walshii*   TL 41–52, HW 25–34

Description Small northern emerald with prominently clubbed abdomen and distinctive brushy appendages. *Male*: Face black with yellowish-brown on sides. Thorax metallic green and brown with prominent yellow spots on sides, front one longer and narrower than rear one. Abdomen black with dull yellowish ventrolateral spots and pale yellow apical ring on S2. Pale basal spots or half-rings on sides of S5–7. Cerci curved slightly downward, broader toward tip, then abruptly swept up; "brush" of hairs at tip obscures

232.1
Brush-tipped Emerald
male—Skamania Co., WA,
August 2005 (posed)

232.2
Brush-tipped Emerald
female—Sussex Co., NJ, June
2004, Allen Barlow (posed)

shape. *Female*: Colored as male but orange dorsolateral spots at base of S3, larger pale spots or half-rings at bases of S5–7. Subgenital plate projecting obliquely, scoop-shaped with rounded tip, about as long as height of S9.

Identification Small size and pronounced wasp-waisted look distinguish it from most other striped emeralds in its range. **Ocellated Emerald** equally small but has two rounded spots on each side of thorax rather than two unequal-sized ones and lacks any markings on midabdomen. Appendages of both sexes quite different; note especially short epiproct of **Brush-tipped**, leaving gap in appendages not notable in **Ocellated**. Breeding habitats separate, but both may be feeding anywhere away from water. Otherwise could be mistaken for **American Emerald**, with no yellow on thorax, shorter appendages, and different habitat (lakes and ponds). **Racket-tailed Emerald** of Manitoba still smaller and with wider abdominal club.

Natural History Males fly slowly back and forth or hover over narrow stream or marsh, usually at knee to waist height. May also cruise fairly rapidly over lengthy beat of 30 to 65 feet. Both sexes fly rapidly around at all heights over forest clearings while foraging. Females usually oviposit by slow flight over water surface, rhythmically tapping water while maneuvering around plant clumps.

Habitat Typically small, slow streams flowing through sedge fens and meadows, sometimes at woodland edge.

Flight Season BC Jun–Aug, WA Jul–Sep, OR Jul–Sep, MT Jun–Aug.

Distribution Also east to Newfoundland, south to Wisconsin and Pennsylvania.

233 Plains Emerald *Somatochlora ensigera* TL 48–51, HW 33–35

Description Mid-sized emerald of prairie region with dramatic yellow spots on sides. *Male* Face yellow, black on top of frons. Thorax dark brown with green highlights; pale yellow stripe on mesothorax, smaller but brighter yellow spot on metathorax. Abdomen black, with large yellow spot on side of S2 and yellow markings on sides of S3. Cerci unique in side view, relatively short and curved downward with short tooth projecting upward near end and epiproct only slightly shorter than cerci. *Female* Colored as male,

233.1
Plains Emerald male—
Cherry Co., NE, July 1998,
Sid Dunkle (posed)

233.2
Plains Emerald female—
Rainy River district, ON,
July 2005, Ilka Milne

face duller. Subgenital plate projecting at right angle, longer than height of S9 and sharply pointed. Cerci relatively short, shorter than S9–10.

Identification No other striped emeralds in southern part of range on plains (Dakotas, western Nebraska, eastern Wyoming, and Colorado) except **Ocellated** in Black Hills; latter similarly patterned but considerably smaller and with quite different appendages on both sexes. Farther east, **Plains Emerald** barely overlaps with a few other species with prominent yellow stripes or spots on sides of thorax. **Clamp-tipped** male has quite different appendages, although females are more similar, but in **Plains** subgenital plate longer than cerci, in **Clamp-tipped** shorter than cerci. May overlap with **Williamson's Emerald** in Manitoba; latter has almost identical subgenital plate but shorter than cerci, also unmarked thorax. **Brush-tipped** considerably smaller, appendages of both sexes quite different. Mostly yellow face diagnostic of **Plains** when it can be seen.

Natural History Males fly beats up and down streams, usually low and with intermittent hovering; quite territorial. Females oviposit by hovering and tapping clay bank or gravel bed above water level or in water itself; may begin in early morning. Both sexes and pairs perch on streamside vegetation.

Habitat Streams and small rivers with pools and riffles, wooded in East but mostly in open areas in plains, where typically lined with band of riparian shrubs.

Flight Season NE Jul–Sep.

Distribution Also east to southern Ontario and western Ohio.

---

234 Mocha Emerald *Somatochlora linearis* TL 58–68, HW 39–47

Description Long, slender southeastern emerald with unmarked thorax. *Male*: Face dark brown with paler sides. Thorax dark brown overlaid with metallic green. Abdomen black with dull yellowish spot on sides of S1–2, slightly indicated pale basal dots on sides of S3–7. Cerci with long, pointed tooth projecting downward near tip in side view; in top view, widened and clublike in terminal half.

Wings become suffused with brown in older individuals. *Female*: Colored as male, but thorax with less obvious metallic green. Abdominal spots often larger. Older individuals de-

234
Mocha Emerald
male—Murray Co.,
GA, September 2005,
Giff Beaton (posed);
female—Floyd Co.,
GA, August 2005,
Giff Beaton (posed)

velop brownish wings. Subgenital plate projecting at right angle, narrowly pointed, about as long as height of S9.

Identification Overlaps with relatively few other striped emeralds in southeastern part of region. Differs from all others in lacking thoracic markings, most distinct from **Ozark** and **Texas** with their brightly marked thorax. More similar to **Fine-lined** and **Clamp-tipped** with their somewhat more obscure markings. As usual, very different appendages will differentiate them for the most part. However, subgenital plates of female **Mocha** and **Clamp-tipped** quite similar, so look at thoracic markings.

Natural History Males patrol a few feet above stream, hovering frequently in sunny spots and cruising over long beats of 60–100 feet. Most active in morning and evening, more likely to be hanging from tree branches in forest during midday. May be found hanging vertically or obliquely from twigs over stream, even quite low. Females oviposit in muddy areas by thrusting eggs into mud or fine gravel at shore or above it, even quite early in morning. Both sexes cruise over open areas to forage, all the way up to treetops.

Habitat Small streams a few yards in width in forest, with rocky riffles or not.

Flight Season TX May–Sep.

Distribution Also from Iowa, Michigan, and Massachusetts south to northern Florida.

---

235 Coppery Emerald *Somatochlora georgiana* TL 44–49, HW 29–33

Description Rare small eastern emerald with red eyes and reddish body. *Male*: Eyes dull red over gray; face golden-brown. Thorax medium brown with two dull creamy-whitish stripes on either side. Abdomen medium brown with cream markings on sides of S2, dark stripe on top of S8–9 when mature. Cerci long, slender, and straight, curled up at flattened tip. *Female*: Colored as male.  Subgenital plate with rear edge at right angle to abdomen, sharply pointed and about as long as height of S9.

Identification No other striped emerald, or emerald for that matter, has eyes red above and a brown body with two pale thoracic stripes. BASKETTAILS are duller brown and shorter-bodied with different appendages in both sexes and no trace of pale stripes on thorax. **Orange Shadowdragon** somewhat similar in color but smaller, also unstriped. Perhaps more likely mistaken for some sort of elongate tropical skimmer, for example,

235
Coppery Emerald
male—Middlesex Co., MA,
August 1997, Blair Nikula
(posed); female—
Rockingham Co., NH,
August 2005,
Stephen Mirick

female **Four-spotted Pennant**, but of course **Coppery Emerald** is a flier that hangs up rather than a percher. Emerald-type appendages in both sexes, with long terminal appendages in male and well-developed subgenital plate in female makes it an emerald, as does abbreviated anal loop in hindwing.

Natural History Males patrol streams at midday but more intensely later in evening. Sometimes in mixed feeding groups over clearings with other striped emeralds. Very seldom encountered in most of range.

Habitat Small, sandy forest streams.

Flight Season East May–Aug.

Distribution Also from southern Mississippi east and north in Coastal Plain to Massachusetts.

236 Fine-lined Emerald *Somatochlora filosa* TL 55–66, HW 36–45

Description Long, slender southeastern emerald with fine white stripes on sides of thorax. *Male*: Face dull brown, metallic green above. Thorax mixed brown and metallic green with two well-defined but narrow white stripes on each side. Abdomen black, white rings on S1–2 continuing thoracic patterning. Cerci long and slender with slightly outcurved and pointed tip. *Female*: Colored as

male but thorax and base of abdomen with more brown. Wings orange-tipped in immature, entirely tinged brown with maturity. Subgenital plate projecting obliquely and upcurved at end, longer than S9 and rather like sled runner.

Identification No other striped emerald in its range in east Texas has fine white vertical stripes on thorax and abdomen base. Both sexes of **Texas Emerald** somewhat like **Fine-lined**, with rather similar abdominal appendages, but **Texas** has wider, bright yellow thoracic stripes and more complicated whitish markings on abdomen base as well as being slightly smaller. **Clamp-tipped Emerald** also similarly marked but with more obscure yel-

236
Fine-lined Emerald
male—Brantley Co.,
GA, October 2005,
Giff Beaton (posed);
female—Long Co.,
GA, November 2005,
Giff Beaton (posed)

lowish thoracic markings and very different appendages in both sexes. **Mocha Emerald** also with very different appendages as well as unmarked thorax. Orange-tipped wings might be clue to overhead females.

Natural History Usually encountered in feeding flight in open pine woods. Males patrol at waist height at edges of swamps and over small, slow streams. One female seen pushing eggs into sandy bank overlaid by silt.

Habitat Poorly known, probably sandy forest streams.

Flight Season TX Aug.

Distribution Also from Missouri and Louisiana east to New Jersey and Florida.

Description Brightly marked emerald of Ozark region. *Male*: Face yellow, top of frons metallic green. Thorax metallic green and brown with two bright yellow stripes on each side, front one narrower. Abdomen black with bright yellow markings on sides of S1–2, small dull yellow dorsolateral spots at base of S3. Cerci straight to half length, then abruptly bent downward at 45° angle; tip widened in side view. *Female*: Colored as male. Subgenital plate very long, reaches end of abdomen; scoop-shaped, pointed.

237.1
Ozark Emerald male—
McCurtain Co., OK, July 2007,
David Arbour (posed)

237.2
Ozark Emerald female—
Van Buren Co., AR, May 2004,
Herschel Raney (posed)

Identification In limited range in eastern Kansas and Oklahoma, could be confused only with **Clamp-tipped Emerald**, a bit larger with more obscure thoracic markings. Abdominal appendages very different in both sexes.

Natural History Has mostly been encountered feeding in open areas; very poorly known and apparently rare.

Habitat Forest streams in hills.

Flight Season MO Jun–Aug.

238 Texas Emerald *Somatochlora margarita* TL 51–59, HW 32–37

Description Slender southeastern emerald with vividly striped thorax and abdomen base. *Male*: Face black, faint yellow on sides. Thorax black with green iridescence, vivid yellow stripes running from top to bottom of meso- and metathorax. Abdomen black, S2 with contrasty pale yellow markings on lower sides and ring at end. Pale yellow markings at base of S3. Cerci long, slender, and fingerlike, parallel in top view and arched downward in side view. *Female*: Colored as male. Subgenital plate very long, extending to end of abdomen, rounded at end, and narrowly scoop-shaped.

Identification In limited range in east Texas, could be confused with no other species because of vivid thoracic markings, visible even on flybys. Green eyes and bright thoracic stripes might invite confusion with RIVER CRUISERS, but considerably smaller and with unmarked abdomen.

Natural History Seemingly rare and poorly known, mostly encountered in high feeding flight over roads in wooded country.

Habitat Breeding habitat unknown, presumably limited forest wetlands. Feeds in open pine forests.

Flight Season TX May–Aug.

238
Texas Emerald male—TX, 1976,
Curtis Williams (posed)

Description Slender eastern emerald with distinctive appendages. *Male*: Face black above, light brown below. Thorax mixed brown and metallic green, two elongate yellow spots on sides that become obscure with age. Abdomen black with yellow ventrolateral spot and pale yellow apical ring on S2. Cerci from side sharply bent downward in middle, overlap strongly curved epiproct like clamp; in top view tips angled inward, then parallel. *Female*: Colored as male but orange dorsolateral spot on S3. Subgenital plate sharply pointed, projecting almost at right angle, longer than height of S9.

Identification Good side view of clamplike appendages of male definitive. Same for long, perpendicular, pointed subgenital plate of female, although much like that of **Mocha Emerald**, which has unmarked thorax. **Ozark** and **Texas Emeralds** similar in shape but slightly smaller and more vividly marked, quite different appendages.

Natural History Males fly up and down streams at knee height or above, at both pools and riffles, with frequent hovering, then may perch at knee height. Female oviposits by tapping water or pushing eggs into wet moss, decaying leaves, or mud. Foraging flight throughout day at woodland edge, rapid and over wide range of altitudes, even persisting until after sunset.

Habitat Small forest streams with rapids and pools, almost always in shade. Also sedgy fens in parts of range.

Flight Season TX Jun.

Distribution Also from Wisconsin to Nova Scotia and south to northwestern Florida.

239
Clamp-tipped
Emerald male—
Murray Co., GA,
September 2004, Giff
Beaton; female—
Cobb Co., GA, August
2003, Giff Beaton

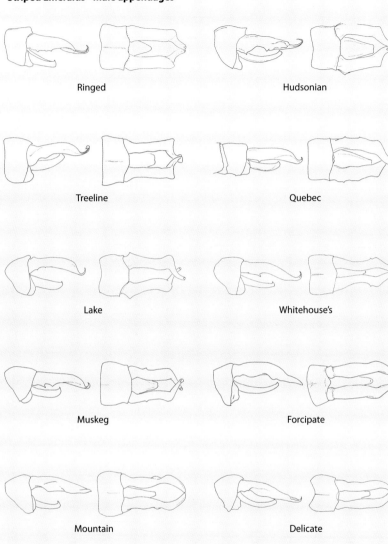

Ringed

Hudsonian

Treeline

Quebec

Lake

Whitehouse's

Muskeg

Forcipate

Mountain

Delicate

Kennedy's

Williamson's

Striped Emeralds - male appendages (*continued*)

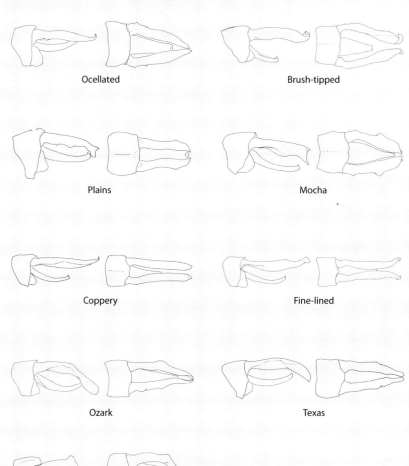

Ocellated

Brush-tipped

Plains

Mocha

Coppery

Fine-lined

Ozark

Texas

Clamp-tipped

Striped Emeralds - female subgenital plates

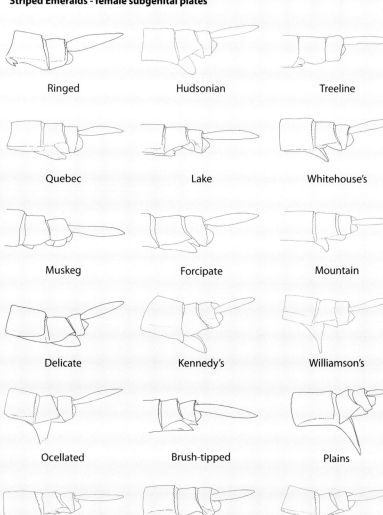

Ringed

Hudsonian

Treeline

Quebec

Lake

Whitehouse's

Muskeg

Forcipate

Mountain

Delicate

Kennedy's

Williamson's

Ocellated

Brush-tipped

Plains

Mocha

Coppery

Fine-lined

Ozark

Texas

Clamp-tipped

Shadowdragons are characterized by dull brown coloration (including the large eyes), entirely pale face, a yellow spot on each side of thorax, and relatively broad wings with dense venation. Dark spots at anterior crossveins are visible in most species. They are indeed dragons of the shadows, their flight time restricted to dusk and into darkness. When warm enough, they also fly at dawn. It is hard to imagine that their flight period is long enough to accomplish either feeding or breeding, but they manage! Because of their time of activity, they are generally very poorly known, but all species may have rather similar habits. Most live on streams and rivers, but a few also patrol the shores of large lakes. They are fast and often erratic flyers, a challenge to the most skillful collector and almost impossible to identify without capture. The large eyes and broad wings are characteristic of crepuscular dragonflies. World 7, NA 7, West 4.

240 Alabama Shadowdragon *Neurocordulia alabamensis* TL 42–46, HW 29–33

Description Small, very plain shadowdragon of Southeast. *Male*: Entirely orange-brown to dull brown, including eyes; sides of thorax become yellowish with maturity. *Female*: Colored as male, recognized by thicker abdomen. Wings with anterior edges heavily dotted with brown in both sexes; entire wings of older individuals suffused with brown.

Identification Distributions poorly known but may overlap with **Cinnamon** and **Smoky Shadowdragons**. In hand, look at dots along front edge of wing. In **Alabama**, dots evenly colored and extend out to nodus and often to stigma. In **Cinnamon**, dots smaller and extend only halfway to nodus. In **Smoky**, dots may extend to nodus but are darker on edge than along crossvein. **Cinnamon** and **Smoky** also usually show bright yellow spot on side of thorax. Perhaps no overlap with **Orange Shadowdragon**, a larger, brighter species with heavily marked wings.

240.1
Alabama
Shadowdragon
male—Bibb Co., AL,
July 2003, Giff
Beaton (posed)

240.2
Alabama
Shadowdragon
female—Alachua Co.,
FL, July 1974, Sid
Dunkle (posed)

Natural History Only seen in dusk flight over breeding habitat; nothing known of behavior otherwise. Scarcely known from West.

Habitat Small to medium-sized slowly flowing streams with sand and muck bottoms, usually in woodland.

Flight Season TX May–Jun.

Distribution Ranges east in Coastal Plain to North Carolina.

241 Cinnamon Shadowdragon *Neurocordulia virginiensis* TL 42–49, HW 32–35

Description Shadowdragon with faint yellow thoracic spots and wing dots. *Male*: Entirely medium brown, including eyes. Thorax with median dark area in front, inconspicuous yellow spot low on sides. Abdomen with faint narrow yellow rings most obvious on basal segments, faint dark markings most obvious on middle segments. *Female*: Colored as male. Both sexes with faint dots on some antenodal crossveins.

Identification Least-marked wings of any shadowdragon, less than **Alabama** and **Smoky** with which it overlaps. Yellow side spot not bordered behind by black, as in **Smoky**.

Natural History Both sexes fly long beats just above water's surface from about sundown until dark (or later?). Males may appear before dark at times. Patrol flights usually above running water, may cruise up and down for short time, then move to a different run, both at shore and out over open water. Day and night roosting apparently in forest near breeding habitat. Scarcely known from West.

Habitat Medium-sized to large rivers with some current in wooded country.

Flight Season East Mar–Aug.

Distribution Also across Southeast to northern Florida, north to Kentucky and Virginia.

241.1
Cinnamon
Shadowdragon
male—Liberty Co.,
FL, April 2005
(posed)

241.2
Cinnamon
Shadowdragon
female—Alachua Co.,
FL, 1973, Sid Dunkle
(posed)

242 Smoky Shadowdragon *Neurocordulia molesta* **TL 45–53, HW 33–38**

Description Shadowdragon with bold yellow thoracic spots and prominent wing dots. *Male*: Entirely light brown, including eyes. Thorax with dark area on front and prominent elongate yellow spot low on sides. Abdomen with yellow basal spot low on sides of S2, narrow yellow rings on about S3–8, and faint dark patterning along middle segments. *Female*: Colored as male. Both sexes with dots at all antenodal crossveins, larger dot at nodus.

Identification Differs from **Cinnamon Shadowdragon** in larger and more extensive wing dots, extending out to nodus. Prominent dark area behind yellow lateral thoracic spot may also distinguish **Smoky** from **Cinnamon**, which usually lacks it. **Alabama** has many dots but lacks obvious yellow spot on side of thorax present in both **Cinnamon** and **Smoky**. **Smoky** may or may not overlap with slightly larger and brighter **Orange Shadowdragon**, with its even more extensive wing markings.

242.1
Smoky Shadowdragon
male—Harrison Co., IN,
June 2005, James Curry

242.2
Smoky Shadowdragon
female—Long Co., GA, June
2005, Giff Beaton (posed)

Natural History Appears at river as dusk is falling, usually last half-hour before darkness. Individuals can be seen coming from forest through clearings to river. Remain at water until scarcely visible and perhaps into full dark. Both sexes fly up and down river very low, males often interacting, and mostly over open water away from shore. Females oviposit by flying rapidly, tapping water every few feet. Roosts in trees and shrubs in forest.

Habitat Large, swift-flowing rivers with sandy shores, wooded banks.

Flight Season NE Jul–Sep, TX May.

Distribution Ranges widely in East from Minnesota and Ohio to Louisiana and northwestern Florida.

---

243 Orange Shadowdragon *Neurocordulia xanthosoma*   TL 48–52, HW 35–40

Description Large orange-brown shadowdragon of southern plains with heavily marked wings. *Male*: Eyes light brown. Body entirely orange-brown. Thorax with faint yellow spot low on sides. Abdomen with faint yellow basal spots on sides of S4–8. *Female*: Colored as male. Both sexes with wings heavily marked with orange or brown, typically dark dots and blotches along anterior

edges but may be heaviest across base; in older individuals, entire wing may become brownish.

Identification Larger and more orange-looking than other shadowdragons, often apparent even at dusk. Extensive wing markings also distinguish it from others of its group. Day-flying habits make it appear when various brown skimmers present, but none of them really looks like it. **Evening Skimmer**, also hanging up in woodland, has round spot in each wing or essentially no wing markings; more likely in pond than river habitats.

Natural History Both sexes fly over rivers beginning at dusk or shortly before and continuing at least until completely dark. Also fly at dawn on warmer mornings and occasionally during cloudy periods at midday. Flight usually just above water but at times a bit higher, from just off shore to well out in open water; beats may be short or much lengthier. Mating seen in dawn flight, with copulating pair flying from water to perching site in tree. Oviposition seen after dawn flight, female scattering eggs in rapid flight. Many roost in forest, but only shadowdragon in West that also regularly roosts in herbaceous and shrubby vegetation adjacent to breeding habitat, perhaps because often in open

243.1
Orange Shadowdragon male—Kimble Co., TX, July 2001, Tom D. Schultz

243.2
Orange Shadowdragon female—McLennan Co., TX, May 1974, Curtis Williams (posed)

country. Thus more easily found in daytime than other shadowdragons, usually hanging below knee height in small shrubs and perhaps showing preference for small islands in rivers.

Habitat Rivers with slow to moderate current in open or wooded landscape.

Flight Season TX Apr–Aug.

Distribution Also east into southern Illinois.

Baskettails *Epitheca*

Among the dullest emeralds, their eyes are less brilliant green and with little or no metallic color on a light brown and black body with spindle-shaped abdomen. Flight season is typically early. Males cruise over water in sexual patrol flight, but females are not often seen, mostly in foraging flights over clearings that may be aggregated when prey is common. Baskettail feeding swarms are a daily occurrence wherever one of the small species is common, much more noticeable in the East than in the West. They perch with abdomen held at around 45° below horizontal except for the long-abdomened Prince Baskettail, which hangs vertically. After mating, the female accumulates a large egg cluster at the end of her abdomen, held in place by very long, forked subgenital plate, which is very conspicuous as she flies low over water. Females seem to spend much time in flight with abdomen conspicuously elevated, looking for optimal oviposition sites. When the right place is found, the abdomen is dragged along water surface, pulling out a long string of eggs, somewhat like egg strings laid by toads but only about 2 1/2 inches long and a few millimeters wide. American species of this genus are often separated in genera *Epicordulia* (Prince Baskettail) and *Tetragoneuria* (all other species). The two Eurasian *Epitheca*, intermediate in size between these two American groups, share the egg strings that appear to unite them all. Body patterns are also rather similar, and all species have at least small brown basal hindwing spots, some considerably more. Some *Tetragoneuria* are frustratingly similar in structure and variable in wing patterns, which vary geographically as well as individually. Field identification is almost impossible for most species, as all are colored essentially the same, and they present difficulties even in hand. Drive slowly along roads that pass through their habitats and look for feeding swarms to capture and identify baskettails. World 12, NA 10, West 8.

244 Common Baskettail *Epitheca cynosura* TL 38–43, HW 26–31

Description May be most common small baskettail on plains, replaced by others farther west. *Male*: Eyes red over gray for much of life but eventually becoming bright emerald green; face dull yellow-orange with black on top of frons varying from small black triangle at base to entire top of frons black (latter perhaps not in West). Thorax brown with darker markings and small yellow spots low on sides. Abdomen with S1–2 brown, black beyond that with elongate yellow spots on sides of S3–8. *Female*: Colored as male; abdomen thicker and cerci much shorter. Populations in southern  part of region (east Texas) with heavily marked hindwing, a triangular brown spot that extends well toward nodus. Farther north (Nebraska), wing markings minimal.

Identification Indeed the common baskettail over much of eastern North America, serving as a standard from which other species must be distinguished. Absent from most of region of this book, however, and presents identification problems only with **Spiny Baskettail**, barely overlapping in the North, and three species on the plains. Male **Spiny** slightly larger, cerci with spine that can only be seen in hand. In fact, most identification of small baskettails not possible except in hand. See **Dot-winged**, **Mantled**, and **Slender** for distinctions from those species in southeast part of region. Female cerci distinctly shorter (shorter than S9) than those in **Dot-winged**, **Slender**, and **Spiny** (longer than S9). Note that wing pattern

244.1
Common Baskettail
male—McCurtain
Co., OK, April 2007,
David Arbour

244.2
Common Baskettail
female—Floyd Co.,
GA, May 2006

alone may distinguish it in some areas, as **Common** only species (outside range of **Mantled**) that has extensive black markings at hindwing bases. But not all **Common** have the wing markings.

Natural History Males fly at knee height over pools in streams as well as along shores of ponds and lakes, often with extended hovering; patrol length 12–30 feet. Copulation in flight, with movements over water at waist height up to hundreds of yards and as long as 5 min. Females at times fly very rapidly across water, stop suddenly to inspect potential oviposition site, apparently find it wanting, and zoom off again. Often survey an area in low flight before even starting to lay eggs. Feeding singly or in small swarms over open areas, from head to well up in trees and each individual moving back and forth rapidly at one height, then pursuing prey and changing height.

Habitat Ponds and lakes with open or wooded margins, also pools in slow streams.

Flight Season NE Apr–Jul, TX Feb–May.

Distribution Range in West may be less than shown because of confusion with Dot-winged Baskettail. Throughout East from southern Ontario and Nova Scotia south.

Description Small southeastern baskettail with broad abdomen and largest dark spots in hindwings. *Male*: Eyes red over gray for much of life but eventually becoming bright emerald green; face dull yellow-orange with black on top of frons varying from small black triangle at base to entire top of frons black (latter perhaps not in West). Thorax brown with darker markings and small yellow spots low on sides. Abdomen with S1–2 brown, black beyond that with elongate yellow spots on sides of S3–8. Hindwings with large dark markings, extending back nearly to wing margin and out almost to nodus. Wing markings even larger in East, reaching nodus and rear of wing. *Female*: Colored as male, thicker abdomen and much shorter appendages evident.

Identification Large hindwing spots, extending almost to nodus and rear edge of wing, should allow easy identification whether on territory or in feeding flight. Confusion most likely with some **Common Baskettails** with especially large hindwing spots, although most **Common** in range of **Mantled** have smaller markings. **Common** a bit larger with more slender abdomen, although difference in females not easily detected. Relatively broad abdomen, middle segments looking wider than long, makes yellow spots on sides

245.1
Mantled Baskettail
male—Travis Co., TX,
April 2007,
Greg Lasley

245.2
Mantled Baskettail
female—Echols Co.,
GA, March 2006,
Giff Beaton

of it more prominent than in other baskettails. Quite different habits, flying around and then hanging up vertically, should distinguish it from small skimmers with dark hindwing patches, for example **Marl Pennant**. **Black Saddlebags** much larger.

Natural History Males fly over water on sexual patrol. Oviposition as in other baskettails, eggs dropped in floating algal mats. One female dragged abdomen tip through 10-inch algal mat quite slowly. Because of large wing spots, flight usually looks fluttery.

Habitat Sandy lakes and ponds with much vegetation.

Flight Season TX Mar–May.

Distribution Also from Georgia to Nova Scotia on Atlantic Coast.

---

246 Slender Baskettail *Epitheca costalis*   TL 38–48, HW 25–32

Description Rather slender southeastern baskettail. *Male*: Eyes red over gray for much of life but eventually becoming emerald green; face dull yellow-orange with small black triangle at base of top of frons. Thorax brown with darker markings and small yellow spots low on sides. Abdomen with S1–2 brown, black beyond that with elongate yellow spots on sides of S3–8. *Female*: Colored as male; abdomen slightly thicker, but very similar, as appendages of both sexes rather long.

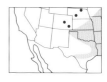

Identification Overlaps extensively with **Common Baskettail** in eastern part of region. Differs in having more slender abdomen, constricted to form suggestion of "wasp waist." Yellow dorsolateral markings on S3 elongate, usually more than twice as long as wide and often tapering sharply to rear, as an indicator of segment shape. In **Common**, these

246.1
Slender Baskettail
male—McCurtain
Co., OK, March 2007,
David Arbour

246.2
Slender Baskettail female—
Laurens Co., GA, April 2005,
Giff Beaton

Emerald Family **363**

markings more rectangular, no more than twice as long as wide. Difficult to quantify, but differences becomes apparent with experience. Appendages very similar, but cerci of **Slender** slightly longer and with more obvious bump at ventral angle, of **Common** slightly shorter and with more of a narrow keel at that angle. Even with examination in hand, not always easy to distinguish. Females differ by **Slender** having distinctly longer cerci, about three times length of S10, not quite two times in **Common**. Many **Common** have extensive dark markings at hindwing bases, not so in **Slender**. Also difficult to distinguish from **Dot-winged** with no dots in wings, as sometimes is the case, but cerci of that species with even less-pronounced ventral angle. No overlap in range with quite similar-looking **Spiny Baskettail**.

Natural History Males fly patrol flights over water as other baskettails but seem slightly more common in wooded areas, often flying in shade and among trees and shrubs. Forage in open areas from just above ground to treetops, often in swarms and may be mixed with other baskettails.

Habitat Sandy ponds, lakes, and slow streams. More likely to be on streams than other small baskettails.

Flight Season NE Apr–Jun, TX Mar–May.

Distribution Possible that range in West less than shown because of confusion with Dot-winged Baskettail. Also from Iowa to New Jersey and south to northern Florida.

Comments Called Stripe-winged Baskettail in some books.

247 Dot-winged Baskettail   *Epitheca petechialis*   TL 41–43, HW 27–31

Description Slender baskettail, many individuals with dotted wings. *Male*: Eyes red over gray for much of life but eventually becoming gray-green; face dull yellow-orange with small black triangle at base of top of frons. Thorax brown with darker markings and small yellow spots low on sides. Abdomen with S1–2 brown, black beyond that with elongate yellow spots on sides of S3–8 or S3–9. *Female*: Colored as male, with thicker abdomen and shorter cerci. Individuals of most populations with small dark dots along front of wings from base to nodus, largest at either end of series and more evident in hindwings.

Identification Individuals with dotted wings safely distinguished from other baskettails and other dragonflies, although note that **Orange Shadowdragon**, differently colored, is somewhat similar in size and may show prominent wing dots. Other SHADOWDRAGONS mostly out of range of **Dot-winged Baskettail**. Individuals without wing dots look exactly like **Slender Baskettail**, even difficult to distinguish in hand; **Dot-winged** male cerci with slightly less prominent angle on ventral surface. Also similar to **Common Baskettail**, although that species often with sizable hindwing marking in Texas. Otherwise, abdomen a bit longer and more slender, cerci slightly different. Easily distinguished from **Mantled Baskettail** by longer abdomen, smaller hindwing marking. Apparently never develops bright green eyes like most other baskettails.

Natural History Apparently like that of other baskettails, although poorly known. Often seen feeding at edges of clearings and in open woodland.

Habitat Lakes, ponds, and slow streams.

Flight Season NM Mar–Jul, NE May–Aug, TX Mar–May.

248 Spiny Baskettail *Epitheca spinigera* — TL 43–47, HW 29–34

Description Slender northern baskettail with dark body and clear wings. *Male*: Eyes bright emerald green at maturity; face dull yellow-orange with black on top of frons as broad-based T-spot. Thorax brown with darker markings and small yellow spots low on sides. Abdomen with S1–2 brown, black beyond that with elongate yellow spots on sides of S3–8. Cerci with small spine under base visible from side, rarely reduced or even lacking. *Female*: Colored as male, eyes duller, dark green and brown at maturity. Cerci rather long.

Identification Wide overlap with **Beaverpond Baskettail**, distinguished by differently shaped cerci in male, longer cerci in female. Also in hand, T-spot on top of frons and black rear of head can be seen in **Spiny**, frons mostly dark and back of head pale in **Beaverpond**. Probably not distinguishable in flight, although appendages might be visible in binoculars in slow or hovering flight. Barely overlaps with slightly smaller **Common Baskettail**. In hand, male **Spiny** distinguished from **Common** by small downward-pointing spine at base of cerci (very rarely lacking), very difficult to see in field. Pattern of top of frons different, prominent dark T-spot in **Spiny** and usually only stem of that marking in **Common**; again, most readily seen in hand. Females more easily distinguished by length of cerci, longer than S9 in **Spiny** and shorter than S9 in **Common**; this could be seen in perched individuals.

248.1
Spiny Baskettail male—
Addison Co., VT, June 2005,
Michael Blust

248.2
Spiny Baskettail female—
King Co., WA, June 2005
(posed)

Natural History Males fly low over water, often far out over lake, in beats 30–65 feet in length. Also males in seeming sexual patrol flight over clearings near water. At some lakes, not along lake shore like most other baskettails. Females fly with raised abdomen, then tap water at intervals, releasing short egg string, as late as dusk. Favored oviposition sites may be inundated with egg clusters, looking much like amphibian eggs. Both sexes feed over clearings, knee height to treetops, sometimes far from breeding localities. From high density of exuviae during concentrated spring emergence, must be locally abundant, but adults not usually seen in large numbers.

Habitat Ponds and lakes of all sizes and kinds, including beaver ponds and bog ponds, more rarely at slow streams. Usually in wooded country but some sites in open grassland or desert.

Flight Season BC Apr–Jul, WA Apr–Aug, OR May–Jul, CA Jun–Jul, MT Jun–Jul.

Distribution Ranges across southern Canada to Nova Scotia and south to Michigan and New Jersey.

249 Beaverpond Baskettail *Epitheca canis* TL 43–48, HW 30–31

Description Northern baskettail with distinctive male appendages, otherwise much like other species. *Male*: Eyes either bright emerald green or gray-blue (perhaps younger individuals), duller below; face dull yellow-orange with black triangle on top of frons. Thorax brown with darker markings and small yellow spots low on sides. Abdomen with S1–2 brown, black beyond that with elongate yellow spots on sides of S3–8 or S3–9. Cerci strongly angled downward near tips. *Female*: Colored as male, eyes more likely to be red over gray. Mature females in eastern populations with entirely brown wings, very distinctive in flight; so far, no such females observed west of Prairie provinces.

Identification Overlaps widely with **Spiny Baskettail**. **Beaverpond** distinguished by head pattern (**Spiny** has more black on top of frons, usually black rear of head), very different male appendages (see drawings), and shorter female cerci (about twice the length of S10 in

Beaverpond, slightly more than that in **Spiny**). Barely overlaps with **Common Baskettail** at east edge of region, differs by pale rear of head (black in **Common**), slightly longer abdomen, very different male appendages. **Common** often has prominent dark hindwing patches, **Beaverpond** only tiny ones. In area of overlap, female **Beaverpond** with brown wings. In hand, ventral view of subgenital plate diagnostic, two lobes almost parallel rather than diverging as in **Common** and **Spiny**.

Natural History Alternate hovering and continuous flight along shore or over open water, knee to waist height. Territories may be as small as 12×6 feet or up to 100 feet long depending on water area. Males at times in what looks like sexual patrol over clearings near water. Ovipositing female may fly around with abdomen elevated for long time, even with 2- to 3-inch egg string hanging from abdomen, before finally dragging it through water to release eggs. Male once found tangled in female egg string. Both sexes hunt by cruising over clearings away from water, sometimes well up in trees.

Habitat Rivers, slow streams, and sloughs seem preferred over ponds and lakes in Northwest, but latter also used. Aquatic vegetation usually prominent. Males more likely at moving

249.1
Beaverpond
Basskettail male—
Columbia Co., WI,
May 1997, David
Westover

249.2
Beaverpond
Basskettail female—
Siskiyou Co., CA, July
2006, Paul Johnson

water than other western baskettails, sometimes even over swift streams. More likely over bog ponds in eastern part of region.

Flight Season BC May–Jul, WA May–Jul, OR May–Jul, CA Mar–Jul.

Distribution Also east across southern Canada to Nova Scotia and south to Wisconsin and West Virginia.

---

250 Robust Baskettail   *Epitheca spinosa*   TL 42–47, HW 29–33

Description Heavy-bodied baskettail, quite rare at southeastern edge of region. Very small brown spots at base of hindwings. *Male:* Eyes bright emerald green at maturity; face dull yellow-orange with black basal triangle on top of frons. Thorax brown with darker markings and small yellow spots low on sides. Abdomen with S1–2 brown, black beyond that with elongate yellow spots on sides of S3–8. Cerci strongly angled downward near tips. *Female:* Colored as male, abdomen thicker at base and appendages slender.

250.1
Robust Baskettail
male—Tuscaloosa
Co., AL, March 2006,
Giff Beaton (posed)

250.2
Robust Baskettail
female—Tuscaloosa
Co., AL, April 2006,
Steve Krotzer (posed)

Identification Very unlikely to be seen in West, but distinctly bulkier than other baskettails; can be distinguished in flight by this. Both sexes have broader-looking abdomen than **Common Baskettail**, more like **Mantled**, and lack extensive dark markings in hindwings as **Mantled** and many **Common** have. Relatively broad abdomen makes yellow spots on sides of it more prominent than in other baskettails. Capture would be necessary for certainty; uniquely shaped cerci in male diagnostic, with upward-pointing spine near tip.

Natural History Males fly along shore and over water as other baskettails, hovering frequently, perhaps a bit harder to approach. Often fly among trees. Much less often seen than other species but occasionally common.

Habitat Swampy woodland ponds.

Flight Season TX Mar.

Distribution Ranges east in Coastal Plain to New Jersey.

---

251 Prince Baskettail *Epitheca princeps* TL 59–65, HW 36–43

Description More like darner than emerald, large and with long abdomen, but note green eyes and big dark spots on wings. *Male:* Eyes bright emerald green at maturity; face light brown, darker on top of frons. Thorax brown with darker markings and small yellow spots low on sides. Abdomen with S1–3 mostly brown, thereafter black above with elongate light brown spots on sides of S4–7 or S4–8 (may be larger and whitish on S3). Fine white ring borders each abdominal segment. Large irregular dark spots at base, middle, and tip of each wing. Geographic variation striking in East, individuals of southern populations much larger and with larger wing markings. Some northeastern populations almost completely lack wing markings. Smaller range of size in plains states, but many with reduced markings in north Texas and farther north. *Female:* Colored as male, eyes more likely to be red over gray; abdomen slightly thicker, cerci straight instead of diverging at ends.

Identification When wing spots can be seen, nothing else like it; easy when overhead, sometimes difficult against dark water. Much larger than female **Common Whitetail** and **Twelve-spotted Skimmer**, with somewhat similar wing patterns, and, as flyer and

251.1
Prince Baskettail
male—Lubbock Co.,
TX, June 2007, Jerry
Hatfield

251.2
Prince Baskettail
male—Williamson
Co., TX, April 2006,
Elizabeth Moon

251.3
Prince Baskettail
female—Travis Co.,
TX, April 2006, Bob
Thomas

hanger, **Prince Baskettail** shows quite different behavior. If wing spots not evident, note it is brown rather than green as in the large STRIPED EMERALDS with which it might be confused. Only common all-brown DARNER in its range and habitat is **Fawn Darner**, with bright yellow spots on thorax.

Natural History Males fly long beats along lake shores or over pools in streams, typically from waist to head height, alternating fluttering and gliding and occasionally hovering. Also fly far out over water of large lakes. Much chasing of other dragonfly species while on sexual patrol. Land in shrubs or trees, even at treetop level, and hang straight down like darner, not with abdomen extended outward as do other baskettails. Both sexes may hold abdomen curved strongly upward when perched, especially exaggerated in females holding ball of eggs. Feeding in flight in open areas from near ground to treetop level, in swarms with its own species or darners. Prey mostly quite small. Feeding seems to peak in early morning and evening and extends at least until dusk. Mating also peaks in morning and evening; pairs couple over water and then fly back and forth for considerable periods, perhaps all of copulation in flight. Females oviposit as smaller baskettails by forming egg mass, flying over water with abdomen strongly curled upward, then dropping to water to drag out egg string. Egg string much longer those of smaller baskettails, up to a foot more.

Habitat Moderate-sized streams to fair-sized rivers, usually slow-flowing, and large ponds and lakes. Most have sandy to muddy bottoms. Tendency to be more at lakes in northern part of range.

Flight Season NE Jun–Sep, TX Mar–Sep.

Distribution Also throughout East from southern Ontario and Nova Scotia south.

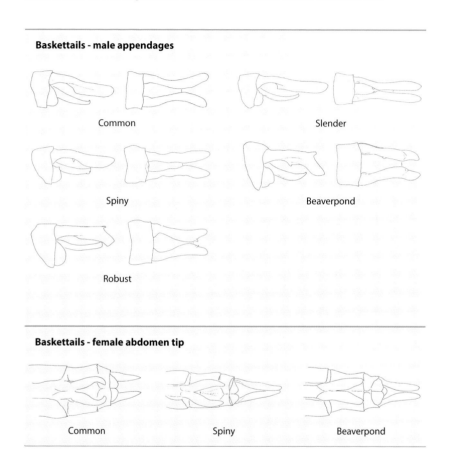

Baskettails - male appendages

Common

Slender

Spiny

Beaverpond

Robust

Baskettails - female abdomen tip

Common

Spiny

Beaverpond

Large eyes touching in center of the head and a foot-shaped anal loop in hindwings are characteristic of this family. Hindwings are similar in males and females, unlike other dragonfly families, which have an angled base on each hindwing and auricles that project on either side of S2. Skimmers comprise the largest odonate family and are the species most likely to be seen by a casual dragonfly watcher. It is both appropriate and wise for beginners to become familiar with common skimmers before attempting to tie down more difficult species of other dragonfly families. The entire array of dragonfly colors is represented among skimmers, although few of ours have the metallic finish of cruisers and emeralds. Some species are strikingly patterned, recognizable as far as you can see them, and many have distinctive color patterns on their wings. On the other hand, some groups of skimmers consist of a set of similar-looking species and present real identification challenges. Most are perchers, but several genera of tropical origin are fliers. One group of fliers includes rainpool gliders, saddlebags, pasture gliders, hyacinth gliders, and evening skimmers; another group clubskimmers and sylphs. Perchers often spend much time in flight when on sexual patrol, but they feed from a perch, as flycatchers rather than swallows. Most species are sexually dimorphic in coloration, sometimes dramatically so, even in wing pattern (for example, whitetails). In many cases, males become pruinose, females not; pruinosity is much less common in other dragonfly families. In most species, immature males are colored like immature females, and usually immature females are colored like mature females. Thus, the sex of many individuals you see, especially away from water, will have to be determined by shape. Males typically have a longer, more tapered and pointed abdomen than females, which have a slightly thicker, blunt or truncate abdomen. Keep in mind that *dragon males are taper-tailed*. Of course, the characteristic bump of genitalia under a male's abdomen base is always diagnostic of sex, and males usually have basal segments more swollen in side view than do females. World 968, NA 108, West 97.

Whitetails   *Plathemis*

These are stocky skimmers of muddy ponds that spend much of their time perching on the ground and other flat substrates. Males have spectacularly pruinose abdomens, and both sexes display striking wing markings. One species is common continent-wide; the other is restricted to the Southwest. Corporals and king skimmers are close relatives, and *Plathemis* has often been combined with *Libellula*. World 2, NA 2, West 2.

252 Desert Whitetail   *Plathemis subornata*   TL 40–51, HW 32–38

Description Vividly black and white broad-bodied southwestern skimmer. *Male*: Eyes dark brown; face blue-black, with yellow markings on side. Thorax dark brown, faintly indicated pale spots low on side. Abdomen entirely pruinose whitish. Wings with dark basal streaks and wide dark subterminal bands, mostly whitish pruinose bases. *Female*: Eyes brown over gray; face yellow. Thorax dark brown with two broad yellow stripes on each side. Abdomen dark brown with dorsolateral series of yellow spots, becoming smaller posteriorly, on S2–8. Wings with dark basal streaks and pair of  brown crossbands beyond nodus on each wing. Immature male with body and wing pattern of female, pruinosity develops at wing base, and area between pair of dark stripes on each wing fills in with black. Note that immatures of both sexes have same wing pattern, quite different from situation in Common Whitetail.

Identification Male most like male **Common Whitetail**, but extensively white inner wings, visible from above or below, distinguish it easily. Pruinosity develops on wings early enough

252.1
Desert Whitetail male—Jeff Davis Co., TX, August 2004, Martin Reid

252.2
Desert Whitetail female—Dallam Co., TX, July 2004, Martin Reid

so that even immatures distinguishable. Female whitetails quite different in wing pattern, lack of black at tip distinctive for **Desert**. Also, stripes/spots on abdomen near edges on **Common**, closer to midline in **Desert**. Male superficially like male **Widow Skimmer** on quick fly-by, but white on wings inside black in this species, outside it in **Widow**. Body pattern of female **Desert Whitetail** much like those of female **Eight-spotted** and **Hoary Skimmers**, but wing pattern different in each one; **Whitetail** also distinctly smaller. All of these species occur together commonly at one place or another.

Natural History Males perch at edge of water and fly back and forth defending territory much as closely related Common Whitetail. Behavior much less well known, and raising abdomen in aggressive display not seen. Perches on ground and flat objects much as its close relative, also commonly on twigs and sedge stems. Females often encountered near breeding habitat, perhaps easily seen because of openness of surroundings.

Habitat Open ponds and slow streams in desert regions; at northern edge of range may be limited to hot springs. Although larval habitat needs probably similar, only rarely seen at same places as Common Whitetail.

Flight Season OR May–Aug, CA May–Sep, AZ Apr–Oct, NM Apr–Nov, NE May–Oct, TX Apr–Jul.

Distribution Sparsely distributed. Ranges south in western Mexico to Jalisco.

Description One of our most common and widespread dragonflies. Unmistakable pied males always notable but different from spotted females. *Male*: Eyes and face dark brown. Thorax dark brown with two faintly indicated pale stripes on each side. Abdomen white (some see it as pale blue) with pruinosity when mature. Wide black crossband near tip and black streak at base of each wing, white behind basal streak at maturity. Immature with spotted abdomen as female but typical male wing pattern; becomes gradually bluish-white, then white. *Female*: Eyes and face brown. Thorax

brown with two white stripes on side of thorax, turning yellow at lower ends. Abdomen brown with white to pale yellow spots along each side, forming lines on S2–3 and extending diagonally on S4–8. Pale spots outlined in black. Three dark spots on each wing, quite different from male. Just-emerged individuals show only very faint wing spots, but distinctive abdominal pattern obvious.

253.1
Common Whitetail male—Hillsborough Co., NH, July 2006, Netta Smith

253.2
Common Whitetail female—Burnet Co., TX, July 2004

253.3
Common Whitetail
immature
male—Multnomah
Co., OR, June 2005,
Jay Withgott

Identification Sexual differences in wing color pattern greater than in almost any other dragonfly. Male like nothing else except closely related **Desert Whitetail**, which has much white on wing bases. More distant from **Band-winged Dragonlet** and **Four-spotted Pennant**, both of which have similar black bands on each wing but no white on wing bases and abdomen slender and black. Female wing pattern much like that of female **Twelve-spotted Skimmer**, differs in abdomen pattern. Skimmer has yellow squares at outer edge of each segment that make abdomen appear bordered along edges, **Whitetail** paler yellow-white slashes that make abdomen appear spotted along edges. **Whitetail** commonly perches on ground and logs, skimmer usually on branches and leaves, but this is indicative, not definitive. See **Banded Pennant** and **Prince Baskettail**, with similar wing patterns but otherwise very different.

Natural History Most of perching away from water is on ground, rocks, or logs, usually very low and in more open areas than where most skimmers perch. Capture small insects that pass by in flight. Males perch at waterside and fly up and down shore, very aggressive to other male whitetails and other somewhat similar-looking species. Strongly territorial, defending an area 12 feet on either side of an oviposition site. Territorial males remain at a breeding site for several hours each day, not necessarily every day, at maximum for 18 days. Larger males defend larger territories and mate more times. Subordinate males may be allowed in dominant male territories, but dominants mate. Patrol flight typically very low, just above water. Broad white abdomen elevated in face-to-face display in flight: one turns and lowers abdomen (still displayed), or both may fly in parallel flight for up to a minute or more. Females visit breeding sites only every few days, peaking at midday but second peak in late afternoon, presumably fertilized females returning to water. Copulation brief (3 sec) and in flight, followed immediately by oviposition. Female usually guarded by male, often hovering about a foot above her. Where common, especially at small ponds, both guarding and territorial defense ineffective, female often snatched up by second male or leaves water. Some females drop to ground at waterside when harassed, then apparently invisible to males. Females oviposit by tapping water near floating vegetation or on clumps of mud or vegetation with glistening surfaces, with frequent position changes or not. May remain in same spot for up to several minutes with male in attendance. Often flick water drops forward with eggs, laying 25–50 eggs at each tap. Females lay around 1000 eggs total, at around 25/sec. Reproductive adults can live up to 36 days.

Habitat Lakes, ponds, and slow streams, including those constructed by humans. Larvae on mud bottom and tolerant of a wide range of conditions, so adults often at wetlands that deserve the term "yucky," for example, cattle-trodden stock ponds. Sometimes only species present.

Flight Season BC May–Aug, WA May–Oct, OR May–Oct, CA Mar–Oct, MT Jun–Aug, AZ Apr–Oct, NM Jul–Sep, NE May–Oct, TX Mar–Oct.

Distribution Ranges south in eastern Mexico to Nuevo León. Also across East, north to southern Ontario and Nova Scotia.

Corporals *Ladona*

This small North American genus is closely related to king skimmers of genus *Libellula* and has often been combined with them. Its nearest relative may be Scarce Chaser (*Libellula fulva*) of Europe, which some have considered *Ladona*. Two light stripes on the thorax of females and immature males, like those that indicate the military rank, give the group its name. Corporals, like whitetails, perch on the ground and other flat surfaces, whereas king skimmers (*Libellula* and *Orthemis*) generally do not. Larvae are quite distinct from all king skimmers. World 3, NA 3, West 2.

254 Blue Corporal *Ladona deplanata* — TL 34–38, HW 26–29

Description Small, wide-bodied, blue or brown southeastern skimmer of early spring with dark streaks at base of all wings. *Male*: Eyes dark brown; face black. Thorax dark brown, pruinose blue in front. Abdomen entirely pruinose blue above. *Female*: Eyes and face brown. Thorax brown with narrow pale stripe at each side of front, equal width darker stripe behind it. Abdomen brown with black stripe down middle becoming wider to rear.

254.1
Blue Corporal
male—Leon Co., FL,
April 2005

254.2
Blue Corporal
female—Lake Co., FL,
April 2005

Identification Nothing much like this species, especially early in spring when it flies. Much smaller and duller than blue KING SKIMMERS such as **Spangled** and **Yellow-sided Skimmer**. Much duller than **Eastern Pondhawk** males that also perch on ground, wider abdomen than **Pondhawk** or **Blue Dasher**. Brown sides of thorax further distinguish it from all blue species. Female with distinctive stripes on thorax and down middle of broad abdomen. No overlap with **Chalk-fronted Corporal** in West.

Natural History Males perch on logs and shore, much like associated clubtails, and are quite aggressive toward one another. Both sexes commonly perch on ground in clearings, on roads and paths, and on sunny side of tree trunks and buildings. Elevated individuals may droop wings slightly. Copulation brief and in flight. Females oviposit by flying low and slowly to rapidly, tapping water regularly or erratically; may be guarded by male. Can be very common at optimal habitats. One of earliest dragonflies to fly in its range.

Habitat Lakes, ponds, slow streams, and ditches with wooded or open borders and at least some mud.

Flight Season TX Feb–May.

Distribution Also throughout Southeast, north to Indiana and New Hampshire.

---

255 Chalk-fronted Corporal *Ladona julia*   TL 41–45, HW 28–34

Description Chunky dark northern skimmer with dark markings at base of all wings. Individuals of western populations indicated here are considerably larger than those of eastern ones. *Male*: Eyes dark brown; face black. Thorax dark brown, pruinose white on front. Abdomen black, S_{1-4} pruinose white. *Female*: Eyes brown; face tan. Thorax brown, paler front bordered by darker stripe. Abdomen brown with black lines on all keels, indistinct black stripe down middle. Older individuals become darker, eventually become pruinose like males.

Identification No other skimmer colored like males and older females with front of thorax and base of abdomen strikingly pale. Combination of robust skimmer-shaped body, paler front of thorax, and clear wings with markings only at extreme base will suffice to identify younger females. Dull female sometimes mistaken for dull brown **Four-spotted Skimmer**, but wing patterns different.

255.1
Chalk-fronted
Corporal male—Pend
Oreille Co., WA,
August 2004

255.2
Chalk-fronted
Corporal female—
Mason Co., WA,
June 2004

Natural History Males perch on shore or on vegetation over water at edge of lakes and ponds, alternating perching and flying back and forth over short distances, both for lengthy periods. Quite aggressive toward one another and somewhat similar-looking Four-spotted Skimmers. Defended territories may be 30–80 feet of shoreline, but some males are nonterritorial, with long flights along shore. Both sexes perch flat on ground when away from water, often on light-colored surfaces. Also commonly perches on sides of trees facing sun, very likely to land on dragonfly net. Feeding occurs away from water, wide variety of prey up to size of whiteface dragonflies. Copulation brief (averaging about 5 sec) and in flight. One female seen to copulate with three males in quick succession. Females oviposit by tapping water and moving short distances between taps, in open and dense vegetation but usually near shore; even in puddles on roads. Males guard ovipositing females by hovering over them; females finished with oviposition fly rapidly into nearby forest. Male reproductive period averages 10 days, extreme 40 days. Adults at very high densities at times, in appropriate habitats outnumbering all other species put together.

Habitat Wooded and open lakes and ponds, especially associated with acid waters of bog lakes.

Flight Season BC May–Aug, WA May–Aug, OR Jun–Aug, CA Jun–Aug, MT Jun–Jul.

Distribution Also across southern Canada to Nova Scotia, south to Wisconsin, Ohio, and West Virginia.

King Skimmers *Libellula*

Species in this group include some of the most familiar to the general naturalist, as they are large and conspicuous, often with distinctive wing patterns. They range all across the North Temperate Zone and extend south into mainland American tropics, one species throughout much of South America. Typically, they perch with the front pair of legs folded behind the head. Males of some species undergo slow and gradual color change, becoming more and more pruinose with maturity. Copulation is brief and in flight in most species. Females of some species have obvious flange on side of S8 that facilitates splashing eggs on bank during oviposition. In bright sunshine, members of this group and related genera droop their abdomen downward rather than raising it in obelisk position. Most North American species fall in distinct groups of species pairs or trios that are most similar to one another and can present identification problems. World 27, NA 18, West 17.

Description Yellow-marked brown northern skimmer with lightly spotted wings. *Male*: Eyes brown above, greenish below, becoming all brown with age; face tan. Thorax rich brown in front, sides similar but yellow below with black lines along sutures. Abdomen rich brown, even orange-brown, at base with bright yellow stripe down outer edge of segments; ends of S4–5 and all of S6–10 black above. Color becomes duller with age, matte brown and very different-looking from brilliant orange-brown individuals. Air spaces in abdomen, probably important insulation for this northernmost of all king skimmers, show up clearly through thin cuticle. Moderate-sized black spot at hindwing base, tiny spot at nodus of each wing; anterior wing veins yellow to orange in some individuals. *Female*: Colored as male, with same variation. One of few king skimmers in which sexes indistinguishable at some distance.

Identification No other brown skimmer with small wing spots at base and nodus, although those not visible in flight. Not really much like anything else in its range and habitat. Female sometimes mistaken for brown but differently patterned **Chalk-fronted Corporal**. Individuals flying cross country could be mistaken for RAINPOOL GLIDERS, which have either unmarked wings or round hindwing spot; very unlikely to be breeding in same habitat.

256.1
Four-spotted Skimmer
male—Okanogan Co., WA,
June 2005, Netta Smith

256.2
Four-spotted Skimmer
female—Fairbanks, AK,
June 2002, Wim Arp

258.1
Neon Skimmer
male—Sonora,
Mexico, September
2006

258.2
Neon Skimmer
female—Cochise Co.,
AZ, September
2004

Natural History Usually perch with broad abdomen held obliquely downward. Males perch over water on twigs from waist height to above head height or make lengthy flights up and down streams. Defend territories vigorously against other males by raising bright abdomen in flight. At most extreme, two males fly parallel and about a foot apart over long distances, even 20 feet or more above ground, then one chases the other, then resume parallel flight until one leaves. Occasionally interact with somewhat similar Flame Skimmers. Disappears from open sunny areas on hot afternoons. On encountering female, male displays by flying below and behind her with raised abdomen, hovering and whirring wings. Similar display to front or side of ovipositing females. Copulation lasts less than half minute, in flight. Females oviposit by splashing water on bank from as much as 8 inches away, usually with male hovering or circling a few feet above. May spend minutes in one spot laying eggs, but males do not persist in guarding.

Habitat Clean streams and flowing ditches, usually wooded but sometimes in open. Also garden ponds in Texas.

Flight Season CA Jun–Oct, AZ May–Nov, NM Apr–Aug, TX Apr–Nov.

Distribution Sparsely distributed west of Texas. Ranges south in uplands to Costa Rica.

Description Large, robust bright orange skimmer with extensively orange wing bases. *Male*: Eyes red-orange; face orange. Thorax and abdomen orange. Wings with most of base orange-brown; veins bright orange, antenodal veins at front of wing often conspicuously paler; also small dark marks at base of each wing. *Female*: Eyes and face brown. Thorax and abdomen pale brown, sometimes orange tinged; only prominent marking thin white line up front of thorax and between wing bases. Usually less color in wings than male but same patterning, with light antenodal veins.  Occasional females colored much as males, with bright orange body and much orange in wings.

Identification Nothing very much like robust orange males except **Neon Skimmer**, even more spectacular with bright red abdomen and front of thorax. **Neon** also has less orange in wing than **Flame**. Females extremely similar, **Neon** even more robust than **Flame**. **Flame** often duller, brownish, whereas **Neon** more likely to be orange, but much overlap. Best distinction might be wings, typically with amber line down front and/or substantial orange

259.1
Flame Skimmer
male—Maricopa Co.,
AZ, May 2004,
Glenn Corbiere

259.2
Flame Skimmer
female—Owyhee
Co., ID, July 2005

suffusion at base in **Flame**, entirely clear in **Neon**. Wings of both can be clear, however; then color of longitudinal veins from above seems distinctive, orange (costa)-yellow (subcosta)-orange (median) from front in **Flame** and all orange in **Neon**. Similarly orange males of **Golden-winged** and **Needham's Skimmers** much more slender, mostly occur farther east. **Mayan Setwing** somewhat similar, much smaller and less robust, brighter red-orange, and with orange wing patch ending short of nodus. Female **Carmine Skimmer** colored something like **Flame** but abdomen much more slender.

Natural History Males perch over or near water and make frequent brief flights. Some defend territories for at least several hours, chasing other males vigorously, but move from day to day. Unlike Neon Skimmer, males remain at water during afternoon heat but may be sedentary in comparison with longer flights when cooler. Two males may fly parallel flights of attrition for long distances and times, but either one may return to perch site! Copulation brief (usually about 10 sec) and in flight, male often guarding female but only briefly. Oviposition throughout day, common after males leave water, and with frequent flights along stream. Oviposit by flicking water drops and eggs as much as a foot forward into water or onto moist shore, turning after each flick. Not as committed to shore oviposition as Neon Skimmer.

Habitat Ponds, lakes, and slow streams, very broad habitat selection. Sufficiently common to be seen almost everywhere, including some distance from water, in Southwest.

Flight Season OR Jun–Sep, CA Feb–Dec, MT Jun–Sep, AZ Mar–Dec, NM Mar–Nov, TX Apr–Nov.

Distribution Ranges south in uplands of Mexico to Oaxaca.

Description Midsized pale skimmer of desert sinks and springs with lightly but distinctly patterned wings. *Male*: Eyes light brown to whitish; face cream. Thorax brown in front, whitish on sides bisected by black stripe. Abdomen pale yellow with black stripes. Gray pruinosity overlies abdomen and front of thorax in mature individuals. Wings with yellow costal vein, brown base, small to moderate brown spot at nodus, and black stigma. *Female*: Eyes gray or brown above; face cream. Thorax cream with wide brown stripes on front, black humeral stripe, and narrower black side stripe. Abdomen mostly black, with series of narrow, pointed pale yellow spots down either side of midline. Spot at nodus usually absent.

Identification Unmistakable with pale face and striking wing pattern, yellow costa contrasting with brown wing spots and black stigmas. Wing pattern not obvious in flight, however. **Comanche Skimmer,** another white-faced skimmer often found with it, has dark costa, pale stigmas, and no wing markings, also bright blue eyes and body in males. Female distinguished from female **Comanche** by wing colors, same as male, and double dark stripe on each side of front of thorax (single broad stripe in **Comanche**). Female **Eightspotted** and **Hoary Skimmers** a bit bulkier, with dark costa and stigmas, larger black spots in wings. Females looks something like female **Blue Dasher** but larger, with all-white face and pale costal vein. When flying around, might be mistaken for **Wandering Glider**.

Natural History Both sexes perch in shrubs near breeding habitat, also sometimes on ground (unusual for king skimmer). Copulation begins in flight and then shifts to perch, also unusual among king skimmers. Lasts no more than few minutes, then female disengages from male genitalia, and pair takes off in tandem, female legs wrapped around male abdomen. Oviposits in tandem, pair descending slowly until female taps water, then rising and hovering for 10 sec or so or dropping to water immediately to tap again; may remain for lengthy period at one place. Pair may also come in on fast runs, tapping water

260.1
Bleached Skimmer
pair—Harney Co.,
OR, July 2005

260.2
Bleached Skimmer
pair—Harney Co.,
OR, July 2006,
Jim Johnson

at intervals like pair of Wandering Gliders. Unique among king skimmers in normally ovi-positing in tandem, although Hoary Skimmer also does so at times. Male may also release female and guard her as she continues. Female typically splashes eggs up in water drop-lets, so eggs not dropped where female hitting water.

Habitat Alkaline marshes, lake borders, and springs, often hot springs in northern part of range. Common at some extremes of water chemistry where few other odonates are found but generally uncommon and local.

Flight Season OR Jun–Aug, CA May–Sep, AZ May–Sep, NM May–Sep.

Distribution Sparsely distributed. Ranges south in Mexico to Coahuila.

Description Large western skimmer with two dark spots in each
wing smaller than in other similar species. *Male*: Eyes dark brown
(gray in parts of Great Basin); face brown, black above, and with
yellow spots on sides. Thorax brown with two pairs of bright yellow spots on each side; front overlaid with pale bluish pruinosity.
Abdomen dull bluish pruinose. Wings with pale bluish pruinosity
surrounding elongate dark basal spots; dark spots at nodus small
and variable in size. *Female*: Eyes variable, as male; face mostly
yellowish-brown. Thorax patterned as male; abdomen dark brown

with rows of bright yellow spots down S2–8, diverging and becoming smaller toward rear.
Very old individuals become pruinose all over. Wing markings as male but, exceptionally,
basal spot scarcely developed.

261.1
Hoary Skimmer
male—Apache Co.,
AZ, July 2007

261.2
Hoary Skimmer
female—Harney Co.,
OR, July 2006,
Netta Smith

Identification Pattern of wing spots differentiates **Hoary Skimmer** from most other species. Much like **Eight-spotted Skimmer** in color pattern of body and wings, but wing spots in both sexes, especially those at midwing, always smaller; spot at midwing extends width of wing in **Eight-spotted**. White pruinosity in male confined to base of wing and not beyond nodal spots in **Hoary**; two big white spots in each wing in **Eight-spotted**. **Four-spotted Skimmer** with somewhat similar wing spots but body color very different, much lighter, and often orange-brown. Body pattern somewhat like female **Desert Whitetail** but note quite different wing pattern. Males might be mistaken for **Bleached Skimmer**, which has much paler face and eyes, lighter brown wing spots, and yellow costal vein.

Natural History Males perch at tips of sedges at breeding habitats or fly back and forth. Often on fences in arid country where they live. In aggressive interactions, males approach others from below and behind. Pairs sometimes fly in tandem while ovipositing; only king skimmer other than Bleached to behave in this way. Male swings female down to water in and at edge of dense grass and sedges in shallow water. Females also oviposit alone.

Habitat Marshy slow streams, ponds, or seeps, often associated with springs, especially hot springs in northern part of range. May breed in same seepage areas as Black Petaltail.

Flight Season OR May–Aug, CA Apr–Sep, AZ Jun–Aug, NM Jun–Aug.

Distribution Ranges south in uplands of Mexico to Michoacan, but rare in that country.

262 Eight-spotted Skimmer *Libellula forensis* TL 49–51, HW 38–40

Description Large western skimmer with two big dark spots on each wing. *Male*: Eyes dark brown; face brown. Thorax dark brown with pair of elongate yellow spots low on sides, front whitish pruinose. Abdomen entirely whitish pruinose with maturity. Each wing with two big black spots, long basal one and transverse one just past nodus; white patches beyond each dark marking develop with age. *Female*: Eyes dark brown; face brown. Thorax brown with spots as male. Abdomen dark brown to black with line of elongate yellow spots on each side of middle, almost forming

262.1
Eight-spotted Skimmer
male—Lewis Co., WA,
August 2004

262.2
Eight-spotted Skimmer
female—Harney Co., OR,
July 2005

stripes, getting lower on sides of segment to rear and ending on S8. Old females, perhaps not all of them, become pruinose on abdomen. Wings of females polymorphic, typically developing white spots as male in drier regions, without them in wetter regions, but both types in some populations.

Identification Most similar species **Twelve-spotted Skimmer**, which differs in both sexes in having dark spots at tips of all wings, as does somewhat similar female **Common Whitetail**. **Twelve-spotted** also has wider yellow stripes down abdomen. Note male **Twelve-spotted** with dirtier-looking abdomen, not glistening white as in **Eight-spotted** (however, **Eight-spotted** in moister regions more like **Twelve-spotted**), and white patches in wings smaller and distinctly less flashy in **Twelve-spotted**. **Hoary Skimmer** also similar, has much smaller wing spots (especially at nodus) and pruinosity only on wing bases. Female **Desert Whitetail** also somewhat similar but has two bands across midwing instead of one and better developed pale stripes on thorax.

Natural History Males perch on exposed perches at shore but spend much time flying up and down; move constantly, not defending fixed territories. Show much aggression to other males of their own and other similar-sized species, especially Twelve-spotted Skimmers, and fly in front of and slightly below other males they encounter; females approached from above and behind. Copulation brief (less than 5 sec) and in flight, female sometimes guarded by male during oviposition. Egg laying varies from lengthy stay in one spot to eggs spread over large area by individual female. May feed in flight like glider on windy days in lee of tree clumps. Hangs vertically under stems in hot, sunny conditions. Away from water, may perch on twigs high in trees.

Habitat All kinds of ponds and lakes, usually with both shore vegetation and open water.

Flight Season BC May–Sep, WA Apr–Oct, OR May–Oct, CA Apr–Oct, MT Jun–Aug, AZ Jun–Aug, NE May–Sep.

Comments Has hybridized with Widow Skimmer.

263 Twelve-spotted Skimmer *Libellula pulchella*   TL 52–57, HW 42–46

Description Large, striking skimmer with three big black spots on each wing. *Male:* Eyes dark brown; face brown, paler on sides. Thorax brown with two pale stripes on each side, gray above and bright yellow below. Abdomen pruinose gray, underlying femalelike pattern often showing through. Wings with three large dark spots, at base, nodus, and tip; pruinose white spots almost fill in clear areas between dark spots. *Female:* As male but thoracic stripes all yellow. Abdomen broadly brown above, continuous

263.1
Twelve-spotted
Skimmer male—
Skamania Co., WA,
July 2005

263.2
Twelve-spotted
Skimmer female—
Hamilton Co., NY,
July 1999

yellow stripe down each side. No pruinosity on wings. Rare individuals of both sexes lack apical wing spots.

Identification Distinguished from all other king skimmers by three dark wing spots in each wing. Similar **Eight-spotted Skimmer** lacks black wingtips. Females have different abdominal pattern, **Twelve-spotted** striped and **Eight-spotted** spotted. Rare **Twelve-spotted** lacking tip spots look much like **Eight-spotted** but dark spots at nodus fall well short of wing margin, almost reach margin in **Eight-spotted**; last white spot in from stigma in **Twelve-spotted**, behind stigma in **Eight-spotted**; females identifiable by abdominal pattern. Although males quite different, female **Common Whitetail** also poses identification problem, as she has same 12 wing spots. In addition to being distinctly smaller, female **Whitetail** has series of yellowish-whitish spots not contacting edge of each segment, while female **Twelve-spotted** has continuous parallel yellow stripes on either edge of abdomen. In side view of thorax, female **Whitetail** has additional yellow dot in front of anterior yellow stripe, lacking in **Twelve-spotted**. Note that much larger **Prince Baskettail** has similarly spotted wings.

Natural History Males fly back and forth along shore and over water with regular hovering, also perch on twigs and other prominent perches at edge of open water and may return to preferred ones. Seem more likely to stay perched than Eight-spotted Skimmer and fly over shorter beats, definitely more territorial. However, territories changed daily. Quite aggressive to other species as well as its own. Territories up to 1000 square feet in area. Copulation fairly brief and in flight. Females oviposit, with peak visitation at midday, by splashing eggs into water with much aquatic vegetation; also at floating logs. Stay in place or move up to a few feet between series of taps, remaining at water for about 3 min. Guarded by males, guarding effective only at low population densities.

Habitat Lakes and ponds with emergent vegetation, in Northwest commonly smaller and marshier ponds than those preferred by Eight-spotted Skimmers.

Flight Season BC May–Aug, WA May–Sep, OR May–Sep, CA Apr–Nov, MT Jun–Aug, AZ Apr–Sep, NM Jun–Oct, NE May–Sep, TX May–Nov.

Distribution Sparsely distributed in Southwest and not recorded from Mexico. Ranges widely in East from southern Ontario and Nova Scotia south to Gulf Coast and northern Florida.

Description Big, showy skimmer with dark wing bases. *Male*: Eyes and face dark brown. Thorax dark brown, pruinose white on front. Abdomen entirely pruinose white above. Wings dark brown from base to nodus, paler at base, then pruinose white from nodus almost to stigma. In populations around Colorado River in California and Arizona, thorax becomes entirely pruinose. *Female*: Eyes brown; face light brown. Thorax brown with large tan spots low on sides. Abdomen with black central stripe widening toward rear, bordered by wide yellow stripes on each side; brown low on sides  at base. Wing bases become dark but not as dark as male and without pruinosity; extreme tips also dark.

Identification No other king skimmer patterned like this, with large dark basal wing patches. Most similar to **Black Saddlebags**, which has smaller patches only in hindwing and no

264.1
Widow Skimmer
male—Cowlitz Co.,
WA, September 2006

264.2
Widow Skimmer
female—Johnson
Co., IA, July 2004

white beyond them. Similarly showy and white-tailed **Common** and **Desert Whitetails** have black bands toward tips of wings, smaller black markings at bases. **Eight-spotted** and **Twelve-spotted Skimmers** also showy black and white wings but very differently patterned. Female **Widow** just as distinctive as male because of wing pattern.

Natural History Both sexes may be common in meadows and on roadsides well away from water, where they perch on herbs and shrubs. Males at water most of day, perching at water's edge or flying conspicuously back and forth over short beats at ankle to head height, aggressive to males of their own and other species. May hover and flip abdomen up and down during patrol flights. May maintain expansive territory or defend group territory in which dominant male most likely to mate. However, territories move and dominance relationships change regularly. Copulation brief (10–20 sec) and in flight, sometimes briefly at rest. Females oviposit, briefly guarded by male or unattended, by tapping open water steadily or flicking eggs onto surface from just above it. Oviposition brief, no more than a few minutes. Longevity after maturation up to 43 days.

Habitat Lakes, ponds, and pools in slow streams of all kinds with mud bottoms and usually much vegetation, in open and wooded habitats. Common at farm ponds and other created habitats.

Flight Season WA Jun–Oct, OR Jun–Sep, CA May–Oct, AZ May–Nov, NM May–Nov, NE May–Oct, TX Apr–Sep.

Distribution Much of range along Pacific coast from recent expansion. Ranges south in western Mexico to Baja California and Durango, also throughout East from far southern Canada to Louisiana and Georgia.

Comments Has hybridized with Eight-spotted Skimmer.

---

265 Yellow-sided Skimmer   *Libellula flavida*   TL 48–51, HW 38–40

Description Southeastern skimmer with bright yellow sides. *Male*: Eyes dark blue-green; face blackish. Thorax mostly pruinose blue, sides obviously yellow under pruinosity. Abdomen pruinose blue, bright yellow under base. Wings with amber stripe on front, stigma dark brown. *Female*: Eyes brown; face tan. Thorax dark brown in front, yellow on sides divided by narrow dark brown stripe. Abdomen

265.1
Yellow-sided Skimmer male—Gadsden Co., FL, June 2005, Giff Beaton

265.2
Yellow-sided Skimmer female—Angelina Co., TX, July 2004

yellow with fairly wide brown to black central stripe, also dark basal side stripe. Wings colored as male but with wide dark tips and paler, usually yellow, stigma.

Identification Both sexes distinguished from other similarly patterned king skimmers and all other skimmers by yellow sides of thorax and golden stripe along wing. Superficially like **Spangled Skimmer** but with dark stigmas. Females somewhat like **Golden-winged** and **Needham's Skimmers**, which have much duller thoracic pattern and wingtips unmarked or with narrower dark markings. Dark abdominal stripe on **Yellow-sided** female wide throughout its length (similar in **Spangled**), in **Golden-winged** and **Needham's** narrow and then suddenly widening at abdomen tip.

Natural History Weakly territorial males perch on stems and twigs at tops of low vegetation and fly back and forth over small area with frequent hovering for a few seconds; more flying than perching. Females in nearby open woodland. Copulation at rest for about 30 sec, then female perches for 10–15 sec before ovipositing by hovering 6 inches above the water, dropping to the water to tap once, then rising again, repeating process for up to several hundred times with brief perching at intervals. Male guards at first but not throughout process.

Habitat Boggy ponds, seeps, and slow streams.

Flight Season TX Mar–Aug.

Distribution Also in Southeast from Missouri, Ohio, and New York south to Gulf Coast; absent from most of Florida.

---

266 Comanche Skimmer *Libellula comanche* TL 47–55, HW 35–44

Description Large southwestern skimmer with mostly white stigmas. *Male*: Eyes blue; face white. Mostly powdery blue, thorax whitish on sides. *Female*: Eyes reddish above, becoming pale blue; face white. Thorax brown in front, white on sides bisected by narrow dark stripe. Abdomen yellow to orange-brown, with three black stripes down its length (lateral stripes may be sufficiently narrow that abdomen looks mostly yellow outside dorsal stripe). Old females become pruinose blue, colored much as males.

Identification Almost no overlap with most similar species, **Spangled Skimmer**, male of which has dark blue-green eyes and dark face. Also, **Spangled** has dark streaks at wing bases; **Comanche** lacks them. Male **Comanche** distinguished from

266.1
Comanche
Skimmer
male—Travis
Co., TX, June
2005

266.2
Comanche
Skimmer
female—
Harney Co.,
OR, July 2005

all other blue skimmers (**Yellow-sided Skimmer**, **Eastern** and **Western Pondhawks**) by mostly white stigmas. Also much larger than blue PONDHAWKS and **Blue Dasher**. Rather like **Bleached Skimmer**, which has dark stigmas, yellow costal vein, and brown in wing bases. Less like **Hoary Skimmer**, which is gray rather than blue and has dark stigmas and prominent wing markings. Occurs with both species at springs. Female **Comanche** differs from **Spangled Skimmer** by blue-gray eyes (brown in **Spangled**) and only an indication of dark color at wingtips (usually widely dark in **Spangled**), from all other similar species by mostly white stigmas. Female and immature **Comanche** sometimes with narrower yellow stripes down abdomen than **Spangled** because of broad black stripe along each side.

Natural History Males perch at head height in tall vegetation and fly back and forth over water, chasing other males rapidly and vigorously. Females in open areas, even open desert, near breeding habitats. Copulation in flight.

Habitat Springs, seeps, and pools in clear streams.

Flight Season OR May–Sep, CA May–Oct, AZ May–Nov, NM May–Oct, TX May–Oct.

Distribution Ranges south in uplands of Mexico to Sonora and Coahuila.

---

267 Spangled Skimmer   *Libellula cyanea*   **TL 41–46, HW 33–35**

Description Midsized eastern skimmer with mostly white stigmas. *Male*: Eyes dark, brown to blue-green; face metallic black. Thorax and abdomen entirely pruinose blue. Stigmas white proximal two-thirds, black distal two-thirds. *Female*: Eyes brown; face tan. Thorax dark brown in front, cream on sides with brown down middle. Abdomen yellow to orange-brown with dark brown central  stripe widening toward rear. Dark stripe on lateral carina sometimes expanded upward. Wings suffused with amber, especially in front, and with broad dark tips.

Identification See **Comanche Skimmer** for most similar species; rarely if ever occur together. No other large skimmers with mostly white stigmas. **Yellow-sided Skimmer** rather similar but has dark stigmas, amber line down wings, and thorax with yellowish sides in male, bright yellow sides in female. Male **Spangled** unlikely to be confused with other all-blue species in range, including smaller **Eastern Pondhawk** and **Blue Dasher**. Female patterned like several other king skimmers, but white stigma always diagnostic.

Natural History Males perch in shrubs and marsh vegetation over water or fly back and forth along shore at waist height or below. Female oviposits on water or wet mud, sometimes in dense vegetation. Both sexes can be very common in roadside weeds and meadows well away from breeding habitat.

267.1
Spangled Skimmer male—
Natchitoches Par., LA, May
2005, Netta Smith

267.2
Spangled Skimmer
female—Grant Par.,
LA, May 2005

Habitat Lakes and ponds, primarily those with silty/muddy bottoms, in open or woodland.
Flight Season TX Apr–Aug.
Distribution Ranges through eastern United States north to Wisconsin, Michigan, southern New York, and Maine; absent from much of southeastern Coastal Plain.

268 Bar-winged Skimmer *Libellula axilena*	TL 60–62, HW 43–49

Description Large southeastern skimmer with black bars along wings. *Male:* Eyes dark reddish brown; face metallic black. Overall blackish, front of thorax and base of abdomen pruinose bluish-white. In older individuals, sides of thorax and sides of abdomen become somewhat pruinose; top of abdomen remains black. Wings typically with well-defined basal streak, dark spot at

nodus, interrupted dark line along front edge between nodus and entirely dark stigma, and narrow dark tip. Sometimes nodal spot and front-edge stripe connected. Extreme base of hindwing becomes pruinose white with age. Immature colored as female. *Female:* Eyes dark reddish brown above; face brown in center, white on edges. Thorax brown in front, mostly pale gray on sides, with dark stripe extending below hindwing conspicuous, wider toward lower end. Abdomen yellow to light brown on sides, paler at base, with black dorsal stripe becoming wider toward rear, covering much of top by S8–10. Wings as in male, dark tip slightly more extensive.
Identification Mature males differ from rather similar **Great Blue** and **Slaty Skimmers** by blue-gray and black pattern (**Great Blue** all blue, **Slaty** all black), wing markings (usually best developed in **Bar-winged**), and face (white in **Great Blue**, black in **Bar-winged** and **Slaty**).

268.1
Bar-winged Skimmer
male—Osceola Co.,
FL, July 1977; inset
immature
male—Washington
Co., FL, June 2004

268.2
Bar-winged Skimmer
female—Washington Co.,
FL, June 2004, Netta Smith

268.3
Bar-winged Skimmer
immature male—Washington
Co., FL, June 2004

Black stigma and duller color distinguish from **Spangled** and **Yellow-sided Skimmers**. Female and immature male very similar to female **Great Blue** and **Slaty**. Differ from **Slaty** by dark face with sharply defined white edges, and dark marking on side of thorax below hindwing very conspicuous, wider near lower end. Differ from **Great Blue** in same ways and also by entirely black femora. Wing markings usually diagnostic, but **Slaty** can be as heavily marked.

Natural History Males typically perch fairly high over water on tree branches but can be seen in wide variety of situations. Both sexes, along with other skimmers, have been found perched in numbers all along fence wires.

Habitat Wooded slow streams and sloughs, forest pools. Occurs with both Slaty and Great Blue Skimmers.

Flight Season TX May–Jul.

Distribution Also widely in Southeast, north to Kentucky, Pennsylvania, and Massachusetts.

Description Large, black eastern skimmer with dark legs, dark stigma, and dark face at maturity. *Male*: Eyes dark brown; face metallic blue-black. Thorax and abdomen black, overlaid with dark slaty-blue pruinosity. Wings usually clear but may have narrow dark tips and tiny spot at nodus; small number with stripe on anterior edge from nodus to stigma. *Female*: Eyes brown; face dull brown, paler below. Thorax brown in front, yellow to gray on sides. Dark marking below hindwing base elongate, varying from quite narrow to slightly more conspicuous; if ex-

269.1
Slaty Skimmer
male—Marion Co.,
FL, June 2004

269.2
Slaty Skimmer
female—Essex Co.,
MA, August 2004,
Tom Murray

269.3
Slaty Skimmer immature
female—Liberty Co., FL,
June 2004

panded, widest toward base. Abdomen yellow to light brown with wide black central stripe becoming wider to rear. Dark side stripe on basal segments narrowing to rear. Old females can become entirely dark gray pruinose. Wings always with dark tips more extensive than in males, also often with dark basal streak, usually with small spot at nodus, often with dark streak along front edge between nodus and stigma. At most heavily patterned, broad dark stripe along front edge of wing from nodus to extensive dark tip.

Identification No other large, all black skimmer. Male **Black Setwing** similarly colored, with brown eyes and black face, but much smaller and more slender, lives on southwestern rivers rather than eastern lakes; slight overlap in range in central Texas. Females and immature males very similar to female **Bar-winged** and **Great Blue Skimmers**. Differs from **Bar-winged** by face whitish to light tan, never black except in mature male; and dark marking on side of thorax less conspicuous, not wider near lower end. Males distinguished from **Bar-winged** by less extensive wing markings, never with basal streak. Females much more difficult, some with identical wings and some female **Slaty** with even more heavily marked wings than **Bar-winged**. Differs from **Great Blue** by entirely black femora (pale at base in **Great Blue**) and somewhat darker face, especially during maturation. Also perceptibly smaller.

Natural History Males perch on shrubs and other vegetation, from knee height to well above water, and fly back and forth along shorelines. Territorial aggression involves circling around one another, then rapid chase. Copulation for 30–60 sec, both perched and in flight. Females oviposit by splashing eggs into vegetation, alternating brief hovering a few inches up and dropping to water for several taps. Male may guard by flying back and forth above her.

Habitat Wide variety of lakes, ponds, and long pools in streams, mostly in wooded areas; sometimes in swamps with Great Blue Skimmers.

Flight Season TX May–Sep.

Distribution Ranges throughout East, north to Wisconsin, southern Ontario, and Nova Scotia.

270 Great Blue Skimmer *Libellula vibrans* TL 56–63, HW 48–51

Description Very large blue skimmer with white face. *Male*: Eyes bright blue or turquoise; face white. Thorax and abdomen entirely pruinose blue with maturity. Wings with dark streak at base, dark spot at nodus, and faint dark tip. Immature colored as female, thorax becoming pruinose first. *Female*: Eyes dark reddish-brown over dull greenish; face dirty white. Body pattern typical of skimmers.

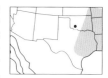

Only dark marking under wings on side of thorax, a faint line under hindwing, barely visible at any distance; if expanded at all, widest toward upper end. Femora mostly pale, dark only at tips (pale color always extends beyond half length). Females become duller and darker with age; eyes may become blue. Wings as males but with dark tips slightly more extensive.

Identification Size alone distinctive in this, our largest skimmer, but some similar species approach it. No other similar king skimmers have blue or blue-green eyes. With strikingly marked dark-and-light thorax and abdomen, immatures and females very similar to **Bar-winged** and **Slaty Skimmers** but femora extensively pale-based. Differ further from **Bar-winged** in white face and much less conspicuous marking on side of thorax below hindwing base, further from **Slaty** in white face (**Slaty** may have very pale face) and slightly less conspicuous marking on side of thorax. Only one of three with no black under forewing, lacking pale point (ear of "wolf's head") of other two.

Natural History Immatures and females may be common in sunny clearings or at forest edge, perching from low to high. Males defend small territories over woodland pools and in dense swamps, usually perching in sun from waist height to well above head height. Few other skimmers are such confirmed swamp dwellers, and males often seen at places lacking most other odonates. Copulation at rest for about 25 sec, then male hover-guards female. Female taps water vigorously, splashing water drops and eggs up to 8 inches away, sometimes on bank.

270.1
Great Blue Skimmer
male—Dixie Co.,
FL, June 2004

270.2
Great Blue Skimmer
female—Newton Co.,
TX, July 2004

270.3
Great Blue Skimmer
immature
female—Putnam Co.,
FL, April 2005

Habitat Wooded swamps and slow streams with dark, mucky water. May breed in pools isolated from stream. Most restricted swamp-dweller of king skimmers.
Flight Season TX May–Sep.
Distribution Also throughout East, north to Wisconsin, southern Ontario, and New Hampshire.

Description Yellow-orange southeastern skimmer with mostly golden-tinted wings. *Male*: Eyes orange over brown; face red-orange. Thorax brown, sides paler. Abdomen orange, paler at base, and with narrow brown stripe down center, becoming darker to rear and widest on S9. Wings amber-tinted, almost all veins and stigma orange, brightest on front edge. *Female*: Eyes brown over pale

greenish; face tan, paler at edges. Thorax brown in front, whitish on sides. Abdomen yellow with narrow black stripe down center, widening on S8–9. Wings less colored than male but anteriormost (costal) veins yellow, faint dusky tips. Some females almost as bright as males.

Identification Few North American species are as bright golden-orange as males of this one, with similarly bright wings. Males really look orange, not the red to scarlet of some other species in their range, and that together with colored wings makes identifying them easy. Look-alike **Needham's Skimmer** poses only difficulty (see that species). Female and immature **Needham's** have basal part of costal vein darker than other wing veins, pale on sides of thorax extends as "thumb" onto brown front. Female **Golden-winged** with thorax and abdomen patterned much as females of other king skimmers, for example **Slaty Skimmer** and its near relatives, but yellow anterior wing veins and orange stigma are sufficient distinction. Females of golden-winged pair of species do not typically show prominent dark wingtips as in clear-winged group, but occasional individuals show distinctly darkened tips and might be mistaken for **Yellow-sided Skimmer**, which has bright yellow sides of thorax and dark abdominal stripe wide throughout its length.

Natural History Males perch on twigs and leaves near and at shore or in tall grass over water, even quite high; also much cruising between perches. Females and immature males

271.1
Golden-winged
Skimmer—Angelina
Co., TX, June 2003,
Greg Lasley

271.2
Golden-winged
Skimmer female—
Lake Co., FL, June
2004

feed in sunny clearings near breeding habitat, perching from ground level to well up in shrubs. Male guards ovipositing female.

Habitat Open ponds and lakes with much shore vegetation, also shallow ponds with tall grasses scattered throughout.

Flight Season NE Jun–Aug, TX Apr–Sep.

Distribution Also widespread in Southeast, north to Ohio, New York, and Massachusetts; scarce and local in northern part of range.

---

272 Needham's Skimmer *Libellula needhami* **TL 53–56, HW 39–44**

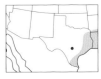

Description Red-orange skimmer with golden-tinted wings, mostly of coastal regions. *Male*: Eyes red over greenish-brown; face red-orange. Thorax orange-brown in front, tan on sides, with extension of tan forward just onto lower part of front. Abdomen red-orange with narrow brown stripe down center, becoming black at rear and widest on S8–9. Costa between base and nodus slightly duller than other veins. In brightest individuals, entire body bright red-orange, stigma and wing veins, especially anterior ones, bright orange; entire wings suffused with orange. *Female*: Eyes brown over pale greenish; face tan, paler at edges. Thorax brown in front, whitish on sides. Abdomen yellow with narrow black stripe down center, widening

272.1
Needham's Skimmer
male—Wakulla Co.,
FL, June 2004

272.2
Needham's Skimmer
female—Collier Co.,
FL, April 2005

on S8–9. Pattern on thorax much more evident than in mature male. Wings with veins fairly dark, stigma and anteriormost veins yellow except basal part of costa dark before nodus.

Identification Very much like **Golden-winged Skimmer**. Males look identical in field, with same degree of variation in brightness, but some **Needham's** have costal vein darker before nodus, contrasting slightly with bright orange veins behind it (same color as rest of wing in **Golden-winged**). Smaller wing veins behind middle of wings tend to be orange in **Golden-winged**, browner in **Needham's**. Some males unidentifiable in field, not easily in hand; look for females, which are easier, with difference in costal vein color more pronounced, difference in pattern on side of thorax obvious. **Needham's** has thumblike extension of pale sides onto front; this shows in younger males and sometimes in mature ones. Hind tibiae black in **Golden-winged**, brown in **Needham's**, but difference not easy to see. Learn which species occurs where; they seem to coexist rarely, and **Needham's** is most common on and near the coast and in far southern Texas. Other large orange-to-red skimmers much more different, for example **Flame Skimmer** (barely overlaps with **Needham's** in central Texas) bulkier, with orange confined to inner half of wings. Male TROPICAL KING SKIMMERS (**Carmine** and **Roseate**) much redder, with uncolored wings. No females other than **Golden-winged** very similar, although various king skimmers without yellow in wings have similar body patterns. Dark brown front and bright yellow sides of thorax, as well as wide dark abdominal stripe, distinguish female **Yellow-sided Skimmer**.

Natural History Males perch along shore and in tall marsh vegetation, typically at waist to head height on stems or tips of grass stalks and cattails. Both sexes forage in sunny clearings in woodland or in open areas. Can be very abundant locally.

Habitat Large ponds and lakes and open marshland, including coastal marshes. Typically larger water bodies than preferred by Golden-winged Skimmer and most common near coast.

Flight Season TX May–Nov.

Distribution Ranges south in Mexico to Jalisco and Tamaulipas, also Quintana Roo, but nowhere common in that country. Also east and north, mostly near coast, to Maine, and in Bahamas and Cuba.

These large dragonflies are shaped and sized much like the king skimmers of temperate latitudes that they replace in New World tropics (and much like *Orthetrum* skimmers in Old World tropics). Wing venation is also similar in these genera, wings with many antenodals and wavy distal veins. Males of most but not all tropical king skimmers are red, females brown. Their presence at ponds at lower latitudes in the West show tropical affinities of those latitudes. They typically perch with front legs folded up behind the head, usually with wings and abdomen level, like king skimmers. Copulation is brief and occurs in flight, so mating pairs are uncommonly seen, but watch for males guarding ovipositing females. World 18, NA 3, West 2.

273 Carmine Skimmer   *Orthemis discolor*   TL 47–53, HW 37–43

Description Large red skimmer of southern Texas. *Male*: Eyes and face bright red. Brilliant carmine-red all over, thorax and abdomen slightly toward purplish from pruinosity (thorax usually more so). Wings unmarked, veins blackish. *Female*: Eyes red above, tan below; face brownish-red. Thorax brown in front, paling to golden-brown on sides, some with faint paler lateral stripes. Con-  spicuous pale median stripe on thorax extends onto abdomen base. Abdomen reddish-brown with very little pattern; conspicuous black flaps on sides of S8.

273.1
Carmine Skimmer
male—Hidalgo Co.,
TX, November 2005

273.2
Carmine Skimmer
female—Puntarenas
Prov., Costa Rica,
January 2005

273.3
Carmine Skimmer
immature
male—Gonzales Co.,
TX, July 2004

Identification Mature males similar in size, shape, and behavior to **Roseate Skimmer**, identifiable by redder color. Face and eyes of **Carmine** red, those of **Roseate** much darker, purplish. Thorax and abdomen usually redder in **Carmine**, but this difference varies, and some males could not be distinguished without seeing head. In hand, **Carmine** males have plain, unmarked thorax, whereas **Roseate** males retain traces of dark spots on lower part of thorax that characterize females and immature males. Females and immature males differentiated readily by thoracic pattern, plain or lightly marked with parallel stripes in **Carmine** and conspicuously striped with somewhat irregular stripes in **Roseate**. **Roseate** also has dark markings low on thorax flanking pale stripe at midlength. Female **Carmine** has black flaps on sides of S8, slightly larger than brown flaps of **Roseate**. Also, **Roseate** females often have dusky spot on top of S8, lacking in **Carmine**. Female **Carmine** somewhat like female **Flame** and **Neon Skimmers**, which are larger, with distinctly broader abdomen, and show orange color at wing bases or at least orange basal veins. Check also other skimmers such as **Mayan Setwing, Red Rock Skimmer, Red-tailed Pennant**, and very rare **Claret Pondhawk,** which are bright red. All either smaller than **Carmine**, with color in wings, or both.

Natural History Males perch and actively patrol territories along water's edge, very aggressive toward conspecific males and sometimes male Roseate and other skimmers. Territory usually well defined, may be about 30 feet long. Copulation brief (about 10 sec), in flight. Females oviposit by tapping water and dropping eggs, more commonly by splashing up water droplets that land either on water or on adjacent bank while facing it. Very methodical, often working sheltered area under overhanging vegetation or bank for several minutes, often with male guarding in flight or perched nearby.

Habitat Ponds, sloughs, and slow streams with mud bottoms, commonly with wooded borders; more characteristic of forested habitats than Roseate Skimmer. Roseate probably found everywhere Carmine occurs, but not the reverse, although Carmine has been found at muddy stock ponds.

Flight Season TX May–Nov.

Distribution Appears to be increasing rapidly in Texas. Ranges south, mostly in lowlands, to Argentina.

Comments Called Orange-bellied Skimmer in earlier publications and confused with Roseate Skimmer in all older literature.

---

274 Roseate Skimmer *Orthemis ferruginea* TL 48–53, HW 36–44

Description Large rosy-purple skimmer of southern states. *Male*: Eyes dark red; face metallic red-violet above, dull reddish below. Thorax purplish, overlaid by matte pruinosity. Abdomen red, pruinosity gives it pinkish cast. Wings unmarked, veins reddish. *Female*: Eyes reddish-brown above, tan below; face dull brownish. Thorax brown with complex pattern on sides of pale yellowish

274.1
Roseate Skimmer male—Hidalgo Co., TX, November 2005

274.2
Roseate Skimmer female—Hidalgo Co., TX, November 2005

274.3
Roseate Skimmer immature male—Galveston Co., TX, July 2004

stripes, black markings (below as well as on lower sides). Contrasty whitish median stripe on thorax and abdomen base. Abdomen reddish-brown; conspicuous dark flaps with pale center on sides of S8. Old females can become dull and even somewhat purplish-pruinose.

Identification Nothing else like it in North America except **Carmine Skimmer,** which see. Female superficially like other large skimmers, but thoracic pattern distinctive. Entirely clear wings (rarely tiny brown area at bases) also distinguish it from some other large reddish skimmers. Other red, clear-winged species in its range include much smaller **Red-tailed Pennant**, without pruinosity. Female **Neon Skimmer** much bulkier, without characteristic thorax pattern of **Roseate**.

Natural History Males perch at and actively patrol territories along water's edge, often with much interaction with adjacent males, also with male Carmine and other large skimmers. Copulation brief (usually less than 10 sec) and in flight, then egg-laying female guarded by male. Females oviposit by tapping water and releasing eggs, more commonly by splashing up water droplets that land either on water or on adjacent bank. Female may remain in small circumscribed area for surprisingly long periods. Both sexes perch on twigs in sunny clearings with abdomen usually below horizontal (obelisking very rarely seen), sometimes several in small area. Long flight season but often most common in fall.

Habitat Very broad habitat tolerance, prefers mud bottoms for larval habitat. Ponds, ditches of all size, and open marshes with mud bottoms, also scummy stock tanks and slow-flowing streams. May be tall emergent vegetation or virtually none, but open water seems necessary. Completely open or wooded country. May breed in very small, including artificial, water bodies.

Flight Season CA May–Jan, AZ all year, NM Mar–Oct, TX all year.

Distribution Ranges south to Costa Rica, east and north to Arkansas and Virginia; also Cuba.

Narrow-Winged Skimmers   *Cannaphila*

This small neotropical genus is related to king and tropical king skimmers by wing venation and flap on S8 of the female, but the hindwing is relatively narrow at base. As do near relatives, members usually perch with four legs, front pair folded behind the head. One species ranges south to Argentina. World 3, NA 1, West 1.

275 Gray-waisted Skimmer   *Cannaphila insularis*   TL 37–44, HW 30–31

Description Medium-sized forest-dwelling skimmer with narrow wings, locally common in south Texas. *Male*: Eyes green; face metallic blue-black above, white below. Thorax conspicuously and evenly striped very dark brown and pale gray to white. Abdomen black with pruinose gray base to S4 or S5. *Female*: Eyes dark reddish-brown over blue-gray; face dark brown above, cream below.  Thorax heavily striped brown and yellow. Abdomen vividly striped brown and yellow, sutures and carinas darker and tip of S8–10 becoming blackish. Darkens with age until finally thorax and abdomen black as in male; some individuals even acquire gray abdomen base. Conspicuous dark flaps on sides of S8. Wingtips with more dark color than in males. Immature of both sexes with orange face, striped thorax, abdomen largely orange to orange-brown with black tip, becoming striped; pale stripes most evident on basal segments.

Identification Distinctive perching style, abdomen along branch. Male superficially like **Blue Dasher** because of head and eye color and striped thorax, but abdomen gray and black rather than blue. Could be mistaken for **Black Setwing** if gray base of abdomen not seen, but very different perching behavior and general habits. Almost overlaps in range with **Bar-winged Skimmer**, larger species with gray-based, black-tipped abdomen but brown eyes and dark face. Immature (striped) female much more brightly marked, with

275.1
Gray-waisted Skimmer
male—Kinney Co., TX,
July 2004

275.2
Gray-waisted Skimmer
female—Bexar Co., TX,
August 2004, Greg Lasley

275.3
Gray-waisted Skimmer
immature female—Puntarenas
Prov., Costa Rica, February 2005

much orange or yellow on abdomen, not like any other species. Look for rather conspicuous flaps on abdomen and recall that considerably larger female TROPICAL KING SKIMMERS also show these.

Natural History Males at water and both sexes away from it may perch at knee height in tall grass or well up on vines or dead twigs in trees and shrubs, usually in shade. Holds abdomen aligned along perch, often almost vertical; does not droop wings or obelisk in sun. Females and immatures stay under canopy, and all retreat up into trees when disturbed. Female oviposits in dense vegetation.

Habitat Shaded swamps and slow muddy streams in forest.

Flight Season TX Jun–Sep.

Distribution Ranges south in lowlands to Panama.

Amberwings are very small, unmistakable skimmers, males with yellow to orange wings and females usually with complex and variable wing patterns. Their abdomen is spindle-shaped, narrower at front and back than in the middle. Legs are long and support the small body well above substrate. Males are found low over water, but both sexes perch from ground level to well up in trees when away from it, where they appear to be wasp mimics. Most species are tropical. World 13, NA 3, West 3.

Table 10   Amberwing (*Perithemis*) Identification

	A	B	C	D
Slough	1	12	3	3
Mexican	2	2	12	12
Eastern	2	1	2	2

A, leg color: 1, with dark markings; 2, all pale.
B, male wing spot: 1, present; 2, absent.
C, thorax: 1, unmarked; 2, pale spots; 3, dark stripes.
D, abdomen markings: 1, none; 2, chevrons/triangles; 3, stripes.

276 Slough Amberwing *Perithemis domitia* TL 22–25, HW 16–19

Description Tiny tropical orange-winged skimmer with dark legs. *Male*: Eyes red-brown above, tan below; face brownish-orange. Thorax evenly striped with dark brown and tan. Abdomen yellow-orange with narrow median and wide conspicuous dorsolateral black or brown stripes. Cerci vary from dull yellow to black. Legs mostly black or yellow with black stripes on outer surfaces. Wing veins and stigmas orange, membrane equally dark. *Female*: Eyes brown over greenish. Body color as male. Wings with orange bases, bordered and spotted with brown, and clear tips (may be brown on extreme tips).

Identification Easily distinguished by dark legs, strongly striped thorax and abdomen; wing color is darker than in **Eastern** and **Mexican Amberwings**. **Mexican Amberwing** quite plain, without stripes on body, and with much paler, golden wings. **Eastern Amberwing** more like **Slough**, with moderate striping on thorax and abdomen, not quite like that on **Slough** but could be confusing. In **Eastern**, pair of pale lateral thoracic stripes wide and wavy, color more intense at lower end, and usually separated by wide brown stripe. In **Slough**, side of thorax pale with two narrow brown stripes. Stripes on abdomen straight and parallel in **Slough**, with diagonal component forming triangles on top in **Eastern**. Dark leg color of **Slough** good close-range mark. Wing veins look orange in **Slough**, yellow in **Mexican**. Female wing pattern distinctive, basal half orange without pronounced darker areas as in other two species. Sufficient variation, however, that occasional **Eastern** or **Mexican** female may look like this; check other characters.

Natural History Males perch low over water, under tree or shrub canopy, typically in shade. Often patrol limited areas, probably because of limited space in preferred habitats, but defend them vigorously against other amberwings. They examine potential oviposition sites and lead females to them, as in other amberwings. Less often seen away from water than other amberwings; females and pairs rarely seen, perhaps mating well away from water. Females perch from near ground to fairly high in trees.

Habitat Wooded pools in streams and sloughs, less often ponds, but always where trees present. May be at very small and isolated water bodies or narrow channels.

Flight Season AZ Jul–Oct, TX May–Nov.

Distribution Ranges south, mostly in lowlands, to Brazil.

276.1
Slough Amberwing male—Puntarenas
Prov., Costa Rica, January 2005

276.2
Slough Amberwing female—Sonora,
Mexico, August 2006, Netta Smith

277 Mexican Amberwing   *Perithemis intensa*   TL 23–26, HW 20–23

Description Tiny, bright yellow-winged skimmer of Southwest with orange legs. *Male*: Eyes brownish red above, pale greenish below; face yellow. Thorax yellow with faintly indicated darker pattern of stripes. Abdomen yellow-orange, often with pair of short and faint black diagonal lines on some abdominal segments. Cerci yellow, often with dark tips. Wing veins and stigma yellow to yellow-orange,

membrane a bit darker orange. Some males with faint brown spot near base of each wing. *Female*: Body colored as male. Wings clear with orange crossbands, one between base and nodus and another just past nodus. Either or both can contain dark brown spot or, less commonly, be entirely brown. Some females with entire wings washed with orange.

Identification Pale overall color, not very brightly marked, distinguishes this species from other amberwings in both sexes. Stigmas and wing veins look yellow-orange in male **Mexican**, red-orange in **Eastern** and **Slough**; especially apparent in sun over dark water. Typically female **Mexican** wings lack larger brown patches typical of other two species, but some have them, and some female **Eastern** and **Mexican** look alike except for thoracic pattern. With larger wings than other amberwings, **Mexican** has crossveins in triangles, visible at close range and in photos.

277.1
Mexican Amberwing
male—Sonora,
Mexico, August 2006

277.2
Mexican Amberwing
female—Sonora,
Mexico, September
2004

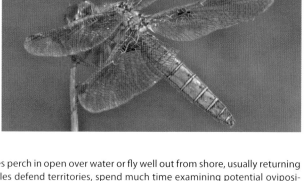

Natural History Males perch in open over water or fly well out from shore, usually returning quickly to perch. Males defend territories, spend much time examining potential oviposition sites and chasing intruding males of their own species or other amberwings. Approaching females quickly led to egg sites, then brief copulation in flight, then at rest, may last a minute. Females oviposit by hovering in one spot, splashing eggs onto floating branches, plant roots, and other objects at surface, even mats of algae; male guards. Both sexes encountered perched on twig tips away from water, from knee to head height and above, with wings and abdomen elevated, sometimes greatly so in midday sun.

Habitat Mud-bottomed lakes, ponds, ditches, and pools in streams and rivers, mostly in open areas but also with wooded banks. All three amberwings may occur at same site.

Flight Season CA Apr–Nov, AZ Apr–Nov, NM Jul–Sep.

Distribution Ranges south in uplands of Mexico to Guerrero, also Yucatan (perhaps in error).

---

278 Eastern Amberwing *Perithemis tenera*   TL 20–25, HW 16–19

Description Tiny eastern skimmer with orange wings and legs. *Male*: Eyes reddish-brown above, pale greenish below; face orange-brown. Thorax brown with faint yellowish anterior stripes, two prominent yellow lateral stripes. Abdomen orange with brown to black dorsolateral streaks slanted toward midline at rear. Series of yellow triangles down middle of abdomen in more brightly marked individuals. Cerci yellow. Wing veins orange, stigma usually darker orange, membrane about same darkness. Just-emerged males with faintly amber-tinged wings. *Female*: Wings clear but with irregular black spot between base and nodus and another, larger one between nodus and stigma; typically both surrounded by orange. Also orange costal stripe; stigma orange-brown. Very rare andromorph females may have entirely yellow-orange wings as males, with some dark smudging.

Identification Pattern on thorax allows distinction from other two regional amberwings. Abdominal pattern also usually distinctive, with series of yellow triangles. **Mexican Amberwing** can show similar triangles but never as contrasty with brown of abdomen; **Mexican** abdomen typically uniformly pale. Yellow cerci contrast with dark abdomen tip in **Eastern**, about same color as abdomen tip in **Mexican** and **Slough**. **Eastern** distinguished from **Slough** where they cohabit by lighter-colored abdomen, yellow-orange versus orange. Many **Eastern Amberwings** have tiny dark spots on each wing between body and nodus, usually lacking in **Mexican** and not present in **Slough**. Wing pattern in female **Eastern** quite different from dark-based wings of **Slough** but overlaps with **Mexican**. Most **Eastern** have extensive brown markings, **Mexican** only small brown spots within yellow areas, but lightest-marked **Eastern** much like **Mexican** and must be distinguished by body pattern.

Natural History Roosts in trees at night. Males stay at water through much of day, perching and patrolling small areas (10–15 feet in diameter) just above open water, and may remain at

278.1
Eastern Amberwing male—Hidalgo Co., TX, May 2005

278.2
Eastern Amberwing female—Hidalgo Co., TX, May 2005

one site for at least several days. Typically perch on flat or elevated water-lily leaves or projecting twigs. Quite aggressive toward other insects of their own size and color, probably mistaken identity. Males examine potential oviposition sites at great length and then defend a small territory that includes a good one. Territories may be as tight as one per 100 square feet. They choose lookout perches farther from shore than their oviposition sites, perhaps to avoid conflicts with neighboring males, but much conflict with wandering males. After quick copulation (averaging 17 sec) at rest, female led to oviposition site, typically something wet at water level to which eggs will adhere. Common sites are floating algal mats, wet logs or projecting twigs, or cups of water in water-lily leaves. Females may tap water or wet substrate 100 times or more in one spot, changing direction at intervals. Both sexes feed away from water, especially females, which may be found widely in weedy fields or perched well up in shrubs or low tree branches. Often perch with wings, or hindwings only, elevated and may point abdomen toward or away from sun at midday. Both sexes, especially females, may be effective wasp mimics. Wing-waving when perched seems to enhance this similarity, and female in flight, with dark wing bases and spindle-shaped abdomen, is sufficiently wasplike to fool an odonate enthusiast and presumably an insect-eating bird.

Habitat Wide variety of ponds, slow streams, and lake shores, usually with mud bottom; in open areas but sometimes associated with woodland.

Flight Season AZ Mar–Oct, NM May–Oct, NE May–Oct, TX Mar–Nov.

Distribution Ranges south in Mexico to Durango and Tamaulipas. Widespread in East, north to Minnesota, southern Ontario, and Maine.

Comments May merge with widespread tropical Pallid Amberwing, *Perithemis mooma*, in northern Mexico, but no definitive study published. Pallid lacks pale spots on thorax, and males lack dark spots in wings that characterize Eastern. Female wings either with contrasty dark bands near base without orange borders that characterize Eastern Amberwing or with broad orange crossbands farther from base. This species could stray north across the border.

Tropical Pennants *Brachymesia*

Two species of tropical pennants are often seen perching at tips of vertical twigs and reed stems, males over open water. With abdomens held horizontal, they look like little flags and are often whipped about by the wind. Often, they hold their wings elevated, perhaps an adaptation to stabilize them in the breeze, and all species are likely to obelisk at midday. Only somewhat smaller metallic pennants, marl pennants, and small pennants normally perch over water in the same way. The Red-tailed Pennant is more varied in perching habits, more like other skimmers in perching on twigs of trees and shrubs and often nearer the ground in vegetation. All of them commonly perch on fence wires. When away from water, they may be found perching on twigs of shrubs and trees, even well above ground. Abdomen shape, with basal segments dramatically expanded in side view, is a good mark for all three species (except female Red-tailed). World 3, NA 3, West 3.

279 Red-tailed Pennant *Brachymesia furcata* TL 41–46, HW 32–36

Description Small tropical skimmer with unpatterned wings, red male and brown female. *Male*: Eyes red over blue-gray; face red, varying to paler reddish-tan. Thorax reddish-brown. Abdomen bright red, often with limited black markings on top of S8–9. *Female*: Eyes dark reddish-brown over blue-gray. Thorax tan, abdomen light reddish-brown, marked as male. Some individuals (oldest?) become about as red as males on abdomen.

Identification Much larger **Carmine** and **Roseate Skimmers** also common at tropical latitudes; males fly back and forth over water at same ponds and lakes but are larger and do not show contrast of browner thorax and redder abdomen. These larger species also slightly more pruinose purplish looking, especially on thorax. Male **Autumn Meadowhawk** smaller and more slender than **Red-tailed**, thorax and abdomen same color, unlikely to be perched over open water. Female **Red-tailed** could be mistaken for **Tawny Pennant** but much more compact, with shorter abdomen. Much smaller than females of **Carmine** and **Neon Skimmers**, also plain brown. Female **Band-winged** and **Black-winged Dragonlets** also rather plain but with more slender abdomen, usually dark rectangles visible along sides; also front of thorax usually darker than sides (not so in **Red-tailed Pennant**). Female **Autumn Meadowhawk** smaller and more delicate with scoop-shaped subgenital plate. Female of rather similarly plain brown but very rare **Claret Pondhawk** has good-sized patch of color at base of hindwing, also much shorter cerci and prominently projecting subgenital plate.

Natural History Males perch on exposed branches over water or fly back and forth along shore and over water, sometimes for lengthy periods with some hovering. Wings held flat or slightly lowered, not raised as is common in two larger species. Patrol beat up to 50 feet in length. Usually perch horizontally but may elevate abdomen in obelisk. Very aggressive to other species as well as its own. Females seldom seen, usually in clearings in wooded areas, may be high in trees. Copulation brief (about 10 sec, can be up to a minute or more), in flight, then pair separates after landing. Ovipositing female dips to water every few feet

279.1
Red-tailed Pennant
male—Monroe Co.,
FL, April 2005

279.2
Red-tailed Pennant
female—Hidalgo Co.,
TX, November 2005

in rapid flight, at times stopping to tap in one place on floating vegetation, usually with guarding male keeping up with her or flying round and round her.

Habitat Lakes and ponds, typically with open shores. Common at coast and may inhabit brackish waters.

Flight Season CA Mar–Nov, AZ Jun–Nov, TX May–Nov.

Distribution Ranges south to Argentina and Chile, also in southern Florida and West Indies.

280 Tawny Pennant *Brachymesia herbida* TL 43–46, HW 34–37

Description Long-bodied midsized skimmer with brown body and brownish-washed wings. *Male*: Eyes reddish-brown over blue-gray; face white with brown markings. Thorax brown. Abdomen light brown with black central line becoming wider toward rear of each segment and almost covering segment on S8–9. *Female*: Colored just as male, distinguished by structure.

Identification Only rather plain brown dragonfly with brownish wings. Immature **Four-spotted Pennants** look very much like this species but always have white stigmas, **Tawny** brown stigmas. Also, **Tawny** lacks black lines on sutures of thorax present in **Four-spotted**, and **Tawny** usually has brown legs, **Four-spotted** black. Color in wings develops between base and nodus in **Tawny**, at and beyond nodus in **Four-spotted**. Female **Band-winged** and **Black-winged Dragonlets**, somewhat similar in size and shape, have different abdominal pattern. **Evening Skimmer** and **Straw-colored Sylph**, similarly brown, hang up to perch. Neither has expanding black line down abdomen like **Tawny Pennant**.

280.1
Tawny Pennant
male—Hidalgo Co.,
TX, November 2006,
Robert A. Behrstock

280.2
Tawny Pennant
female—Hidalgo Co.,
TX, November 2005

Natural History Males perch on tips of plants over water, usually in smaller numbers and on smaller water bodies than Four-spotted Pennant. Alternate perching with patrolling along shore. Both sexes often perch with wings up. Habits probably generally similar to Four-spotted but less well known. One female oviposited by tapping water surface at long intervals.

Habitat Lakes, ponds, and marshes, usually with more vegetation than typical of other tropical pennants.

Flight Season TX Apr–Nov.

Distribution Ranges south in lowlands to Argentina; also Galapagos, southern Florida, and West Indies.

Description Conspicuous slender spotted-winged skimmer of open wetlands. *Male*: Eyes dark brown; face metallic black. Thorax black, thinly overlaid with gray pruinosity. Abdomen dark brown, darker along top, may become black. Wings with large brown spots just beyond nodus, stigmas shining white. *Female*: Eyes brown over blue-green; face white, heavily marked with black.

281.1
Four-spotted
Pennant male—
Chambers Co., TX,
July 2004

281.2
Four-spotted
Pennant female—
Dade Co., FL,
April 2005

Thorax medium brown with dark lines in sutures. Abdomen light reddish brown with black lines along all sutures, black central line widening toward rear of segment on S7–9. Wing spots more diffuse than in male, but old females may become almost black, with large wing spots, and look like male, only distinguished by shape and structure.

Identification Only other dragonflies with large black mark on each wing are male **Common Whitetail** and male **Band-winged Dragonlet**, both of which have black mark larger and rectangular (most extreme **Four-spotted Pennants** almost this prominent but never perfect rectangles). **Common Whitetail** has abdomen short and broad, white in mature males, and perching habits entirely different. **Band-winged Dragonlet** perches and flies low in marshes, stigmas black instead of white. In south Texas, see **Tawny Pennant**, always with brown wings and dark stigmas.

Natural History Males perch on tips of twigs, cattail leaves, and grass inflorescences, often waving in the breeze. May fly far out over water of large lakes and sometimes fly back and forth low over water very fast, with black wing spots conspicuous. Females perch similarly, more likely away from water but sometimes in mixed-sex groups. Both sexes may perch high in trees or on transmission lines. Both abdomen and wings usually much elevated with high temperatures and overhead sun. Copulation brief and in flight, then immediate oviposition with male in attendance (although more typically solo). Oviposition quite varied, flight from just above water to knee height, tapping water at very short to moderately long intervals, varying from regular and unidirectional to scattered with much course change. One of few perching skimmers that commonly feeds like flier, even in small swarms. Swarming is more common in morning and late afternoon. Tiny flies most common prey in these circumstances, although larger prey occasionally taken. Reaches great densities in large open lakes, especially near coast.

Habitat Mostly lakes, often those with relatively low odonate diversity. Also at large coastal marshes in Texas and may be on large drainage canals and artificial lakes. Tolerates high alkalinity and perhaps some salinity.

Flight Season AZ Jul–Aug, NM Apr–Sep, TX Mar–Oct.

Distribution Also east and north, mostly in Coastal Plain, to New Jersey.

Small Pennants *Celithemis*

Small skimmers of marshy lakes and ponds, these are almost restricted to eastern North America north of Mexico and should be especially treasured as part of the rich odonate fauna of that region. Go to the eastern edge of the western region to see them. With patterned thorax and central red or yellow spots on the abdomen, they are somewhat similar to whitefaces but have colored faces, and most species have prominent wing markings. As do other pennants and very different from whitefaces, they perch almost invariably at tips of vertical stalks, males over water and females away from it. All species characteristically elevate their wings, forewings higher than hindwings; at most extreme, forewings vertical, hindwings horizontal. They could be considered the butterflies of the dragonfly world, and in flight, those species with most heavily marked wings are indeed reminiscent of butterflies. They share that distinction with distantly related tropical Old World genus *Rhyothemis* (flutterers), with amazing convergence in wing pattern between certain species in the two genera. Small pennants are among relatively few skimmers that usually oviposit in tandem, but like meadowhawks and others, a pair may separate and the female continue on her own. World 8, NA 8, West 6.

282 Amanda's Pennant *Celithemis amanda* TL 27–31, HW 23–26

Description Small red or yellow pennant with large hindwing patches. *Male*: Eyes dark red over gray-brown; face dark red-brown. Thorax reddish-brown with black markings. Abdomen black with red-orange triangular or oval spots on S3–7, becoming smaller toward rear. Cerci red-orange (pale in immature). Hindwing patches dark brown, edged in orange; veins running through them bright red; stigmas orange. *Female*: Eyes reddish-brown over greenish; face tan to yellow. Front of thorax with broad black stripe bordered by brown, sides yellow with very sparse brown to black markings. Abdomen black, S2–3 mostly yellow, S4–7 with large oval yellow spots. Cerci pale yellow. Wing patches orange with dark brown markings fore and aft; stigmas light brown.

Identification Only pennant with large, complex hindwing marking but otherwise clear wings. No other regional dragonfly exactly like this. Female **Marl Pennant** perhaps most similar but with much smaller markings, also larger and heavier-bodied. Body color rather like **Ornate Pennant**, but wing markings much larger, nearly reaching rear edge of hindwing, going back only about halfway in **Ornate**. For satisfying close-range confirmation, male **Amanda's** has red cerci (pale in immature), **Ornate** black. Female **Amanda's** has sides of thorax almost entirely yellow, female **Ornate** with conspicuous black markings.

Natural History Males in grass near breeding habitats or over open water like other pennants. Pairs oviposit in tandem or female by herself. Tends to fly later in season than other small pennants.

Habitat Open ponds with band of emergent vegetation at shore or ponds with dense grass, usually in open habitats.

Flight Season TX Jun–Aug.

Distribution Ranges east in Coastal Plain to North Carolina.

282.2
Amanda's Pennant female—
Polk Co., FL, August 2002, David
McShaffrey

282.1
Amanda's Pennant male—
Washington Co., FL, August
1998, Robert A. Behrstock

Description Orange-brown or yellow pennant with moderate-sized hindwing marking. *Male*: Eyes bright red over dull brown; face brown, top of frons metallic blue-black. Thorax orange-brown with wide black stripes on front and sides. Abdomen black with orange base and elongate orange triangles on S4–7. Cerci black. Hindwing patch of moderate size, extending back a bit more than halfway across wing and out to end of triangle; mostly dark brown with red veins. Stigma orange. *Female*: Eyes dark reddish-brown over pale greenish; face tan. Thorax brown in front, yellow mixed with brown on sides, with black stripes as in male. Markings in abdomen as in male but dull yellow-orange. Immatures of both sexes strikingly black and yellow.

Identification Distinguished from other pennants by clear wings with midsized hindwing markings smaller than **Amanda's**. Thorax more heavily marked with black on sides than **Amanda's**, spots on abdomen narrower, cerci always black. Other small dragonflies with red or yellow coloration and conspicuous basal wing markings include female **Marl Pennant**, heavier-bodied and with mostly pale abdomen; **Checkered Setwing**, larger and

283.1
Ornate Pennant
male—Clay Co.,
FL, April 2005

283.2
Ornate Pennant
female—Lake Co.,
FL, April 2005

with white markings on thorax and abdomen; and **Red-faced Dragonlet**, smaller and with solid red thorax and pruinose blue abdomen in male, all plain yellow in female. None of these species common where **Ornate Pennants** occur.

Natural History Both sexes, males mostly immature, can be common in weedy fields well away from water. Males perch on tips of grasses and sedges over water, often in fairly dense vegetation. Pairs oviposit in tandem flight among emergent and floating vegetation.

Habitat Grass and sedge beds at shores of sand-bottomed lakes and large ponds.

Flight Season TX May.

Distribution East and north in Coastal Plain to New Jersey.

Comments Formerly called Faded Pennant.

284 Calico Pennant *Celithemis elisa*   TL 29–34, HW 27–28

Description Small red-brown or yellow pennant with prominently brown-spotted wings. *Male*: Eyes dark red; face red. Thorax red with black markings. Abdomen black with bright red dorsal triangles or heart-shaped spots on S3–7. Cerci red (pale in immature). Forewing reddish at extreme base, with brown spot just past nodus and dark brown tip. Hindwing with large patterned basal patch extending just past triangles, filled with red veins, and dark postnodal spot and tip as in forewing. Stigma bright red, contrasting with dark wing spots. *Female*: Eyes red over pale greenish; face

284.1
Calico Pennant male—St. Tammany Par., LA, April 2004, Gayle and Jeanell Strickland

284.2
Calico Pennant female—
Burnett Co.,
WI, June 2007

yellow-brown. Thorax brown on front, with black central stripe and bordered by black; yellow on sides, with two prominent black spots. Abdomen black with large yellow spots from base to S7. Veins in hindwing spot yellow, stigma orange.

Identification Wing pattern diagnostic; no other pennant with clear wings, large basal patch and dots at midwing and tip. Basal patch much as that of other species, but markings beyond nodus distinctive. Abdominal spots, although not dramatically different from other pennants, more often heart-shaped and memorable because of that. Female **Eastern Amberwing** has heavily patterned wings but otherwise very different.

Natural History Often found in numbers feeding in weedy fields near water. Breeding males perch up on the tallest herbaceous or shrubby vegetation over water and fly actively back and forth along shore and over the water at knee height. Also perch on tips of herbaceous plants in nearby fields to watch for females; much pairing occurs away from water. Copulation for about 5 min; then pair flies to water to oviposit. Tandem oviposition for about 3 min, rarely followed by another few minutes of solo oviposition. About 700–800 eggs laid.

Habitat Ponds and lakes with vegetated margins, including bog ponds.

Flight Season NE May–Sep, TX Apr–Sep.

Distribution Also throughout East from southern Ontario and Nova Scotia south to northern Florida.

---

285 Banded Pennant *Celithemis fasciata* TL 30–38, HW 25–32

Description Dark pennant with dark-banded wings. *Male*: Eyes dark brown; face brown, metallic blue-black above. Thorax black with faint paler markings. Abdomen black, may develop slight pruinosity at base. Wing markings quite variable in size but typically include tip, incomplete band between nodus and stigma, and complex pattern across wing bases black. Commonly yellow wash

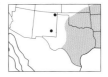

285.1
Banded Pennant male—Kinney Co., TX, July 2004

285.2
Banded Pennant male—Angelina Co., TX, July 2004

285.3
Banded Pennant
female—Angelina
Co., TX, July 2004,
Netta Smith

across wing base enclosed by black markings. Dark markings reduced in some individuals, but basic set always present. *Female*: Eyes reddish-brown over pale greenish; face cream, metallic blue-black above. Thorax bright black and yellow striped. Abdomen black with yellow spots, longer on S4–6 than on S3 or S7.

Identification Easily distinguishable from **Halloween Pennant** by lack of color between dark wing markings except for wash at base. In very poor lighting could still be distinguished by greater extent of basal markings, continuous from base to nodus (interrupted in **Halloween**, basal spots separated by nodal spots). Mature male **Banded** all black, **Halloween** reddish. Female **Banded** with more black than yellow on abdomen, spots discontinuous (forming line in **Halloween**). Some KING SKIMMERS, WHITETAILS, and **Prince Baskettail** patterned somewhat like **Banded Pennant** but much larger and without complicated patterning of basal spots of pennant.

Natural History Both sexes roost and even feed well up in trees, also spend time in weedy fields near water. Males perch on leaf and branch tips at open-water shore or among marsh vegetation, often on grass stems. Abdomen and wings may be well elevated, forewings more than hindwings. Also spend much time flying low over water. Pairs oviposit in tandem, often flying around wildly covering large area. Then pair may separate and female continues by herself, more sedately in smaller area and with or without male guarding. Females oviposit by hovering, dropping to water and tapping once, then rising and moving to do it again.

Habitat Sandy lakes and ponds with emergent vegetation, also slow streams.

Flight Season NM Aug–Sep, TX May–Oct.

Distribution Ranges in East north to Michigan, southern Ontario, and Massachusetts.

286 Halloween Pennant *Celithemis eponina*   TL 36–42, HW 32–34

Description Medium-sized pennant with dark-spotted colored wings. *Male*: Eyes dull red over orange-brown; face red-orange. Thorax orange-brown with black stripes below each wing and third, incomplete, stripe between them. Abdomen orange at base, otherwise black with full-length orange spots forming stripe from base through S7. Cerci black, some orange at base. Wings entirely golden-tan with black band just in from stigma and another on inner side of nodus, one black spot near base of forewing and two near base of hindwing. Most wing veins and stigma red-orange.

Female: Eyes red over gray-green; face yellow-orange. Thorax brown in front, yellow on sides, stripes as in male. Abdomen black with pale yellow base and yellow stripe extending to S7. Wings as male, most veins and stigma yellow.

Identification Unmistakable because of wing color pattern. **Banded Pennant** most like it but has pale part of wings clear rather than tan, abdomen all black in males. **Calico Pennant** similar but more restricted markings. KING SKIMMERS with spotted wings much larger and without overall tan color. Much smaller AMBERWINGS have unmarked or differently marked wings as well as characteristic short, spindle-shaped abdomen.

Natural History Males perch on grass and sedge tips over water and fly at one another. Copulation while perched, lasting 3–6 min, oviposition immediately thereafter. Pairs and solo females oviposit in open water or among floating vegetation, more commonly in morning and even quite early, often over open water in windy weather. Both sexes perch similarly away from water, sometimes in large numbers, with wings elevated. Fluttery flight and colored wings recall butterfly, prompting speculation whether they look enough like Monarch to gain some protection from bird predators while away from water. Sometimes in great numbers away from water along Gulf coast when blown out of extensive breeding habitat.

Habitat Open lakes and marshes of all kinds, with at least some emergent vegetation. Very common coastally, also in slightly brackish marshes.

286.1
Halloween Pennant
male—Fort Bend Co.,
TX, May 1998, Robert A.
Behrstock

286.2
Halloween Pennant
female—Collier Co., FL,
April 2005

286.3
Halloween Pennant
immature male—Clay Co.,
FL, June 2004

Flight Season NM Jun–Oct, NE Jun–Sep, TX Apr–Oct.

Distribution Ranges south in Mexico to Coahuila and Nuevo León, also throughout East from southern Ontario and Maine south.

287 Double-ringed Pennant *Celithemis verna* TL 32–35, HW 25–28

Description Pennant with black or black and yellow body and tiny dark hindwing markings. *Male*: Eyes dark brown; face dark glossy brown. Thorax and abdomen black, overlaid with dark blue pruinosity. Small black spots at base of hindwing. *Female*: Eyes reddish over gray; face yellow, metallic blue above. Thorax yellow with wide median black stripe and lateral brown stripes on front, two black stripes on side. Abdomen yellow at base with black on top of S1–2, yellow ring on base of S3–4, remainder black. Veins within black wing markings yellow.

Identification Male unlike other pennants in being entirely black, with very small wing markings. Might be mistaken for other small black dragonflies with brown eyes that overlap in range although not in habitat. **Seaside Dragonlet** never with bluish tinge from pruinosity, lacks dark marking at wing base, at coast or alkaline lakes. **Black Setwing** larger, pruinosity gray rather than blue, most often on streams. **Marl Pennant** chunkier, with large hindwing spots, more common at coast. Mostly black abdomen of female **Double-ringed** with two pale rings distinguishes from other female pennants and any other dragonfly. Female with sides of thorax somewhat like other female pennants but with distinctive pattern of one stripe with lightning-bolt effect followed by narrow straight stripe.

Natural History Males perch on overhanging herbaceous vegetation at knee height and below and fly short beats along shore over open water. Pairs oviposit in tandem.

Habitat Open ponds and small lakes with band of emergent vegetation along shore.

Flight Season TX Jan–Apr.

Distribution Ranges in East from southern Missouri, Indiana, and New Jersey south to northern Florida.

287.1
Double-ringed Pennant male—
Cumberland Co., NJ, July 1999,
Robert A. Behrstock

287.2
Double-ringed Pennant
female—Laurens Co., GA,
May 2005, Giff Beaton

Whitefaces *Leucorrhinia*

Whitefaces constitute a north-temperate genus of small black dragonflies frequenting lakes, ponds, and marshes with conspicuous white face and yellow and/or red markings. Females of all species have a dull morph and a bright, male-like morph. Usually, there is a small brown triangular spot at the hindwing base. They perch flat on leaves or ground or on plant stems, their wings flat or sometimes drooped forward. Male whitefaces seem more likely than other dragonflies to attempt tandem with males of pairs already in tandem. Well-adapted to northern latitudes, they are almost always seen perched flat on light-colored rocks, logs, and tree trunks in the morning, where dark coloration allows quick warming in sun. Old females become quite pruinose on underside of the abdomen. They are sometimes very abundant with pairs commonly seen. Identification is often tricky, especially of females. Examination of male hamules and the female subgenital plate may be necessary for the last word. World 16, NA 7, West 7.

288 Frosted Whiteface *Leucorrhinia frigida* TL 28–32, HW 21–24

Description Small white-faced skimmer with pale-based black abdomen, barely enters region from east. *Male*: Eyes dark brown; face white. Entirely dark brown to black, mature males with abdomen base (typically S2–5) heavily white pruinose. *Female*: Eyes brown; face white. Thorax ochre, black in front and scattered spots on sides. Abdomen black, yellow at base and with elongate spots on S4–7 becoming narrower to rear, that on S7 always a fine streak. Subgenital plate with two large, pointed lobes. Immature thorax almost entirely unpatterned golden with square black patch on front.

Identification Mature male unlike anything else in region. Populations of mature male **Belted Whiteface** farther east look like this, but probably not in range of this book. If so, **Belted** slightly larger than **Frosted**, with more vividly marked thorax and shows red at wing bases if not elsewhere. Female **Frosted** much like other female whitefaces, but most of abdomen lacking light markings or present only as fine streaks. Thorax also with reduced black markings in comparison with most other species. Immature **Dottailed** may have pale thorax but pale carina makes fine streak down front of thorax (all black in immature **Frosted**) and usually more heavily spotted abdomen.

Natural History Males perch low in emergent vegetation or on lily pads at water, sometimes defending small territories but not aggressive at other times. Females usually on ground or in herbaceous vegetation away from water. Copulation for 10–20 min on ground or vegetation, may fly briefly over water. Female then perches for few minutes, then oviposits with male guarding. One female oviposited by tapping water twice, hovering briefly, and double-tapping again six times, then perching again.

288.1
Frosted Whiteface male—Nipissing Co., ON, July 2005, Greg Lasley

288.2
Frosted Whiteface female—Hampden Co., MA, August 2003, Glenn Corbiere

Habitat Mud-bottomed lakes and ponds with abundant emergent vegetation, especially pools in fens and bogs.

Flight Season WI May–Aug.

Distribution Also through Northeast from southern Ontario and Quebec and Nova Scotia south to Illinois, Ohio, and New Jersey.

289 Belted Whiteface *Leucorrhinia proxima*   TL 33–36, HW 24–27

Description Small white-faced skimmer with pale-based black abdomen. *Male:* Eyes dark brown; face white. Thorax brightly marked red and black, becoming darker and duller with age. Abdomen black with red base (S1–3); otherwise all black or may have narrow red streaks down center of middle abdominal segments. *Female:* Polymorphic, andromorph colored like male but heteromorph with light areas yellow. Subgenital plate with two short, rounded lobes.

Identification Other whitefaces with red at base of abdomen and little red beyond that include **Crimson-ringed**, most similar species, and **Canada**. **Crimson-ringed** differs in wing venation (usually two or more cells in radial planate on each wing doubled in **Crimson-ringed**, usually not so in **Belted**) and hamule shape (inner branch straight and then hooked at tip in **Crimson-ringed**, smoothly curved in **Belted**). Female **Belted** looks like females of most other whitefaces, exactly like female **Crimson-ringed**; capture and scrutiny of subgenital plate may be essential for identification. Lobes of plate barely evident in **Crimson-ringed**, short but quite obvious in **Belted**, still longer in other species. Wing venation also good separation from female **Crimson-ringed**, sometimes visible in field. **Canada** smaller and more slender and lacks pale spot on S7.

Natural History Males perch on twigs over and near water and often flat on lily pads, defending small territories. Can be abundant in optimal habitats. Away from water, perch from ground to well up into trees. Pairs couple at water, immediately fly in wheel away from

289.1
Belted Whiteface male—
Okanogan Co., WA, June 2005

289.2
Belted Whiteface yellow
female—150 Mile House,
BC, July 2006

289.3
Belted Whiteface red
female—70 Mile
House, BC, July 2006

water into shrubs and trees, where copulation quite lengthy. Mating pairs, often in numbers, seen throughout day.

Habitat Lake shores and ponds with much aquatic vegetation, often boggy margins; usually in forested landscapes.

Flight Season YT Jun–Jul, BC May–Aug, WA Jun–Sep, CA Jun–Aug, MT May–Aug.

Distribution Ranges east across Canada to Newfoundland, south to Michigan, West Virginia, and New Jersey. In mountains in southern part of range.

Comments Called Red-waisted Whiteface in earlier publications.

290 Dot-tailed Whiteface *Leucorrhinia intacta* TL 29–33, HW 23–25

Description Small all-black dragonfly with white face, yellow dot on abdomen. *Male*: Eyes dark brown; face white. Thorax and abdomen black, abdomen with paired contiguous yellow dots on S7. Immatures brown with spotted abdomen as female, males at water may retain some pale markings. *Female*: Eyes brown, face white. Thorax brown with dark spots below base of forewings and scattered smaller spots. Abdomen yellow at base, yellow extending down sides of S3–4. Large yellow spots on S3–7, shortest on S7. Thorax and abdominal spots become darker with age, and oldest females have dark brown thorax and black

Skimmer Family **425**

red replaced by yellow; S4–6 may or may not have narrow yellow spots on basal half of segment. Subgenital plate with two short, rounded lobes.

Identification Smallest whiteface in range, more slender than others and perhaps recognizable by this alone. Both sexes like small, slender-bodied **Belted** or **Crimson-ringed Whitefaces**, abdominal spots smaller than in **Boreal** and **Hudsonian**. At close range, both sexes distinguished from other whitefaces by creamy-yellow tint on face; all others have chalk-white faces.

291.1
Canada Whiteface
male—Klondike, YT,
July 2006, Cameron
Eckert

291.2
Canada Whiteface
yellow female—
Deadman Lake, AK,
June 2003, Robert
Armstrong

Natural History Males perch low in breeding habitat, often right on moss; otherwise seen on ground, rocks, and logs as in other whitefaces.

Habitat Bog ponds and fens with much low emergent vegetation and especially mats of floating mosses, more restricted to this habitat than any other whiteface; not in dense sedges with Hudsonian Whiteface.

Flight Season YT Jun–Aug, BC Jun–Aug.

Distribution Also widely in eastern Canada east to Nova Scotia.

---

292 Hudsonian Whiteface *Leucorrhinia hudsonica* TL 27–32, HW 21–27

Description Small marsh-dwelling whiteface with prominently spotted abdomen. *Male*: Eyes dark brown; face white. Thorax red with black triangle on front and extensive black markings on sides. Abdomen black, most of S1–2 and base of S3 red, and rather narrow red spots extending about halfway down top of S4–7. *Female*: Polymorphic, andromorph colored exactly as male and heteromorph with light areas ochre on thorax and brighter yellow on abdomen; spots on abdomen longer, extending much of length of segments. Subgenital plate with two prominent lobes, their inner edges touching.

Identification Red spots along most of abdomen distinguish male from other whitefaces except **Boreal**; considerably smaller than **Boreal** and with no spot on S8. Superficially like **Belted**, **Canada**, and **Crimson-ringed**, but all of them have little or no red on middle abdominal segments, and only **Canada** as small as **Hudsonian**. Female similar to female **Belted** and **Crimson-ringed** but smaller, and spots on abdomen

292.1
Hudsonian Whiteface
male—Skamania Co.,
WA, July 2005

292.2
Hudsonian Whiteface
yellow female—
Okanogan Co., WA,
June 2005

292.3
Hudsonian Whiteface red
female—Hampden Co., MA, June
2004, Glenn Corbiere

usually wider and longer, reaching two-thirds length or more of each middle segment (usually halfway in other species); also like **Canada** but somewhat more robust. Look at subgenital plate for definitive identification; large in **Hudsonian**, with two prominent lobes or projections touching one another; in other species projections well separated (**Boreal, Canada, Dot-tailed, Frosted**) or not much more than bumps (**Belted, Crimson-ringed**).

Natural History Males perch on sedge and grass stems and defend small territories. Females away from water, often perch on ground or light-colored logs in woodland clearings. Pairs leave water for lengthy copulation at rest. Females oviposit in open water in sedge beds, often guarded by male hovering nearby.

Habitat Marshes, sedge meadows, coldwater fens, and bog ponds, occupying denser vegetation than most whitefaces. Only whiteface in bog habitats in southern parts of range, together with others in north.

Flight Season YT May–Aug, BC Apr–Aug, WA Apr–Oct, OR May–Sep, CA May–Sep, MT May–Aug, NE May–Sep.

Distribution Also east across Canada, south to Michigan and New Jersey, in mountains to West Virginia.

Description Small white-faced skimmer with pale-based black abdomen. *Male*: Eyes dark brown; face white. Thorax brightly marked red and black, becoming darker and duller with age. Abdomen black, red at base (S1–2) and sometimes with fine red streaks down center of middle segments. *Female*: Polymorphic, andromorph colored as male; heteromorph thorax varies from brown and dull yellowish to rather contrasty black and yellow. Subgenital plate barely visible.

Identification Looks exactly like western populations of **Belted Whiteface**, and capture would be necessary for identification unless wing venation could be clearly seen or photographed. Presence of doubled cell rows in radial planate on all wings (rarely not doubled on one or two wings) often visible with close-range binoculars. Most **Belted** have single cell row, rarely one or more doubled rows in one or two wings, but a few just like **Crimson-ringed** and identification only by venation would be flawed. Slight difference in

293.1
Crimson-ringed
Whiteface
male—Skamania Co.,
WA, July 2005

293.2
Crimson-ringed
Whiteface yellow
female—Sussex Co.,
NJ, June 2004,
Allen Barlow

male appendages (epiproct extending beyond half of rather slender cerci in **Belted**, about half of slightly thicker cerci in **Crimson-ringed**) might also be visible in binoculars, but distinct difference in hamules impossible to see except in hand. Female identical in coloration to female **Belted** although dark area on front of thorax more likely to narrow smoothly to point above, same dark area usually remains wider and angled before end in **Belted**, but with overlap. Wing venation in both sexes, hamule shape in male, and/or subgenital plate in female furnish definitive identification. Female distinctly larger than similar-looking female **Hudsonian Whiteface**.

Natural History Males often flat on water lily leaves but also commonly on stems of emergent vegetation as well as waterside trees and shrubs. Perch flat on ground away from water but also very often in trees. Females rarely seen except in pairs in wheel, seen as they approach lake from surrounding woodland. Lengthy copulation at rest, as long as 30 min or more.

Habitat Lakes and ponds in forested regions, often with boggy margins; vary from very little to abundant aquatic vegetation.

Flight Season BC May–Aug, WA Jun–Aug, OR Jun–Aug, CA May–Sep, MT Jun–Jul.

Distribution Ranges east across Canada to Newfoundland, south to Michigan and West Virginia. In mountains in southern parts of range.

294 Boreal Whiteface *Leucorrhinia borealis* TL 44–46, HW 29–32

Description Largest whiteface, with heavily red-spotted abdomen. *Male*: Eyes dark brown; face white. Thorax with red-brown and black pattern. Abdomen black, S1–2 mostly red, S3–8 with large red spots almost covering top of each segment; less often, little or no red on S8. *Female*: Polymorphic; colored just like male or pale color yellow to ochre; S8 usually all black but may show pale basal marking. Subgenital plate with two large, rounded lobes.

Identification Only whiteface distinctly larger than others of group. Similarly colored male **Hudsonian Whiteface** conspicuously smaller and lacks obvious spot on S8 (rarely also absent in **Boreal**). Larger size and full-length pale spot on S7 will distinguish female

from females of other whitefaces such as **Belted** and **Crimson-ringed** that occur with it, but subgenital plate shape distinctive for each species in hand.

Natural History Males perch on grasses and sedges over water, also flat on algal mats and ground. Near optimal open-marsh habitats may be extremely abundant. As many as 10 per square foot seen perching on sunny sides of aspen trees and other pale surfaces in early morning. Only whiteface restricted to short, early flight season.

Habitat Marshy ponds and lake margins, bogs and fens, typically with much upright emergent vegetation, even tall grasses. Most common in open country.

Flight Season YT Jun–Jul, BC Apr–Jul, WA Jun–Jul, MT Jul.

294.1
Boreal Whiteface
male—Takhini Valley,
YT, July 2006,
Cameron Eckert

294.2
Boreal Whiteface red
female—Fairbanks,
AK, July 2002,
Wim Arp

Whitefaces - male hamules

Frosted

Belted

Dot-tailed

Canada

Hudsonian

Crimson-ringed

Boreal

Whitefaces - female subgenital plates

Frosted

Belted

Dot-tailed

Canada

Hudsonian

Crimson-ringed

Boreal

Appropriately named, these medium to large skimmers are voracious predators of other insects, including dragonflies, up to their own size. They capture them by flycatching. Walking along the waterside, you may flush teneral odonates, and if there are pondhawks in the vicinity, you may see a swift attack and can watch the process of a meal being consumed. The genus is characterized by some extra-large spines on its hind femur, surely an adaptation to take large prey. Most species perch and fly low, often on the ground or broad leaves or logs. Perhaps low perching facilitates capture of larger prey, as this behavior is common in skimmers of Old World genus *Orthetrum*, similar in numerous ways, and also in many clubtails. Wings usually are held flat, but they may be drooped when on an elevated perch. Raising wings or pointing abdomen upward is not noted in this group. The species fall into two categories, with either excessively slender and elongate or normally shaped abdomen. Pin-tailed and Great are the "slender" species of our fauna; the others belong to the "normal" group. World 10, NA 7, West 7.

295 Flame-tailed Pondhawk *Erythemis peruviana* TL 36–43, HW 28–34

Description Medium-sized red and blue or brown skimmer with fairly short abdomen. *Male*: Eyes dark purple-brown; face dark blue-violet. Thorax and S1–3 dark pruinose blue, S4–10 entirely bright scarlet red. Immature colored as female, gets increasingly red abdomen while still in immature color, then pruinosity develops. *Female*: Eyes dark brown; face tan. Sides of thorax brown, bordered by dark brown in front, then front of thorax pale brown. Abdomen pale brown with dark brown rectangles on side of each segment. Pale yellow stripe between wings contrasts strongly with rest of upper surface.

Identification Nothing else in West colored as mature male. Males attaining maturity, with brown thorax and bright red abdomen, also unique in our fauna, although most like

295.1
Flame-tailed Pondhawk male—Puntarenas Prov., Costa Rica, February 2005

295.2
Flame-tailed
Pondhawk
female—Bolívar,
Venezuela,
February 2006,
Wulf and Eva
Kappes

mature male **Red-tailed Pennant,** which perches differently and has entire abdomen, including base, red. Pale front and dark sides of thorax good mark for females and immatures, shared by female and immature **Pin-tailed,** also **Plateau** and **Red-fronted Dragonlets**, which are much smaller and have more heavily patterned abdomens.

Natural History Perches low in grasses and other dense aquatic vegetation. Male patrol flights low, usually over vegetation.

Habitat Open marshes and pond margins, usually heavily vegetated.

Flight Season TX Jul.

Distribution Ranges south in lowlands to Argentina, also Jamaica. So far only a single record from the United States, male collected at Llano River State Park, Texas, in 2001.

296 Black Pondhawk *Erythemis attala*   TL 42–44, HW 32–38

Description Medium-sized black or black and yellow skimmer of south Texas, abdomen slender but not excessively so. *Male*: Eyes dark brown; face black. Thorax and abdomen entirely black; appendages white. Some mature males retain pale bands of immaturity on abdomen. Basal dark hindwing markings extend out to triangle. *Female*: Eyes dark brown; face tan. Thorax brown. Abdomen black,  yellow at base, and with basal yellow bands on S4–7, bands increasingly restricted with age to paired yellow blotches as abdomen darkens, then may become entirely black like male. Yellow on S4 and S7 may remain conspicuous as abdomen darkens. Subgenital plate projects below abdomen, visible from side.

Identification Very few other dragonflies in its range in North America are entirely black like mature males and fully black females. Male **Marl Pennant** superficially most like it, slightly smaller and with very different habits, perching on plant tips in open areas. **Marl Pennant** also has black appendages, pondhawk conspicuously pale cerci. **Pin-tailed Pondhawk** colored similarly but with much more slender, longer abdomen (longer than wing), and smaller brown spot at wing base not conspicuous in most views (conspicuous in **Black Pondhawk** and **Marl Pennant**). Some female **Black** have reduced wing spot and must be distinguished from **Pin-tailed** entirely by abdomen shape; also, when younger, have

296.1
Black Pondhawk
male—Canal Zone,
Panama, February
2006, Dan and
Kay Wade

296.2
Black Pondhawk
female—Cameron
Co., TX, October 2007,
Terry Fuller

entirely brown thorax rather than pale-fronted thorax of **Pin-tailed**. **Seaside Dragonlet**, usually on coast, much smaller and with dark appendages. **Black Setwing** pruinose rather than shiny black, on elevated perch in open with wings depressed. Blackish KING SKIMMERS such as **Slaty Skimmer** are much larger than pondhawks. Dark thorax and brightly yellow-banded abdomen of female and immature **Blacks** not duplicated by any other species.

Natural History Perches on ground or low in vegetation usually, on twigs or flat surfaces. Sexual patrol flight low, usually over floating vegetation. Females at small clearings or woodland edge, perched up to chest height.

Habitat Ponds and swamps with densely vegetated surface, typically water lettuce or duckweed.

Flight Season TX May–Nov.

Distribution So far known in West only from population on King Ranch in 2004–2005 and additional individuals in Lower Rio Grande Valley in 2007–2008. Ranges south in lowlands to Argentina, also in West Indies.

Description Medium-sized red or reddish-brown skimmer with fairly short abdomen; narrow dark brown patch at base of hindwing extends width of wing. *Male*: Eyes and face red. Brilliant scarlet red all over. *Female*: Light reddish-brown, virtually no pattern. Subgenital plate projects below abdomen, visible from side.

Identification Of several entirely bright red species found in south Texas, this may be reddest, further distinguished by dark brown contrasty wing patch. With typical pondhawk behavior, unlikely to be mistaken for bright red SADDLEBAGS; basal hindwing marking a bit less extensive than in **Antillean** or **Vermilion Saddlebags** and much less extensive than in **Carolina** or **Red Saddlebags**. **Carmine Skimmer**, also red all over, is more carmine than scarlet and has no wing markings. **Carmine** has pale legs, **Claret** dark. **Red-tailed Pennant** similar in having abdomen brighter than thorax but a bit smaller, perches differently, and has much smaller patch of color at hindwing base. Brilliant scarlet **Mayan Setwing** of upland streams has larger hindwing patch. Female **Claret Pondhawk** superficially like other light brown to reddish females, for example **Red-tailed Pennant** (smaller hindwing patch, subgenital plate not visible from side), **Band-winged Dragonlet** (no hint of reddish, sides of abdomen patterned with dark rectangles), **Carmine Skimmer** (distinctly larger, dark flange on S8, narrow pale stripe on front of thorax), and similarly rare **Flame-tailed Pondhawk** (smaller hindwing patch, front of thorax pale).

Natural History Males on patrol flight fly low over water, usually over vegetation. Females oviposit among floating plants, including duckweed.

Habitat Swamps with well-vegetated water surface, typically water lettuce or duckweed. Only U.S. records in woodland away from water.

Flight Season TX May–Oct.

Distribution Only U.S. records are from Lower Rio Grande Valley, Texas. Ranges south in lowlands to Bolivia and Argentina.

Comments Northern populations (Mexico and Guatemala) now considered this species actually represent an undescribed species, listed as *mithroides* pending species description.

297
Claret Pondhawk
male—Cameron Co.,
TX, October 2007,
Terry Fuller

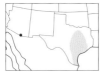

Description Medium-sized dark skimmer with long, slender abdomen making normal-sized base look swollen. *Male*: Eyes dark brown; face purple-black. Thorax and abdomen dark brown to black. Appendages cream-colored. Immature colored as female, thorax becomes entirely black while abdomen still banded. Small dark brown patch at hindwing base. *Female*: Eyes dark brown; face brown, paler at edges. Thorax medium brown on sides, pale brown on front, with darker stripe between them emphasizing contrast. Pale front of thorax continued as pale stripe between wings. Abdomen light brown at base, rest black, with pale yellow crossbands at tip of S3 and base of S4–7 interrupted by fine black line on dorsal carina, all duller and darker with age.

Identification Shape alone distinguishes this outrageously slender species, especially contrast between pinlike part of abdomen so much narrower than vertically expanded base. No other black dragonflies have such a slender abdomen, although much rarer **Black Pondhawk** could be mistaken for it, as its abdomen is rather slender as well. **Black** shorter with less contrast between abdomen base and tip and larger dark spot on hindwing base, usually conspicuous in any view. Both species have pale appendages, unlike most other

298.1
Pin-tailed Pondhawk
male—Hidalgo Co.,
TX, November 2005

298.2
Pin-tailed Pondhawk
female—Sucre,
Venezuela, February
2006, Wulf and
Eva Kappes

black dragonflies, for example **Black Setwing**, somewhat similar but with purple face and different perching habits. Female and immature **Black Pondhawk** have dark thorax, without pale front characteristic of **Pin-tailed**. Female **Band-winged Dragonlet** smaller, without such swollen abdomen and with front of thorax as dark as or darker than sides.

Natural History Perches low, on ground and other flat objects, also up into shrubbery but usually below head height. Males quite aggressive toward each other, seem to detect approaching male no farther away than about 6 feet. Males arrive at water late morning, mating activity highest early afternoon. Pairs copulate for about a minute at rest near water, then male flies to oviposition site, apparently to show it to female, then back to make contact with female, then both fly to water where female oviposits and male guards by flying close around her. Females often oviposit without males, especially later in day, moving over large areas.

Habitat Ponds and open marshes of all sizes, edges of canals and lakes, even temporary pools, usually with emergent and/or floating vegetation.

Flight Season AZ Sep, TX Apr–Nov.

Distribution Ranges south in lowlands to Argentina, also peninsular Florida and Greater Antilles.

---

299 Great Pondhawk *Erythemis vesiculosa* TL 56–59, HW 39–40

Description Large green skimmer with very long, slender abdomen making normal-sized base look swollen. *Male*: Eyes brown (immature) to blue (mature); face green. Thorax and abdomen grass green with black bands (brown in immatures) across rear of S4–7; S8–10 black; cerci white to pale chartreuse. Note green stigma, especially in younger individuals. *Female*: Almost identical 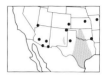 to male, even in shape. Probably goes through same color changes with maturity; female eyes more likely to be brown. Cerci cream-colored, shorter than those of male.

Identification Much larger and longer-bodied than female and young male **Eastern Pondhawk** of same color. Abdomen of **Great** a bit longer than hindwing, very slender, and dramatically swollen at base; obviously shorter than hindwing and not such swollen base in **Eastern**. Dark markings on abdomen transverse in **Great**, pointed in front in **Eastern**. S1–3 entirely green in **Great**, all carinas marked with heavy black lines in **Eastern**. Abdominal markings always green, never white, as they sometimes are in **Eastern**. Femora pale in **Great**, entire leg black in **Eastern**. Female subgenital plate projects in **Eastern**, not in **Great**. In flight, easily mistaken for DARNER, in particular species with green thorax such as **Amazon, Blue-faced,** or **Common Green**, but green bands across abdomen distinctive, as well as perching habits.

Natural History As other pondhawks, perches on ground, but much more likely to perch higher, to head height and above. Males usually perch below waist height at waterside or fly back and forth over open water, hovering from time to time with abdomen elevated and looking a bit like a darner. Mating rarely observed, copulation at rest. Oviposition by regular taps in one spot, then moving on to another area to repeat. Regularly takes large prey, including dragonflies and butterflies. Usually more common in fall than earlier in season.

Habitat Marshy ponds, including temporary ones; also fly over pools in streams but may not breed there.

Flight Season AZ Aug–Sep, NM Jul–Aug, TX Feb–Nov.

Distribution Ranges south in lowlands to Argentina, also southern Mississippi, southern Florida, West Indies, and Galapagos.

299.1
Great Pondhawk
male—Collier Co.,
FL, June 2004

299.2
Great Pondhawk
female—Cojedes,
Venezuela,
December 2000

Description Medium blue or green skimmer with medium-length slender abdomen. *Male*: Eyes blue-green, paler than those of Western, with dark pseudopupils much more in evidence; face green. Thorax and abdomen entirely pruinose blue; appendages white. Immature colored as female, becomes pruinose first on abdomen and then thorax. Color change begins at about a week of age and takes about 2–3 weeks, fastest at higher temperatures. *Female*: Eyes olive-brown to yellowish-green to fairly bright green; face green. Thorax and abdomen base bright green, rest of abdomen banded

black and white or black and pale green, tip black; appendages white. Subgenital plate projects downward from abdomen at right angle. Considerable geographic variation, with black markings reduced in northern plains females, from Colorado and New Mexico east at least to Iowa and Minnesota in North but not in southern plains (Texas). Plains and northern populations also have slightly broader abdomens, in both color and shape approaching condition of Western Pondhawk, and the two may be conspecific. See Western Pondhawk.

Identification Green females and immatures unmistakable, only thing similar are female and immature **Western Pondhawk** and both sexes of **Great Pondhawk**, considerably larger with much thinner abdomen (see those species). Blue males most likely mistaken for **Blue Dasher,** but that species has white face, bright green eyes, and a bit of color at hindwing base. Perching habits distinguish them, **Pondhawk** near and often on ground, **Dasher** up in vegetation, and males on territory above waist height. **Blue Dashers** droop their wings, **Pondhawks** do not. Two of most common species within their ranges, these two can be compared again and again. In some parts of plains and Southwest, both

Eastern and **Western Pondhawks** and perhaps intermediates between them occur, so look closely at pondhawks in those areas. Green individuals may be mistaken for CLUBTAILS, especially as they fly swiftly and land on ground. Note very different abdominal pattern.

Natural History Often superabundant, clearly favored in wetlands of today. Perches low, usually below waist height and with four legs; on ground and other flat surfaces, also leaves and branches. One of few skimmers that habitually chooses ground perching. Voracious predator, especially females, eating odonates of all kinds smaller than themselves, up to Blue Dashers, other pondhawks, and, rarely, immature king skimmers. Males return to water at about 2 weeks of age, perch along shores and in marsh vegetation, and spend relatively little time in patrol flights. May remain on territory for up to 5 hr in a day. Interacting males fly in parallel flight and then have mutual display, "vertical circling," seeming to leapfrog but in

300.1
Eastern Pondhawk
male—Marion Co.,
FL, June 2004

300.2
Eastern Pondhawk
southern female—
Hays Co., TX, July
2004

300.3
Eastern Pondhawk
northern plains
female—Black Hawk
Co., IA, July 2004

300.4
Eastern Pondhawk immature male—Lafayette Co., FL, June 2004

fact rear one moving *under* the other, and pair usually repeating action at least several times. Copulation brief and aerial, may be followed by resting period, lasting about 20 sec to 2 min. Female usually continues to perch after separation, quite inconspicuous to males. Her mate, remaining on territory, may guard her once she starts laying eggs. Most oviposition without male in attendance, usually in beds of vegetation and fairly brief, often under a minute spent at water (but returning repeatedly). Quite variable in distance moved and taps made per site, but up to several hundred when undisturbed. Often abdomen conspicuously raised above horizontal before hitting water. Clutch size averages 900 eggs, released at 8–10/sec.

Habitat Almost any body of water—lakes, ponds, slow streams, ditches, canals—with emergent vegetation. Seems especially to prefer carpets of floating plants such as water hyacinth and water lettuce. Shuns unvegetated wetlands.

Flight Season NE May–Oct, TX all year.

Distribution Ranges south in eastern Mexican lowlands, then to Costa Rica, although very scarce south of Mexico. Also throughout East from Minnesota, southern Ontario, and Maine south.

Comments See Western Pondhawk.

301 Western Pondhawk *Erythemis collocata* TL 40–42, HW 31–32

Description Medium blue or green ground-perching skimmer with fairly short abdomen. *Male*: Eyes dark turquoise blue; face green. Entire body pruinose blue (S10 may remain black), appendages black. Immatures look like females, with pale appendages; abdomen becomes pruinose before thorax. *Female*: Eyes light green, yellowish-brown when immature, may become turquoise in older individuals; face green. Thorax and abdomen green, may shade to yellowish-green, yellowish, or light brown on posterior or most abdominal segments. Fine black lines on all abdominal  carinas, most pronounced down midline of abdomen, where they become wider toward rear. Appendages pale. Subgenital plate projects downward from abdomen at right angle. Some local variation, those at hot springs at Yellowstone National Park with heavy black markings on thorax and abdomen. Tenerals of both sexes entirely brilliant lime green.

Identification Most like **Eastern Pondhawk**. Typical male **Western** differs from **Eastern** in having shorter, broader abdomen, parallel-sided rather than slightly constricted behind

301.1
Western Pondhawk
male—Catron Co.,
NM, July 2007

301.2
Western Pondhawk
female—Harney Co.,
OR, July 2005

301.3
Western Pondhawk
Yellowstone
female—Teton Co.,
WY, September 2007

swollen base, with dark cerci (white in **Eastern**). Eyes of mature male distinctly darker and bluer than those of **Eastern** and pseudopupils less apparent. Females also with shorter abdomen, in most individuals lightly marked except for black lines along carinas (**Eastern** with striking black crossbands, line on dorsal carina not at all prominent). Immatures even more differently colored, **Western** with middle abdominal segments green with tan edges, whereas in **Eastern** those segments are vividly black and white banded (white changes to green with maturity). Femora typically partially pale in **Western** females, entirely black in **Eastern** females. Unlike **Eastern Pondhawk**, not likely to be mistaken for CLUBTAIL, because abdomen shorter and wider.

Natural History Perches low, often on ground. Males fly patrol low over open water. Females often common in grass and weeds away from water. Males rarely and only briefly perform "vertical circling" characteristic of Eastern Pondhawk. Copulation at rest, fairly brief (20–30 sec), then female often perches for another 10–20 sec before ovipositing; then male guards her, staying very close and flying around and around rather than hovering. Females oviposit in circumscribed area, tapping somewhat irregularly or several times in one spot, then moving to another, in algal mats or open water near vegetation. Not as oriented toward large prey as Eastern Pondhawk, only rarely eating dragonflies, but damselflies commonly taken.

Habitat Wide variety of still waters, from lake shores and ponds to canals, slow streams, and springs. Often at hot springs in northern part of interior range.

Flight Season BC May–Jul, WA May–Sep, OR Apr–Oct, CA Feb–Nov, AZ Feb–Nov, NM Mar–Nov.

Distribution Ranges south in uplands of Mexico to Michoacan.

Comments In some areas both types of pondhawk have been found together, but over southern plains and into Southwest they may constitute a hybrid swarm that indicates they may be the same species. They are kept separate for convenience because typical populations of both are quite different, easily distinguishable, and apparently somewhat different in natural history.

Dragonlets *Erythrodiplax*

This large genus of small to medium skimmers of New World tropics has few species reaching North America and one found only there. Most live in marshes with dense vegetation. In our area, there are two large black species with black wing markings, three smaller species with pruinose blue abdomens, and one slender black species of seashore and saline lakes. Females within these three groups are quite difficult to distinguish, unfortunately, as immature males are colored just like females, and immatures are often common. At rest, wings are held level or drooped forward, usually perching parallel to a branch; they are likely to obelisk in midday sun. World 58, NA 6, West 6.

302 Red-faced Dragonlet *Erythrodiplax fusca*   TL 24–30, HW 19–24

Description Small skimmer with red face and blue abdomen. *Male*: Eyes dark red; face red. Thorax dark red, abdomen with S1–2 dark red, S3–7 pruinose blue, tip of S7 and S8–10 black, appendages brown to black. *Female*: Eyes dull red over gray-green; face light brown. Thorax tan to brown with wide dark humeral stripe. Abdo-  men tan with narrow black stripe down middle of segments and black rectangles on side of posterior half of S4–7, covering entire segment on S8–10. Sub- genital plate scooplike, extending to end of S9. Older females become dark brown. Imma- ture male colored as female.

Identification No other small dragonfly colored like this one. See **Plateau Dragonlet**. Fe- male and immature male **Plateau** identical, distinguished by range. Female and immature **Little Blue** very similar, but black on abdomen tip more extensive, covering S7–10, and vir- tually no color at hindwing base.

302.1
Red-faced Dragonlet
male—Madre de
Dios, Peru, July 2002,
Netta Smith

302.2
Red-faced Dragonlet
female—San Luis
Potosí, Mexico,
October 2002, Robert
A. Behrstock

Natural History Males territorial at water, perch low over water or along shore. Females oviposit in dense vegetation. Yellowish immatures can be common at and near breeding sites.
Habitat Shallow marshy ponds, permanent or rainy-season.
Flight Season TX Jul.
Distribution Very scarce in North America but at least a few breeding populations in south-central Texas (recently found in Lower Rio Grande Valley). Ranges south in lowlands to Argentina, also into Lesser Antilles.

303 Plateau Dragonlet   *Erythrodiplax basifusca*   TL 26–30, HW 21–26

Description Small southwestern skimmer with dark head and thorax, blue abdomen, often basal hindwing spots. *Male*: Eyes dark brown; face metallic blue-black. Thorax very dark brown. Abdomen dark brown at base, S3–6 pruinose blue, S7–10 black; appendages brown. Pruinosity sometimes invading base and tip of abdomen. Wings clear or hindwing with small brown basal spot. *Female*: Eyes reddish-brown over tan; face tan, not metallic. Immature yellowish-tan with brown humeral stripes on thorax, brown to blackish rectangles low on sides of abdominal segments, more pronounced toward

rear. Wing spot when present often pale and diffuse. Subgenital plate scooplike, extending to end of S9. When mature, all brown with dark wing spot, may become washed with pruinosity. Immature male like female.

Identification This species could overlap with **Red-faced Dragonlet** in Central Texas, in which case males would be distinguishable by metallic purple face, darker thorax, and less brown at hindwing base. Females indistinguishable, even in hand, although **Plateau** averages slightly larger than **Red-faced**. Could also overlap with **Little Blue Dragonlet** in same area, distinguished by dark thorax (mature **Little Blue** with bluish pruinose thorax), dark appendages (light in **Little Blue**), and larger size. Female distinguished from female **Little Blue** by uniformly patterned abdomen; **Little Blue** has extensive black tip. Male **Blue Dasher** easily distinguished from **Plateau Dragonlet** by striped or light blue thorax, green eyes, and white face.

Natural History Males perch low in marsh vegetation, aggressively defending small territories, sometimes raising blue abdomen in display. May greatly elevate both wings and abdomen in midday sun. Females and immatures anywhere near water, immatures sometimes common at and near breeding sites. Pairs rarely seen. Tandem oviposition reported, may not be normal mode.

Habitat Shallow marshes, often with dense grass and/or sedges. Also similarly vegetated pools in streams.

Flight Season AZ Mar–Nov, NM Jun–Sep, TX Jul–Oct.

Distribution Ranges south in Mexico on west coast and in uplands to Oaxaca.

303.1
Plateau Dragonlet
male—Sonora,
Mexico, August 2006,
Netta Smith

303.2
Plateau Dragonlet
immature
female—Cochise Co.,
AZ, September 2004

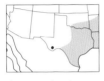

Description Very small clear-winged skimmer with blue male. *Male*: Eyes dark brown, blue-tinged in oldest individuals; face metallic blue. Thorax pruinose blue or blackish, abdomen pruinose blue with S8–10 black, appendages pale. *Female*: Eyes reddish over blue-gray, blue in oldest individuals; face yellowish-tan. Thorax brown in front, yellowish to tan on sides; abdomen yellowish with rectangles on sides, stripe down middle, and tip black. Oldest females may develop entirely blue eyes. Immature similarly patterned but pale colors yellow.

Identification No other skimmer in its range is tiny and blue. Dark face and reddish-brown/gray eyes readily distinguish mature male from male **Eastern Pondhawk** (green face, dark blue-green eyes), male **Blue Dasher** (white face, brilliant green eyes), and male **Blue Corporal** (dark brown eyes, face, and thorax and conspicuous basal wing spots). All of these largely blue dragonflies are obviously larger. Females and immature males drab but recognized by small size. Most similar is barely overlapping **Red-faced Dragonlet** and probably not overlapping **Plateau Dragonlet**, both slightly larger and with more color at hindwing base. **Little Blue** should be distinguishable from both by shining white appendages contrasting with entirely black abdomen tip (others with pale color reaching tip and brownish appendages). Dark stripe down center of abdomen and black tip distinguish female **Little Blue** from female MEADOWHAWKS.

Natural History Males not very conspicuous as they perch below knee height in marsh vegetation, but they fiercely chase others from their small territory. Often perch with wings depressed and abdomen elevated, like miniature Blue Dasher. Copulation brief and in flight, followed immediately by oviposition, male often hovering in attendance. Females oviposit in dense vegetation or in open water at edge of vegetation, either moving rapidly over open water and tapping at irregular intervals or remaining in restricted area and tapping slowly and methodically.

Habitat Shallow marshy ponds and lake margins.

Flight Season TX Feb–Oct.

Distribution Also from Illinois and New York south, very sparse in northern part of range.

304.1
Little Blue Dragonlet
male—Lake Co., FL,
April 2005

304.2
Little Blue Dragonlet
female—Lafayette
Co., FL, June 2004

304.3
Little Blue Dragonlet
immature
male—Richmond
Co., GA, May 2006

305 Seaside Dragonlet *Erythrodiplax berenice*   TL 31–35, HW 21–27

Description Small slender skimmer of Texas coast and southwestern alkaline lakes. *Male*: Eyes dark red-brown over pale greenish; face metallic blue-black. Entirely glossy black at maturity, eventually overlaid by dusky pruinosity. *Female*: Eyes red-brown over greenish-gray; face cream colored, metallic blue-black on top of frons. Thorax yellow to orange, with many narrow black stripes. Abdomen black

on S1–3 and S8–10; S4–7 orange above, black on lower sides. Many individuals with black thorax and bright orange abdomen, eventually (all of them?) becoming entirely black as male. Alternatively, rare individuals have striped thorax, entirely black abdomen. Populations in alkaline lakes in New Mexico include females with intensely colored wings, at extreme almost entirely brown; some coastal individuals may show brown wing spots beyond nodus. Subgenital plate conspicuous—long, pointed, and extending downward from abdomen.

Identification Unique habitat and range makes identification relatively easy; no other similar species occurs with it. **Double-ringed Pennant**, also small and black, becomes more bluish pruinose and has small brown spot at hindwing base. Overlap in habitat very unlikely. Superficially similar **Black Meadowhawk** a northern species of fresh water, no overlap in range. Female **Seaside Dragonlet** distinctive because of bright orange and black or yellow and black in various combinations. Some female SMALL PENNANTS closest in appearance, but most have colored patch at hindwing base and black abdomen with pale spots down middle. When changing from orange to black, thorax in **Seaside Dragonlet** becomes black when only top of abdomen orange, a unique combination. Female **Double-ringed Pennant** has light color only at abdomen base.

Natural History Both sexes often in large numbers in breeding areas; seem less territorial than other dragonlets, but males aggressive to other males at close range. Often at freshwater locations near coast where they may not breed and occasionally disperse well inland, perhaps with prevailing winds. Perch in marsh vegetation, also on shrub and tree branches when in wooded habitats. Tandem oviposition unique in genus, and pairs may fly long distances in tandem. Oviposition slow and methodical, often with repeated dips to water at same spot, especially on algal mats, then moving a few inches to repeat the procedure.

Habitat Only American dragonfly that breeds in salt water. Salt marshes in much of range, mangrove swamps in tropics; also saline ponds and lakes, some of them temporary. In New Mexico and west Texas, breeding in waters saltier than ocean. When ephemeral pools dry out, larvae survive in masses of stonewort until they fill again.

Flight Season NM May–Oct, TX May–Nov.

Distribution On Atlantic coast extends north into Nova Scotia, surprising for tropical genus but probably because of specialized habitat. Ranges south around Gulf and Caribbean to Venezuela, also on Pacific coast of Mexico from Baja California and Sonora to Oaxaca.

305.1
Seaside Dragonlet
male—Cameron Par.,
LA, May 2005

305.2
Seaside Dragonlet
striped female—
Cameron Par., LA,
May 2005

305.3
Seaside Dragonlet
black female—
Cameron Par.,
LA, May 2005

305.4
Seaside Dragonlet
desert lake
female—Chaves Co.,
NM, June 2005,
Dustin Huntington

306 Band-winged Dragonlet *Erythrodiplax umbrata* TL 38–45, HW 25–33

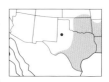

Description Mid-sized slender skimmer, male with black bands across wings. *Male*: Eyes dark brown; face black. Thorax and abdomen black, each wing with rectangular black band past midwing. Immature males fairly quickly develop light brown hint of definitive wing pattern. *Female*: Eyes brown over gray; face brown. Polymorphic, tan to brown with dark markings on front of thorax, dark rectangles along abdomen, and clear wings. Small percentage of mature females andromorphic, with black body and banded wings as males.

Identification Nothing else like mature male except **Four-spotted Pennant**, especially those that have larger wing spots. **Pennant** differs by having edges of spots rounded rather than straight across and white stigmas, also very different habit of perching high on tips of vegetation. Male much like male **Black-winged Dragonlet** in habitat and behavior, but wing patterns diagnostic. Female almost exactly like female **Black-winged** (see that species), but **Black-winged** no more than accidental in range of this book (global warming could change that!). Note that male **Common Whitetail** and female **Filigree Skimmer**, both in Texas, also have black band on each wing, but both very different otherwise, with wide abdomen and different body colors.

Natural History Males perch low in or at edge of tall grasses and fly low over water, often chasing one another. Flight looks fluttery, perhaps a consequence of patterned wings. Females and immatures perch on branches and herbaceous stems, even at some height in shrubs and trees, but mature males stay low. In tropics, spend dry season in woodland.

306.1
Band-winged
Dragonlet
male—Collier Co., FL,
April 2005

306.2
Band-winged
Dragonlet
heteromorph
female—Hidalgo Co.,
TX, June 2005

306.3
Band-winged
Dragonlet
andromorph
female—Veracruz,
Mexico, September
1999, Netta Smith

Mating pairs rarely seen, copulation is brief and in flight. Ovipositing female, guarded by male or not, taps slowly and methodically without rising far from water. Ovipositing andromorph females, with black wing bands, may display to males as if they were also males and thus avoid harassment. Huge numbers of this species seen on central Texas coast in July 2004, estimated to be millions. This magnitude of occurrence remains unexplained, although species probably in part migratory. Immatures often retire to woodland until ready to breed; may spend entire dry season there in tropics. Abundant and widespread in tropics.

Habitat Shallow, marshy ponds, often temporary, with scattered to dense, low to high sedges and grasses.

Flight Season TX all year.

Distribution Ranges south in lowlands to Argentina, sporadically east to Georgia and north to Ohio, where it has bred. Common in southern Florida and West Indies.

---

307 Black-winged Dragonlet *Erythrodiplax funerea* TL 38–42, HW 25–34

Description Midsized slender skimmer with black-winged male, not likely to be seen in United States. *Male*: Eyes dark brown; face blue-black. Entirely black pruinose with forewing and hindwing bases dark brown to halfway between nodus and stigma. Appendages tan. Immature males fairly quickly develop light brown hint of definitive wing pattern. *Female*: Polymorphic. Eyes reddish-brown over gray; face brown. Tan to brown with darker markings on front of thorax and dark rectangles along abdomen. Majority of females with clear wings, small percentage of mature females black-winged as males, body becomes entirely black. Even fewer with dark brown spot at base of hindwing and smaller spot at forewing base, also becoming black. Not known if all females become black eventually, but some very old ones still brown.

Identification Male most like male **Filigree Skimmer** that has different habitat and behavior, black in wings that extends to stigmas, and differently patterned forewing and hindwing. Note also dark terminal appendages in **Filigree**, somewhat pale in **Dragonlet**. Colored somewhat like male **Widow Skimmer**, but that species much larger and heavier, with white pruinosity beyond black wing bases and on abdomen when mature. Heteromorph female virtually identical to female **Band-winged Dragonlet** in coloration but distinguished in hand by some double cells in medial planate. Immature, well-patterned females might be distinguished by abdominal pattern. **Black-winged** has dark markings on sides of middle abdominal segments rectangular and reaching halfway forward on segment.

307.1
Black-winged Dragonlet
male—Sonora, Mexico,
August 2006, Netta Smith

307.2
Black-winged
Dragonlet
heteromorph
female—Puntarenas
Prov., Costa Rica,
January 2005

307.3
Black-winged
Dragonlet
andromorph
female—Nayarit,
Mexico, September
2001, Netta Smith

Band-winged has same markings pointed in front and often reaching slightly less than halfway forward. This average difference might not be diagnostic on every individual, and the pattern is somewhat different on immature males.

Natural History Males perch in tall or short grasses and fly low over water, often chasing one another. Flight looks fluttery, perhaps a consequence of dark wings. Immatures retire to shrubby and wooded areas until ready to breed. Mating pairs rarely seen. Wings sometimes held in characteristic way unusual for North American dragonflies, with forewings elevated well above slightly lowered hindwings. Mass movements have been seen at beginning of rainy season, should be sought at that time in the United States.

Habitat Primarily temporary ponds and marshes of wet season, usually quite open.

Flight Season AZ Aug–Oct, TX Jun.

Distribution Perhaps only vagrant to North America; no known breeding population and no recent records. Any sighting should be documented. Ranges south to Ecuador, most common in Pacific lowlands, where it can be one of the most abundant dragonflies.

This unusual genus inhabiting rocky rivers may be most closely related to pondhawks and dragonlets, but its color pattern, well adapted to its rocky habitat, is distinct from all other skimmers. World 1, NA 1, West 1.

308 Filigree Skimmer   *Pseudoleon superbus*   TL 38–45, HW 30–35

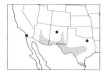

Description Midsized rocky-stream skimmer with exceptionally fancy wing and body patterns, most striking in immatures. *Male*: Eyes dark brown; face black. Thorax and abdomen black with faintly indicated pattern of fine lines. Forewing all black beyond nodus except for clear spot at tip, also black streak at base and other scattered black markings in basal half; hindwing entirely black except for clear tip; stigmas brown. Some individuals with a bit less black. Immature male with head and body colored as female, wings with dark areas as male but brown and patterned rather than black. *Female*: Eyes finely barred with dark brown and whitish; face cross-barred brown and white. Thorax indescribably variegated, mostly brown with scattered black markings and fine whitish and tan lines scrawled all over. Abdomen brown with complex V-pattern of alternating dark blotches and pale wavy fine lines on each segment, strongest on S2–7. Wings with dark bands beyond nodus and extensive dark flecking all over rest of wings, especially rear base of hindwing; stigmas at least half white.

Identification Both sexes unmistakable because of wing pattern alone. Finely lined eyes and body also unique among North American dragonflies. Most likely confused is male **Black-winged Dragonlet**, which has both wings with identical patterns. That species, however, very rare in North America.

Natural History Males alternate perching flat on rocks and sometimes on twigs at water with flying rapidly up- and downstream or over ponds, with much chasing of other males. Wings usually held down as if pressed onto rock surface, but at hot midday, abdomen points to sky and wings raised. Long and rapid chases of males and of females by males are regular. Copulation brief (5–30 sec), in flight. Females oviposit solo, often guarded by male,

308.1
Filigree Skimmer
male—Sonora,
Mexico, September
2005

308.2
Filigree Skimmer
female—Sonora,
Mexico, July 2003

by tapping in one spot, then moving, or slowly moving over water surface and tapping at intervals of a foot or so (several per second), preferring algal mats and collected detritus. Both sexes encountered resting flat on ground away from water, often on roads and usually on light-colored substrates. Superbly camouflaged on patterned ground but probably do not know it, as they often perch on white substrates.

Habitat Rocky rivers and streams with slow to moderate current, usually in open but also on some narrow streams within riparian belt. Also ponds adjacent to such streams.

Flight Season AZ Feb–Nov, NM Mar–Aug, TX Mar–Jan.

Distribution Ranges south, mostly in uplands, to Costa Rica.

Meadowhawks *Sympetrum*

Meadowhawks are the only small red dragonflies seen over most of North America and all across Eurasia. The only other small red dragonfly in the West is the Red-tailed Pennant, of limited distribution in our region, and there are few meadowhawks in the same range and habitat. Other red dragonflies in the region are larger or have much more black patterning mixed with the red. Black Meadowhawk is an exception to the red coloration prevalent in this genus and could be mistaken for several species of whitefaces except for its dark face. Female meadowhawks are more problematical, as most are brown, and there are other small, brown dragonflies, for example, dragonlets. Note dark markings on abdomen of female dragonlets are usually rectangular, not so for meadowhawks; also, dragonlets are of tropical origin, meadowhawks temperate, so there is relatively little overlap in range. Immature meadowhawks may be everywhere in meadows at high density and often more than one species. Immature males are colored just like females but are easily distinguished by a more slender abdomen. Some female meadowhawks, perhaps the oldest individuals, turn entirely red as in males. Both sexes may be seen away from water, sometimes far away, especially in the migratory Variegated Meadowhawk. Perching with wings drooped forward is common, even when on tips of grass inflorescences where pennants perch with wings level or raised. All meadowhawks oviposit in tandem, and some of them fly long distances in tandem between mating and egg laying. Other red dragonflies that oviposit in tandem, including saddlebags and small pennants, look quite different. Some species oviposit in water, but others lay eggs on dry ground with hatching after winter rains fill temporary ponds. Larvae develop quickly during summer, and the cycle begins again. These species lay relatively large eggs, up to 3 mm in diameter. This breeding cycle is characteristic of species that fly in late summer, and most meadowhawks are seen in late summer and well into fall. Cardinal and Variegated are exceptions, both appearing fairly early in spring. Variegated are further noteworthy for migratory and wandering tendencies. World 62, NA 14, West 13.

Table 11   Meadowhawk (*Sympetrum*) Identification

	A	B	C	D	E
Variegated	21	25	14	1	2
Cardinal	2	5	1	2	3
Red-veined	2	12	1	13	1
Blue-faced	4	2	3	1	3
White-faced	3	1	2	1	1
Ruby	1	1	2	13	1
Striped	1	2	21	1	123
Cherry-faced	2	1	2	13	1
Saffron-winged	2	1	12	1	12
Band-winged	1	3	2	3	1
Black	5	4	12	1	1
Autumn	2	1	1	1	3
Spot-winged	2	1	3	2	3

A, mature male face color: 1, tan; 2, red; 3, white; 4, blue; 5, black.

B, thoracic markings: 1, none; 2, pale stripes; 3, black stripes; 4, black chain; 5, pale spots below.

C, lateral abdominal markings: 1, none; 2, black triangles or lines; 3, black spots or rings; 4, white spots.

D, wing markings: 1, none; 2, dark basal spot in hindwing; 3, orange-brown base.

E, femur color: 1, all black; 2, black and pale striped; 3, all brown or paler.

309 Variegated Meadowhawk   *Sympetrum corruptum*   TL 39–42, HW 29–30

Description Robust migratory meadowhawk with fancy abdominal pattern and pair of wide white stripes or yellow spots on either side of thorax. *Male:* Eyes red above, pinkish to gray below; face red in front, brownish on sides. Thorax brown with pair of small bright yellow spots on either side. Abdomen overall reddish, but complexly patterned brown and red at close range. Legs black with brown stripe on outside of femur and tibia. Anterior wing veins red. Younger males (common away from breeding habitats) patterned as females, with white lateral spots persisting, but with much red on abdomen. *Female:* Eyes and face brownish-tan. Thorax brown with yellow spots as in male. Abdomen brown with  whitish lateral spots on S3–8. Sutures between segments richer orange-brown. Anterior wing veins yellow. Immature of both sexes more vividly marked, with wide white stripes on sides of thorax leading down to yellow spots; bright white spots low on sides of S2–8, bordered by black above and below; and black dorsal markings on S8–9, all producing variegated look that gives species its name. Stigma at all stages darker in center than at ends. Legs prominently black and yellow striped.

Identification Yellow spots low on sides of thorax diagnostic in combination with complexly patterned abdomen in immatures; no other dragonfly so patterned. Mature males and females, much duller except for yellow thoracic stripes, confused most readily with **Cardinal Meadowhawk**, of similar size but with dark markings at hindwing bases and no trace of pattern on abdomen. **Cardinal** much brighter scarlet, **Variegated** always dull red and often showing whitish abdomen spots. Other meadowhawks with pale stripes on thorax at any time of life include **Red-veined** and **Striped**, neither of which have stripes prominently yellow at lower ends and both of which have abdomen plain except for dark ventrolateral markings. **Saffron-winged Meadowhawk** has similarly brighter anterior wing veins but has stigma uniformly colored, **Variegated** darker in center; **Saffron-winged** body also scarcely patterned.

Natural History May feed in flight in swarms with other fliers but more typically a flycatcher like most skimmers. Males alternate perching on twigs, sedges, and grasses and frequent

309.1
Variegated
Meadowhawk
male—Greenlee Co.,
AZ, July 2007

309.2
Variegated
Meadowhawk
female—Hidalgo Co.,
TX, November 2005

309.3
Variegated
Meadowhawk
immature
male—Harney Co.,
OR, July 2006,
Netta Smith

flights low over water. Pairs oviposit in tandem, staying fairly low between descents and often tapping repeatedly before moving a short distance to do it again. Pairs at times very common, even at small pools or flooded parking lots. Away from water, commonly perches on ground; during migration, seen well up in trees. More likely to droop abdomen downward during midday heat than elevate it, as other meadowhawks do. Presumably migrant like Common Green Darner, with mature males appearing in spring in North all across region, from British Columbia to Manitoba, before any sign of emergence. Large numbers emerge in fall in northern part of range and then disappear, and wintering individuals fairly common far from water in deserts of Southwest, typically only odonate present at that time and place. In fact, can be ubiquitous away from water in many parts of Southwest. Winter breeding definitely known in southern Arizona and Texas. Not a simple pattern, as both spring and fall breeding noted in all parts of range. Substantial fall movements high along western mountains (documented from Cascades, Blue Mountains, and Sierra Nevada) and through Central Valley of California, as late as late October. Probably similar migration all across range, numbers seen in western Oklahoma. Also substantial fall flights (Aug–Oct) along Pacific coast, perhaps normal movements drifting westward with east winds until stopped by coastline (but also found over ocean). Maximum numbers in thousands or even tens of thousands, passing coastal points at rates of up to 500/min.

Habitat Breeding habitat shallow open or marshy lakes and ponds, often alkaline. Also pools with shore vegetation in streams and rivers. Oviposition seen in temporary ponds in many areas, even rain puddles on roads; not known if larvae develop to emergence.

Flight Season BC Apr–Sep, WA Apr–Nov, OR Feb–Oct, CA all year, MT May–Aug, AZ all year, NM all year, NE Apr–Oct, TX all year.

Distribution Recorded south to Belize and Honduras, perhaps only as vagrant. Regular south onto the Mexican Plateau; southern limits of breeding not known. Scattered records throughout East, north to southern Ontario and Nova Scotia, but breeding status poorly known.

310 Cardinal Meadowhawk *Sympetrum illotum* TL 28–40, HW 26–28

Description Robust bright scarlet meadowhawk of far West with dark streaks at wing bases. *Male*: Eyes red above, tan below; face red-orange above, tan-orange below. Entire body scarlet-red, thorax with two vivid pale yellow spots low on each side. Wings with anterior and basal veins reddish, dark brown basal streaks in all wings, larger in hindwings. *Female*: Patterned as male, but body either brown or reddish, either polymorphic or changes with age. Stout-bodied in comparison with other meadowhawks.

Identification Brilliant red males with yellow spots low on either side of thorax are distinctive. Other bright red meadowhawks lack these spots. **Variegated Meadowhawk** males have poorly defined spots when mature but are much duller, with both dark and light markings along sides of abdomen. Bright mature male **Red-veined Meadowhawk** looks quite similar, but look closely to see if yellow thoracic spots and dark basal wing markings are present or absent. Female distinguished in same ways as male, unmarked except for yellow thoracic spots and dark wing streaks. **Spot-winged Meadowhawk** occurs in southeast Arizona with **Cardinal**, distinguished by black stigmas and lack of spots on thorax; flight season and habitat barely overlap. **Cardinal** abdomen rather broad (thus so impressively colorful) and straight-sided, unlike other meadowhawks that have slightly swollen base and constriction at junction of S3 and S4.

Natural History Males perch on leaves and twigs at waterside or over water and fly out at other male dragonflies, usually their own but also of other species. May obelisk on sunny midday. Females relatively rarely seen, nowhere in numbers as in so many other meadowhawks. Copulation in flight and then at rest, relatively brief. Oviposits in tandem or solo with or without male guarding; usually near shore and typically in beds of low vegetation

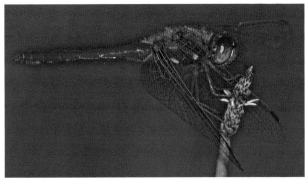

310.1
Cardinal
Meadowhawk
male—Sonora,
Mexico, September
2005

310.2
Cardinal
Meadowhawk red
female—Whatcom
Co., WA, August 2007

or on algal mats. Pair or single female taps water in one spot for some time, hovering just above water; male swings female when slightly higher, only female moves abdomen when lower. Ovipositing pairs attracted to one another.

Habitat Lakes and ponds in northern part of range, even very small ones such as backyard pools. More common on pools in rocky streams in southern areas. Often in same breeding habitat as Variegated Meadowhawk in Northwest but, unlike that species, rarely seen at any distance from water.

Flight Season BC Apr–Sep, WA May–Oct, OR Mar–Oct, CA Jan–Dec, AZ Apr–Nov, TX Apr–May.

Distribution Ranges south in uplands to Panama. Oddly distributed in West, considering its wide range to south.

---

311 Red-veined Meadowhawk   *Sympetrum madidum*   TL 42–45, HW 28–31

Description Fairly large meadowhawk with conspicuous thoracic stripes, red-veined wings. *Male*: Eyes and face red. Thorax red, faint indication of black lines on sutures and trace of two pale spots low on each side. Abdomen red, in some with black specks on posterior segments corresponding to areas where other species have more black. Legs black. Veins on front and base of wings reddish, may be orange suffusion in basal wing membranes. *Female*: Eyes red-brown over pale greenish; face tan. Thorax brown with two broad white stripes on either side, in some individuals fainter whitish stripes on front. Abdomen tan with narrow black lateral stripe

on S4–9, whitish areas below that; narrow black median line on S8–9. Some females become red like males, retaining much of pattern.

Identification   Mature male quite similar to **Cardinal Meadowhawk**, with bright red body and glowing orange wings, but **Cardinal** has dark streaks at wing bases and small yellow spots low on sides of thorax. **Cardinal** usually flies over open water, **Red-veined** over grasses and sedges. Male **Cherry-faced Meadowhawk** also bright red as **Red-veined** but

311.1
Red-veined
Meadowhawk
male—Humboldt
Co., NV, July 2005

311.2
Red-veined
Meadowhawk yellow
female—70 Mile
House, BC, July 2006

311.3
Red-veined
Meadowhawk red
female—Josephine
Co., OR, July 2003,
Chris Heaivilin

461

311.4
Red-veined
Meadowhawk
immature
male—Glenn Co., CA,
June 2004

has prominent black markings along side of abdomen and cerci with long point and prominent ventral tooth (blunt-ended, no tooth in **Red-veined**); also a bit smaller. Immature and female **Red-veined** with striped thorax extremely like **Striped Meadowhawk**, distinguished by much more reddish wings in **Red-veined**; **Striped** may have yellow to reddish anterior veins and orange at wing base but not reddish suffusion over entire wings. Many **Red-veined** lack frontal stripes on thorax, while **Striped** usually have them. **Red-veined** also has longer stigmas than **Striped**; with familiarity, this could be good mark. Stripes on thorax of **Red-veined** more likely to be obscured with age, less likely in **Striped**. Finally, easily seen in hand or photos, **Red-veined** has doubled cell rows in all radial planates, **Striped** not.

Natural History Males perch low in herbaceous vegetation or on ground around marshy areas. Females may be nearby, often perched on rocks or ground when temperature low in morning. Oviposits in tandem or alone, typically with male hovering nearby to guard female. Eggs laid in shallow water about to dry up or in dry pond basins, presumably overwintering and hatching in spring after ponds fill in winter.

Habitat Shallow, open, often somewhat saline ponds with abundant emergent vegetation, usually or perhaps always drying up by mid- or late summer. Also marshy pools in small slow-flowing streams, not much associated with open water.

Flight Season YT Jun–Jul, BC May–Sep, WA Jun–Sep, OR May–Sep, CA Apr–Sep, MT Jul–Aug, NE Jun–Sep.

Distribution Sparsely distributed in Rocky Mountains and Great Plains.

312 Blue-faced Meadowhawk   *Sympetrum ambiguum*   TL 36–38, HW 26–28

Description Brightly marked southeastern meadowhawk with blue face in male. *Male:* Eyes pale blue, brownish-tinged above; face white overlaid by blue to turquoise, sometimes looks greenish. Thorax mostly creamy gray with rather faint brown stripes. Abdomen red, black apical rings on S4–9, widening to rear to cover most of S9. Appendages pale. *Female:* Eyes brown over gray; face tan. Thorax as male. Abdomen tan, black apical dorsolateral spots on S3–9 becoming larger to rear and covering most of S8–9.

Identification Male only red dragonfly with bright blue face and bluish eyes. Gray, brown-striped thorax of both sexes not like thorax of any other meadowhawk or, actually, any other species in its size range. Abdomen also with more ringed look than in other meadowhawks. Overlaps widely with **Autumn Meadowhawk**, a brown-legged species that lacks black rings. Slight overlap at northern edge of range on southern plains with **Cherry-faced**, **Ruby**, **Saffron-winged**, and **White-faced Meadowhawks**, in all of which both sexes have black markings low on sides of most abdominal segments rather than as rings or partial rings. Those species also have dark legs, **Blue-faced** tan.

Natural History Both sexes perch in trees at woodland edge or within forest when away from water. Males perch low around edge of shallow pools or dried-up ponds. Copulation at rest. Females oviposit by dropping eggs in grass or on mud, guarded by male hovering or perched.

312.1
Blue-faced
Meadowhawk
male—McCurtain
Co., OK, August 2006,
David Arbour

312.2
Blue-faced
Meadowhawk
female—Chattooga
Co., GA, August 2005,
Giff Beaton

Habitat Ponds, including temporary ones, in or out of woodland; also on some wooded streams. More characteristic of wooded habitats than any other North American meadowhawk.
Flight Season NE Jul–Oct, TX May–Nov.
Distribution Ranges to east from Michigan and New Jersey south to northern Florida.

313 White-faced Meadowhawk *Sympetrum obtrusum*   TL 31–39, HW 20–29

Description Meadowhawk with snow-white face, black abdominal markings, and black legs. *Male*: Eyes red-brown over pale greenish to grayish; face white. Thorax brown, may look paler on sides than front with faint indication of ventrolateral pale spots. Abdomen bright red, black ventrolateral triangles on S4–8, most of lower surface of S9 black. Appendages reddish. *Female*: Eyes reddish-brown over pale greenish; face pale yellowish. Thorax brown in front, lighter brown on sides with pale yellowish areas below. Abdomen tan with wide black ventrolateral stripe from S4–9, covering larger parts of segments toward rear. Wings may have some orange suffusion at base. Some females may become red as in other meadowhawks but rarely seen in this species.

Identification Mature male only one of three species with vivid black triangles on red abdomen that has white face. **Cherry-faced** has red face, **Ruby** light brown face. Female like **Cherry-faced** but usually paler face, darker wing veins, and more contrasty thoracic pat-

313.1
White-faced
Meadowhawk
male—Fayette Co.,
IA, July 2004

313.2
White-faced
Meadowhawk
female—Kittitas Co.,
WA, September 2007

313.3
White-faced
Meadowhawk
immature
female—Chelan Co.,
WA, June 2004

tern; like **Ruby** but not usually with orange-based wings (this character works in region covered by this book, not farther east). Many individuals will be puzzling. Subgenital plate in female **White-faced** with lobes not much expanded, pointed directly to rear. Female and immature male with brightly marked thorax look much like **Striped Meadowhawk,** but markings not quite so clearly arranged in stripes. Usually a bit brighter yellow than **Striped**, with more extensive black on abdomen.

Natural History Males can be superabundant in extensive sedge meadows, as many as one per 10 square feet. Copulation presumably lengthy, as pairs in wheel can also be abundant at times when scarcely one in tandem is seen. Oviposits in tandem or alone, usually with male hovering nearby to guard female. Male may release female and both land, then he guards her when she flies up to lay eggs again, or she may lay eggs even while perched. Eggs dropped from above in shallow water about to dry up or in dry pond basins, presumably overwintering and hatching in spring after ponds fill in winter.

Habitat Shallow marshes, bogs, and fens that may or may not dry up each summer; also similar habitats at edge of lakes. Often most common meadowhawk in cold, boggy wetlands, and associated with forest over most of its range.

Flight Season YT Jun–Jul, BC Jun–Sep, WA Jun–Oct, OR Jul–Oct, CA May–Oct, MT Jun–Sep, NE Jun–Oct.

Distribution Also in East from Ontario and Nova Scotia south to Kentucky, Virginia, and Maryland. In mountains in southern part of range.

314 Ruby Meadowhawk *Sympetrum rubicundulum* TL 33–34, HW 24–30

Description Red meadowhawk with sharply defined black triangles along abdomen, light brown face, and pale orange wing bases. *Male*: Eyes red over grayish-green; face yellowish. Thorax red. Abdomen bright red with elongate black ventrolateral triangles on S4–9. In most individuals, wings washed with orange across entire base, out to nodus; populations farther east clear-winged. *Female*: Eyes reddish-brown over pale greenish; face yellowish to chartreuse. Thorax brown, paling to yellowish on sides. Abdomen tan with black markings similar to those of male but more like rectangles than triangles. Some females bright red like males.

Identification In West, orange wing bases good field mark in both sexes to distinguish **Ruby** from quite similar **Cherry-faced** and **White-faced Meadowhawks**. However, some females of other species, especially **Cherry-faced**, have much orange at wing base, in which case capture and examination of subgenital plate are necessary for identification. In

314.1
Ruby Meadowhawk
male—Chattooga
Co., GA, August 2005,
Giff Beaton

314.2
Ruby
Meadowhawk
immature
male—Cherry Co.,
NE, July 1998

314.3
Ruby Meadowhawk
female—Winneshiek
Co., IA, July 2004

this species, subgenital plates much expanded, bulging downward, and pointing straight to rear. **Ruby** averages larger than **Cherry-faced** and **White-faced**, so size alone might be clue. See more details under **Cherry-faced**. Orange wing bases superficially like those of **Band-winged Meadowhawk**, which has brown crossband at midwing where it occurs with **Ruby**, also black markings on thorax.

Natural History Males perch in tall grasses and shrubs at breeding locations. Tandem pairs and lone females oviposit by dropping eggs into grass near water; also lay in water, unlike near relatives Cherry-faced, White-faced, and Striped.

Habitat Edges of extensive marshes and marshy ponds and lake shores, typically in open areas in West.

Flight Season NE May–Oct.

Distribution Because of much confusion in literature among similar species, western edge of range may need revision. Also east to Nova Scotia, south to northern Alabama and Georgia.

---

315 Striped Meadowhawk *Sympetrum pallipes*　　　　　　TL 34–38, HW 25–28

Description Red meadowhawk with variable amounts of black on abdomen, clearly striped thorax. *Male*: Eyes red-brown over tan; face tan. Thorax brown with cream-colored stripes, two on each side and two narrower ones in front. Abdomen red, varying with climate from conspicuous black ventrolateral markings on S4–8 in wetter regions to scarcely any black in dry interior. Legs entirely black in wetter climates (*pallipes*, "pale-footed" in Latin, is a misnomer over much of range), femora mostly brown in drier interior regions. *Female*: Eyes red-brown over pale greenish; face tan. Thorax light brown with cream stripes as in male; abdomen tan with markings as those of male, similarly variable with climate. Wings may have much orange suffusion at base, extending along front edge to nodus. Abdomen can become bright red as in male. Thoracic stripes can be obscured in very old individuals of both sexes.

Identification Distinctive as only meadowhawk with conspicuous thoracic stripes throughout adult life, but other species have similar stripes when immature. Most similar is **Red-veined Meadowhawk**, may be difficult to distinguish except in hand, when different hamules, cerci, and/or subgenital plate are evident. **Striped** usually has one row of cells in radial planate, **Red-veined** many doubled cells (occasionally doubled in one or two wings in **Striped**). Fortunately, **Red-veined** has reddish suffusion throughout wings that **Striped** lacks, visible in many lighting situations. Immature **Variegated Meadowhawk**

315.1
Striped Meadowhawk dark male—Thurston Co., WA, September 2005

315.2
Striped Meadowhawk
dark female—
Thurston Co., WA,
September 2005

315.3
Striped Meadowhawk
pale male—Apache
Co., AZ, July 2007

with prominently striped thorax also shows white-spotted abdomen. Immature and female **White-faced Meadowhawk** much like **Striped**, but very slightly smaller and pale markings on sides of thorax not formed into such obvious stripes.

Natural History Males set up territories over dry grassy, weedy basins, commonly over grass lawns near water. Copulation lengthy, at least 5 min, and mostly at rest. Pairs oviposit in tandem (sometimes only female) over dry grassy and sedgy basins, dropping visible white eggs from above (as low as 3 inches) onto ground. Typical effort flicking abdomen down about once per second for several minutes, dropping one or few eggs each time. Has been seen ovipositing in grass lawn near backyard pond and large lake. Pairs apparently strongly attracted to one another, as groups of many pairs in small area often seen, with none in apparently suitable habitat between them. Both copulating and tandem pairs attracted. Single males and copulating pairs sometimes make dipping motions over certain spots, presumably to indicate appropriate oviposition site to female. Perches at and near water more commonly up in shrubs than other meadowhawks of same habitats but also perches on ground, especially on cool, sunny days. Very common in Pacific Northwest coastal lowlands.

Habitat Shallow marshes that flood temporarily or shallow basins adjacent to lakes that do the same. Often in open areas but more regular in woodland.

Flight Season BC May–Oct, WA Jun–Nov, OR Jun–Nov, CA Apr–Nov, MT Jun–Aug, AZ Jun–Oct, NM Jul–Aug, NE Jun–Sep.

Distribution In mountains in southern part of range.

316 Cherry-faced Meadowhawk *Sympetrum internum* TL 31–36, HW 23–27

Description Small meadowhawk with reddish face, colored basal wing veins, prominent black markings on sides of abdomen, black legs. *Male*: Eyes red, shading to dull pale greenish below; face reddish. Thorax and abdomen bright red. Abdomen with elongate black ventrolateral triangles on S4–9, widening to rectangles to rear; may be fine median black line on S7–8. Veins of wing base, or most of wing, may be yellow-orange, and sometimes yellowish suffusion in this area. Some northwestern populations with relatively dark veins, only costa yellowish. *Female*: Eyes red-brown over tan; face tan. Thorax and abdomen tan; latter with continuous black ventrolateral stripe on S4–9, continuous with narrower stripe higher up on S2–3. Wings as in male, more likely to have extensive orange suffusion on hindwing base. Small percentage of females get quite red on upper surface of eyes, front of thorax, and all of abdomen.

316.1
Cherry-faced
Meadowhawk
male—150 Mile
House, BC, July 2006

316.2
Cherry-faced
Meadowhawk
female—70 Mile
House, BC, July 2006,
Netta Smith

Identification One of three small species with plain thorax, black legs, vivid black markings on side of abdomen. Male distinguished from otherwise identical **White-faced Meadowhawk** by face and wing-vein color, from **Ruby Meadowhawk** by slightly smaller size, redder face, and lack of orange wing bases (most **Ruby** in range of this book have orange-based wings, as do some female **Cherry-faced**). Much more black on abdomen than in **Saffron-winged**, which may have somewhat similar wing color but differs also in yellow, black-bordered stigmas. Mature female like **White-faced** but face somewhat redder, wings with yellow costa, and thorax tends to be all one color, whereas thorax of **White-faced** tends to have browner front contrasting with yellower sides. Immature males and all females difficult to distinguish from **Ruby** and **White-faced**; must examine hamules or subgenital plates for certainty. **White-faced** has small notch in hamule, **Cherry-faced** and **Ruby** large notch; this difference visible with hand lens. **Ruby** has slightly larger notch than **Cherry-faced**, with more prominent shelf extending inward. Subgenital plates similar, all narrowing toward tip and then notched, with groove between two halves. Paired tips point outward in **Cherry-faced**, parallel in other two. In side view, each half bulges dramatically in **Ruby**, flatter and more parallel with abdomen in **White-faced**.

Natural History Males set up small territories, often close to one another, over dry grassy, weedy basins. Copulation presumably lengthy, as pairs in wheel common. Male holds wings backward to touch female to signal tandem flight, which may be lengthy. Oviposits in tandem or solo, dropping eggs from below knee height on dry ground of pond basin, often persistent at one spot and usually collecting in clusters. May lay eggs unproductively in grass on wet lawns. Can be spectacularly abundant in good habitat, with many tandem pairs flying cross-country.

Habitat Shallow marshy ponds and lake edges that dry up during summer. Many of these habitats have dried up in regions of West with persistent drought, and the species has disappeared from parts of its former range because of this.

Flight Season YT Jun–Sep, BC Jun–Sep, WA Jul–Oct, OR Jul–Sep, CA Jun–Sep, MT Jun–Sep, NE Jun–Oct.

Distribution Sparsely distributed on Pacific coast. If Jane's Meadowhawk, *Sympetrum janeae*, of East is included with this species, occurs east to Newfoundland, south to Kentucky and North Carolina.

---

317 Saffron-winged Meadowhawk *Sympetrum costiferum* **TL 31–37, HW 25–28**

Description Meadowhawk with brightly colored line down front edge of wings and long, black-bordered stigma; male all dark red. *Male*: Eyes dark red, paler below; face reddish-tan. Thorax red-brown with fine black lines in sutures. Abdomen red, duller at base, and with black ventrolateral lines from S3 to S9, not always present, and narrow black median stripe on S8–9. Legs light brown to mostly black. Anterior wing veins reddish; stigma red, strong black border fore and aft. *Female*: Eyes dull reddish over tan; face tan. Thorax and abdomen tan to reddish, black abdominal markings as male but much reduced. Legs brown on outer surfaces. Anterior wing veins yellowish to orange; stigma clear yellow, bordered fore and aft by black. Small percentage become almost as red as male.

Identification Mature male dark red all over with dull brown face; no other meadowhawk exactly like this, and most a bit brighter red. However, brightest **Saffron-winged** much like **Autumn** because of entirely red body and (in some populations) entirely brown legs; distinguished most readily by wing coloration, **Autumn** lacking saffron band on front and having dark color at hindwing base. **Cherry-faced**, **Ruby**, and **White-faced Meadowhawks** with black legs, much more prominent black abdominal markings. Female colored much like female **Autumn** but legs usually darker, less color at wing bases, and easily

317.1
Saffron-winged
Meadowhawk
male—Okanogan
Co., WA, August 2007

317.2
Saffron-winged
Meadowhawk
female—Benton Co.,
WA, August 2007

317.3
Saffron-winged
Meadowhawk
immature
female—Prescott
and Russell Co., ON,
July 2005,
John C. Abbott

distinguished by end of abdomen (**Autumn** with expanded scoop-shaped subgenital plate). Female **Saffron-winged** wing color also distinctive, with dark-bordered yellow stigmas and saffron band along front edge. Female **Variegated Meadowhawk** with somewhat similar wings but yellow stripes on thorax and patterned abdomen.

Natural History Males on territory at and near shore, perched low over open water on twigs or hovering at knee to waist height offshore, more hovering over water than is typical of meadowhawks (except Cardinal and Variegated). Breeding habitat more open than in most other meadowhawks. Oviposits in tandem, usually in open water but also on wet shore. Some pairs tap water half-dozen times, then move slowly to next nearby spot, very methodical.

Habitat Permanent marshes, ponds, and lakes with much emergent vegetation. Often at alkaline ponds in West.

Flight Season BC Jun–Oct, WA Jun–Nov, OR Jul–Oct, CA Jun–Oct, MT Jul–Oct, NE Jun–Oct.

Distribution Also east across Canada to Newfoundland, south to Illinois and New Jersey. Rare in southern parts of range.

318 Band-winged Meadowhawk *Sympetrum semicinctum* TL 28–40, HW 22–28

Description Meadowhawk with extensive orange-brown patches at wing bases, fine black markings on sides of thorax. *Male*: Eyes red-brown over greenish-tan; face reddish-brown. Thorax reddish-brown with black pattern above legs and two or three narrow black lines extending from legs up toward wing bases, one of them forked. Abdomen bright red with narrow black dorsolateral stripe on S2–3, conspicuous black ventrolateral stripe on S2–10, black median line on S8–9. Bright red when mature, wing patches dark. *Female*: Duller, usually brown on front of thorax and yellow on sides,  black lines more conspicuous. Continuous black on sides of S5–10; wing patches usually paler than those of males. Older females with red on top of abdomen.

Identification Only meadowhawk with extensive orange-brown patches at bases of all wings. Patches much reduced or even lacking in *californicum* subspecies. Then best field mark is black markings on sides of thorax more extensive than in other red meadowhawks. Female **Black Meadowhawk** sometimes has orange wing bases but has even heavier black markings on thorax, lacks upper dark stripe on basal abdominal segments. **Ruby Meadowhawk** also with orange wing bases but less extensive than **Band-winged** and without dark outer marking of co-occurring *fasciatum* subspecies of **Band-winged**, nor dark thoracic lines.

Natural History Individuals of both sexes at times fly back and forth over clearings up to 30 feet above ground, like little saddlebags and with much hovering. Are these individuals feeding or on display? Both sexes tend to perch higher than other meadowhawks, males high on twigs and leaves at breeding sites. Pairs often seen away from water, flying cross-country in tandem. Perhaps conspicuous wing markings, as in saddlebags, facilitate sexes finding one another away from breeding sites, and pairs probably come together away from water. Oviposits in tandem in shallow open water, often among plants.

Habitat Open ponds and marshes, usually permanent but sometimes small seepage areas; in open or wooded country. Also spend much time in grassy meadows.

Flight Season BC Jun–Sep, WA Jun–Oct, OR Jun–Oct, CA Apr–Oct, MT Jul–Oct, AZ Jun–Nov, NM Jun–Oct, NE Jun–Sep.

Distribution In East from southern Ontario and Nova Scotia south to Illinois and New Jersey, in mountains to northern Alabama.

Comments Western populations of this species long considered separate species, Western Meadowhawk, *Sympetrum occidentale*. Subsequently intermediate specimens have been found just east of region, but no one has conducted thorough study to document extent of

318.1
Band-winged Meadowhawk
male—Kittitas Co., WA,
September 2004

318.2
Band-winged Meadowhawk
female—Harney Co., OR, July
2006, Netta Smith

318.3
Band-winged
Meadowhawk
male—Apache Co.,
AZ, August 2004,
Doug Danforth

intermediacy. Three subspecies named from West: *S. s. occidentale* in Northwest, *S. s. californicum* in Great Basin, and *S. s. fasciatum* in Rockies and Great Plains. *S. s. occidentale* has evenly colored dark wing patches; *S. s. californicum* has wing patches reduced and pale, even lacking in some individuals; and *S. s. fasciatum* has conspicuous darker band on outer edge of each wing patch. Much variation, however, and more research needed to define accurately ranges and even validity of these subspecies.

Description Small black or black and yellow meadowhawk of northern regions. *Male*: Entirely dark brown to black from face to appendages. Face becomes metallic. Younger individuals colored as females become increasingly dark with age but retain showy pattern for some time. *Female*: Eyes reddish-brown over pale green; face tan. Thorax rich brown in front, pale yellow on sides, with complex black pattern of anastomosing stripes. Abdomen yellow at base, otherwise tan with continuous black lateral stripe from S2–9 or S2–10, black median stripe on S8–9. Legs black as in male. Costa yellow.

Identification Males unmistakable by all-black coloration, easily distinguished from whitefaces if face can be seen. Also more slender-bodied than typical of whitefaces. Females, with striking pattern on thorax, most like female **Band-winged Meadowhawk**, also with black markings on thorax. **Black** has more complex pattern, with small pale spots contained in black on sides, whereas **Band-winged** has dark lines and no pale spots. Most **Band-winged** have orange basal wing patches, but some female **Black** similarly colored. Pale sides of thorax of **Black** contrast strongly with brown front, differing from most other meadowhawks. No overlap in West with **Seaside Dragonlet**, another small species with black males and black-and-yellow to orange females. Metallic purple frons in dragonlet would distinguish males of that species; differently patterned thorax would distinguish females. See also **Ebony Boghaunter**, a small, mostly black dragonfly that does not overlap seasonally with **Black Meadowhawk**.

Natural History Males scattered through marshy wetlands, very conspicuous against bright green sedges but not particularly territorial. Immatures and females often common nearby, perching in herbaceous and shrubby vegetation or sometimes on ground; females more

Black Meadowhawk male—Kittitas Co., WA, August 2002, Netta Smith

319.2
Black Meadowhawk female—Baker Co., OR, August 2004

319.3
Black Meadowhawk
immature male—
Baker Co., OR,
August 2004

likely higher up in shrubs. Much mating takes place away from water, then pairs in tandem come to water by midday. Lone females also become more common at water after midday. Oviposits in tandem or solo in open water or on moss, less often mud, even dropping eggs from air; usually attracted to one another. Pairs dip toward surface about 1.6 times/sec. Eggs scattered widely in open or among dense vegetation, but most pairs use single site while present. Perches in obelisk position in sun at midday.

Habitat Typically shallow lake borders, marshes, and fens with abundant emergent vegetation, especially sedges; also quaking bogs.

Flight Season YT Jun–Oct, BC Jun–Oct, WA Jul–Oct, OR Jun–Oct, CA Jun–Oct, MT Jul–Nov.

Distribution Ranges east across Canada to Labrador and Newfoundland, south to Wisconsin and New York. In mountains in southern part of range. Also occurs all across northern Eurasia.

320 Autumn Meadowhawk *Sympetrum vicinum* TL 31–35, HW 21–23

Description All red or yellow late-season meadowhawk. *Male*: Entirely red, only black narrow median spot on S9; however, some older individuals with additional dark markings toward end of abdomen. Legs reddish-brown. *Female*: Eyes reddish over tan; thorax and abdomen tan or red, same black on S9 as male but also may be on S8. Legs light brown to yellowish. Subgenital plate prolonged into large scoop-shaped structure.

Identification Pale legs usually distinctive, as is entirely red color, including face, with little or no black. Lacks dark markings on thorax, abdomen, wings, or legs that characterize most other meadowhawks, but dark spot on S9 distinctive. Some **Saffron-winged** almost lacking in black markings but a bit larger, distinctly darker red in male, than **Autumn**, with characteristic wing coloration. Black dorsal markings in male **Saffron-winged** always on S8–9, not just S9 as in **Autumn**. **Cardinal Meadowhawk** about same shade of red as **Autumn**, distinguished by greater bulk, dark basal wing markings. Scoop at end of abdomen of female **Autumn** easily visible from side.

Natural History Adults common in clearings and at forest edge near breeding sites, males not defending territory at water. Usually perch well up in shrubs, also on tree trunks, wood piles, and ground in sun on cool days. Arrive at water already paired by late morning or midday and accumulate in certain spots, in or at edge of dense vegetation, but often fly far and wide around area before laying eggs. Dipping movements at water may in fact precede and stimulate copulation. Pairs oviposit in tandem and seem attracted to one another, show no aggression at all to others within inches. Pair usually oriented toward

320.1
Autumn
Meadowhawk
male—Kitsap Co.,
WA, September 2006

320.2
Autumn
Meadowhawk
female—Kitsap Co.,
WA, September 2005

and close to shore while laying eggs, either in shallow water, on wet vegetation, including moss-covered logs, or on wet mud. Female forms ball of eggs within subgenital plate, pair drops to substrate from 3–6 inches up, rises and hovers, then drops again, often moving a few inches each time. Often alternate tapping shore and water. In West, emerges late in season, whereas emergence may be in midsummer in East with long interval before returning to water. Latest species to fly in northern parts of range, usually until first heavy frosts. Able to maintain activity by constant basking in sun.

Habitat Well-vegetated ponds and lakes, usually permanent and usually associated with woodland and forest.

Flight Season BC Jun–Oct, WA Jul–Nov, OR Aug–Nov, CA Sep–Oct, NM May–Nov, NE Jul–Nov, TX Jun–Dec.

Distribution Also east through southern Canada to Nova Scotia, south to northwestern Florida. Very few records between western and eastern populations.

Comments Called Yellow-legged Meadowhawk in some recent publications.

321 Spot-winged Meadowhawk *Sympetrum signiferum* TL 33–42, HW 25–32

Description Southwestern stream meadowhawk with distinctive wing pattern. *Male*: Eyes red above, blue-gray below; face red. Thorax reddish-brown in front, redder on sides. Abdomen bright red with apical posteroventral black spots on S5–8. Legs reddish-brown. Anterior wing veins golden-orange, stigmas black; orange patch at base of both wings, black spot behind it in hindwings. *Female*: Eyes red-brown above, pale blue-gray below. Thorax brown in front, yellowish-tan on sides. Abdomen orange-tan with same black markings as male. Legs yellowish-tan. Eyes of immature red-brown over pale greenish. Older females become redder.

Identification Black stigma contrasting with golden anterior wing veins diagnostic, as is black spot at base of hindwing. Only other species with somewhat similar wing spot is **Cardinal Meadowhawk**, more robust with redder wing veins, no black on sides of abdomen, and big white spots on sides of thorax. The two species scarcely overlap. Black markings on abdomen also different from any other coexisting species. Strikingly bicolored red and blue-gray eye color of mature individuals distinctive among our dragonflies.

Natural History Males fly around and around over pools or perch on twigs and grass stems from near water surface to head height, often at high density; at other times defend small territories aggressively. Oviposition in tandem in shallow water among low emergent plants. Immatures sometimes common in tall grasses of breeding habitat, which furnishes only moist environment for them.

Habitat Pools of hill streams, typically where there is much tall grass for perching, in open or woodland.

Flight Season AZ Aug–Nov.

Distribution Ranges south in uplands of Mexico to Nayarit.

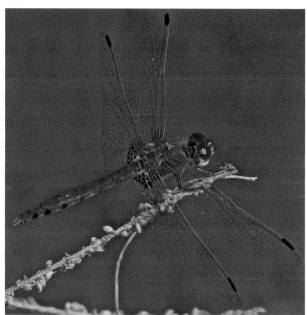

321.1
Spot-winged
Meadowhawk
male—Cochise Co.,
AZ, September 2004

321.2
Spot-winged
Meadowhawk
female—Cochise Co.,
AZ, September 2004

Meadowhawks - male hamules

Red-veined

Blue-faced

White-faced

Ruby

Striped

Cherry-faced

Saffron-winged

Meadowhawks - female subgenital plates

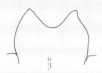
Red-veined

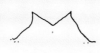
Blue-faced

White-faced

Ruby

Striped

Cherry-faced

Saffron-winged

These neotropical skimmers have glowing green eyes and white faces, metallic above, just like related Blue Dasher. Species vary from tiny to midsized, all with clear wings but variably patterned bodies, typically including a strongly striped thorax. Almost all have pale spots on S7, a common marking in dragonflies, the function of which badly needs explanation. Two bridge crossveins in each wing are diagnostic of the genus, one such vein in all similar genera. They perch quietly at the edge of ponds and swamps, drooping wings and elevating abdomen in the midday sun. Females have very varied egg-laying behavior, including doing so while perched. This is a large genus, extending barely north into North America. World 47, NA 3, West 3.

322 Spot-tailed Dasher *Micrathyria aequalis*   TL 28–33, HW 21–25

Description Small tropical dasher with triangular spots on abdomen. *Male*: Eyes brilliant green; face white, top of frons metallic purple. Thorax and abdomen base pruinose blue, posterior abdomen black with conspicuous triangular yellow spots on S7. *Female*: Eyes dull green over blue-gray, face white. Thorax brown in front with narrow pale green stripe on either side not reaching top; side

more or less evenly striped brown and pale green, IYI-pattern usually clearly indicated with half-height dark stripe after that, behind hindwing. Abdomen dark brown to blackish with

322.1
Spot-tailed Dasher
male—Starr Co., TX,
November 2005

322.2
Spot-tailed Dasher
female—Hidalgo Co.,
TX, November 2005

dorsolateral pale green stripe on S2–4, interrupted to form narrow triangles on S5–7, triangle often wider on S7. Pale green ventrolateral areas as well on S2–3. Old females can become very dull, grayish pruinose all over.

Identification Only other similar smallish skimmer in range slightly larger **Thornbush Dasher**, male of which has complexly marked thorax and entirely spotted abdomen. Still larger male **Three-striped Dasher** with striped thorax, larger square spot on S7. Also larger male **Blue Dasher** with pruinose blue abdomen as well as thorax, or only abdomen blue. Female **Spot-tailed** very much like **Thornbush,** but side of thorax more heavily striped, with one additional dark stripe at rear. Also much like female **Blue Dasher** except distinctive thoracic pattern, with more complex lateral stripes.

Natural History Males perch low in herbaceous vegetation at shore or in grasses and sedges in open marshes, some individuals staying at water all day. Droops wings and raises abdomen when perching in midday sun. Females usually away from water, surprisingly hard to find (but sometimes in numbers where species especially common); often roost up in shrubs and low trees. Females oviposit by hovering over algal mats, then landing on them and applying eggs to substrate, both on surface and curling abdomen beneath leaves and algal mats. Male stays nearby in flight and perched.

Habitat Ponds, ditches, and marshes, often those formed during rainy season and usually with much aquatic vegetation.

Flight Season AZ Nov, TX Mar–Nov.

Distribution Ranges south to Ecuador and Guianas, also southern Florida and West Indies.

323 Thornbush Dasher *Micrathyria hagenii*   TL 33–35, HW 25–29

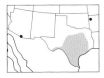

Description Medium-small tropical dasher with multispotted abdomen. *Male*: Eyes brilliant green; face white, top of frons metallic purple. Thorax complexly striped dark brown and pale greenish, with dark YI stripe pattern, with no dark stripe under hindwing. Abdomen black with pale elongate bluish spots on basal segments and square whitish spots on S7, black beyond that. At maturity, pruinose blue between wings. *Female*: Colored much as male, but no pruinosity between wings and abdominal spots larger and those on S1–6 yellow; wings may be orange at base. Old females develop some pruinosity on abdomen.

Identification Lack of blue on thorax and abdomen easily distinguishes from male **Spot-tailed** and **Blue Dashers**, but immatures of latter more similar. Female and immature male **Thornbush** with large spots on abdomen back through S7, just as in **Blue** and **Spot-tailed Dashers**. **Blue** has straight and separated stripes on thorax, **Thornbush** with at least one Y-shaped stripe that joins other stripes at top. **Thornbush** has vivid flattened X at top of front of thorax, formed from yellow line across top and yellow sclerites before wing bases; with pale median line looks like tiny dragonfly in flight! In **Blue Dasher** only lower yellow line evident (sometimes faint upper line evident, but never such an obvious "pair of wings"). Distinguishing **Thornbush** from **Spot-tailed** even more difficult, but **Spot-tailed** has one more dark stripe toward rear of thorax. Spots on midabdomen of **Spot-tailed** tend to be pointed, those on **Thornbush** more irregular, that on S7 squarish. **Thornbush Dasher** considerably smaller than SETWINGS that might be mistaken for it.

Natural History Males on twigs and other prominent perches at water's edge, often higher than smaller Spot-tailed. Both sexes common in shrubs away from water. Wings of males at water typically drooped but away from water may obelisk with raised wings. Copulation brief and in flight, oviposition immediately thereafter. Females drop clusters of eggs from above water or land on floating submergent vegetation and walk along it, extruding eggs.

Habitat Ponds and open marshes, frequent at rain pools.

Flight Season AZ Sep, TX Apr–Dec.

Distribution Ranges south to Panama, also Greater Antilles.

323.1
Thornbush Dasher
male—Hidalgo Co.,
TX, November 2005

323.2
Thornbush Dasher
female—Hidalgo Co.,
TX, November 2005

324 Three-striped Dasher *Micrathyria didyma*   TL 35–41, HW 25–33

Description Slender south Texas dasher with square-spotted abdo-men. *Male*: Eyes brilliant green; face white, top of frons metallic purple. Thorax somewhat evenly striped brown and pale yellowish-green, abdomen black with fine yellow streaks on base and large squarish white spots on S7. Well-developed blue pruin-osity between wings at maturity. *Female*: Essentially as male in color and pattern but abdomen slightly thicker with more color at base and no pruinosity between wings; may have wide dark wingtips.

Identification Larger size, lack of spots on basal abdominal segments, and more evenly striped thorax (without Y-shaped stripes) all distinguish from **Thornbush Dasher**. Dis-tinctly larger than **Spot-tailed Dasher**, with large, square spots that fill segment rather

325.1
Blue Dasher eastern
male—Burnet Co.,
TX, July 2004

325.2
Blue Dasher western
male—Pinal Co., AZ,
July 2007

325.3
Blue Dasher female—Butte
Co., CA, June 2004

325.4
Blue Dasher immature
female—Marion Co., FL, April
2005

Habitat Any body of standing water with some aquatic (floating or emergent) vegetation might be appropriate, including lake shores, ponds, ditches, and even wooded wetlands where some sunlight penetrates. Also at slow streams with abundant vegetation. Although they commonly perch in trees and shrubs, these are not necessarily present. Adults more likely to be in tall (for example, cattails) than in short emergent vegetation, but larvae abundant in floating carpets of water lettuce and water hyacinth. Shuns unvegetated wetlands.

Flight Season BC May–Aug, WA Jun–Oct, OR May–Sep, CA Feb–Nov, AZ Apr–Nov, NM Mar–Oct, NE May–Sep, TX all year.

Distribution Ranges south into upland Mexico, less common in tropical lowlands but rarely to Belize. Also throughout East, north to southern Ontario and New Brunswick, and in Cuba and Bahamas. Spottier in West than East. Appears to be extending range northward.

Setwings   *Dythemis*

This small neotropical genus is related to sylphs, clubskimmers, and rock skimmers. As in near relatives, the female has underside of S9 characteristically elevated, curved, and keeled, plainly visible from the side and different from flat segment in other skimmers. It is named for its characteristic perching with wings depressed and abdomen up, like the "ready, set, go" of a runner at a track meet. However, all species also perch like small pennants, with abdomen and forewings raised substantially, hindwings level or a bit raised, and may assume full vertical obelisk position at midday. Most species have a slender abdomen. Mature male coloration is always distinct, but females of two species are somewhat similar and need close study. Most females have prominent dark wingtips. Mayan Setwing is quite distinct, more likely to be mistaken for dragonflies other than setwings. World 7, NA 4, West 4.

326 Checkered Setwing   *Dythemis fugax*   TL 44–50, HW 36–38

Description Setwing with red-brown, black, and white pattern and dark brown wing patches. *Male:* Eyes red over red and gray; face red. Thorax mostly reddish-brown with indistinct black markings, pale gray on lower sides. Abdomen black with cream streaks at base and pair of long spots on S7. All wings with large brown patches at bases, darker veins running through them. *Female:* Eyes red over blue-gray,

326.1
Checkered Setwing
male—Kinney Co.,
TX, July 2004

326.2
Checkered Setwing
female—Travis Co.,
TX, July 2004

face tan. Thorax heavily striped brown and white. Abdomen with each of S2–7 irregularly marked white at base, black at tip; S8–10 mostly black. Wings narrowly dark-tipped.

Identification Nothing else in range like male, with reddish head and thorax, black and white abdomen, and prominent brown wing patches. Female rather like female **Black** and **Swift Setwings** with which it occurs except for basal wing patches. Female **Marl Pennant** also with brown hindwing patches, but forewings with only tiny patches and abdomen shorter and thicker, mostly pale rather than strongly patterned as in **Checkered Setwing**. Patterned abdomen and different wing pattern distinguishes from vaguely similar SMALL PENNANTS that often perch similarly.

Natural History Males perch on twigs along shore. Much more active than other setwings, floating over water in flash of color much like appearance of small pennants. Fly back and forth slowly over pools, and two males may engage in very rapid and lengthy parallel flights. Copulation brief and in flight, female often leaving water to return later.

Habitat Slow streams and rivers and open lakes, sometimes ponds, usually in open country with a shore belt of low trees or shrubs.

Flight Season AZ May–Nov, NM May–Sep, TX Apr–Dec.

Distribution Ranges south in Mexico to Tamaulipas.

327 Black Setwing *Dythemis nigrescens*　　　TL 40–45, HW 30–35

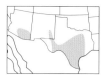

Description Slender setwing of river and lake shorelines, male solid black. *Male*: Eyes purple-brown; face metallic blue-black. Thorax and abdomen pruinose matte blue-black or very dark gray, slightly younger ones showing pale streaks on abdomen. Wingtips usually clear, in some individuals smudged with dark. *Female*: Seems very oddly polymorphic. Dark females with eyes reddish-brown over gray; face light brown. Thorax with complex pattern of dark and light stripes; abdomen blackish with profuse pale markings as two rows of streaks on S2–8 or S2–9, coalescing into larger marking on S7. Lower row disappears with age, leaving mostly black abdomen with pale paired spots above. Paler individuals with pale markings much more extensive than

327.1
Black Setwing
male—Baja California
Sur, Mexico, October
2006, Steve
Mlodinow

327.2
Black Setwing dark
female—Sonora,
Mexico, August 2006,
Netta Smith

327.3
Black Setwing pale
female—Sonora,
Mexico, June 2005,
Doug Danforth

dark, at extreme most of abdomen pale tan. Oldest individuals with gray-green, not reddish-brown, eyes and may develop pruinosity on thorax. Wings with brown marks at base, narrowly dark-tipped.

Identification Mature males unmistakable, not exactly like anything else with which they occur. Other slender species with black males in its range are **Pin-tailed Pondhawk**, brown-black rather than blue-black, its more slender abdomen with greatly expanded base, and with pale appendages; **Seaside Dragonlet**, considerably smaller, solid shiny black, and with red-brown eyes prominently contrasting with black body; and **Double-ringed Pennant**, also smaller and with small but conspicuous dark markings at hindwing base. None of these species characteristically perches with wings drooped. Dark purple face distinctive as it develops in older immature males, but female and immature male much like **Swift Setwing**. Thoracic stripe patterns different: in **Black**, first stripe usually wider, with enclosed pale spot; in **Swift** narrow, without spot. These patterns can be read as HII or HIY (reading from front on left side) in **Black**, YIY in **Swift**. Note that in rather similar **Three-striped Dasher**, pattern forms III. In **Black**, middle abdominal segments typically have two streaks on either side, only one on either side in **Swift**. In **Swift**, pale markings on S7 very large and conspicuous against solid black abdomen tip and much smaller markings out to S6. In **Black Setwing**, basal markings not so dramatically smaller than those on S7, and usually small pale markings on S8. Female **Swift** typically has more extensive dark color at wingtip, reaching stigma; less so in **Black**. Setwings sometimes seen perched above observer, and view from below allows positive identification. In **Black**, markings under thorax ill-defined, whereas same markings in **Swift** vivid black on white. Especially distinctive is short line running fore to aft, not present in **Black**. **Swift** has narrow black line running around margin of top of face, **Black** only tiny black dot in depression on top of frons. Finally, rear of head in immature **Black** pale, with two cross stripes on either side running out to eye margin; black in **Swift**, with three small white dots at eye margin. Check enough of these small differences and you can differentiate these two similar species.

Natural History Males perch on twigs over water and defend small territories against other males. Also perch on rocks in stream or ground, unusual for setwing. Fly back and forth along shore at knee to waist height, with much chasing of their own and other species. Both sexes perch on shrubs away from water, may elevate both wings and abdomen in hot sun. Females oviposit in open water or on algal mats by tapping water frequently in one spot and rising briefly between taps or flying rapidly low over water and tapping at intervals.

Habitat Streams and rivers with moderate to slow current, sand to mud bottom; also ponds and lakes with open shorelines but good perch sites.

Flight Season AZ Apr–Nov, NM Jun–Sep, TX Mar–Jan.

Distribution Ranges south in Mexico to Oaxaca.

---

328 Swift Setwing *Dythemis velox* TL 41–48, HW 32–35

Description Brightly marked slender setwing with conspicuously spotted abdomen and usually obviously dark wingtips. *Male*: Eyes red-brown over blue-gray; face brown, paler on outer edges. Thorax heavily marked with brown and cream in complex stripes. Abdomen mostly black, white basal markings on S1–3 and small spots on S4–7, largest on S7 where they form an oval half the length of  the segment. Very small brown spot at base of hindwing, indication of dark border at wingtips. *Female*: Colored as male but pale abdominal markings larger, spots on S7 cover much of segment. Usually prominent dark wingtips.

Identification Male differs from **Black Setwing** in its bright pattern. Females and immature males much more similar, but female **Swift** has black abdomen tip and conspicuous pale markings on S7, **Black** more pale color all along abdomen. See **Black Setwing** for further distinction from **Swift**. Female **Checkered** has big basal wing patches. Nothing else much

328.1
Swift Setwing
male—Floyd Co.,
GA, July 2006,
Marion Dobbs

328.2
Swift Setwing
female—Gillespie
Co., TX, September
2005, Dan and
Kay Wade

like **Swift**, especially as females, with thicker abdomen, also have dark wingtips to distinguish them from **Blue Dasher** and TROPICAL DASHERS. Female actually rather like rare **Three-striped Dasher**, but **Swift Setwing** with YIY instead of III thoracic pattern.

Natural History Males perch on twigs over water and defend small territories against other males. Spend most time perching, relatively seldom in flight, but males sometimes fly rapidly back and forth in restricted space, then land on tip of branch, or two males fly together in parallel flight. Females in open on elevated perches nearby. Commonly in shade on sunny days.

Habitat Streams and rivers with slow to moderate current, less often pond and lake shores. Usually wooded or shrubby banks.

Flight Season AZ May–Oct, NM May–Oct, TX Mar–Nov.

Distribution Ranges south in Mexico to Durango and Nuevo León, east and north, mostly in Piedmont, to Tennessee and Virginia.

---

329 **Mayan Setwing** *Dythemis maya* TL 43–45, HW 36–40

Description Heavy-bodied setwing of Mexican border, brilliant red-orange male with large wing patches. *Male*: Entirely red-orange, from face to abdomen tip. Abdomen slightly expanded, unlike other setwings. Large basal wing patches orange-brown with contrasting orange veins. *Female*: Eyes red-brown over blue-gray; face brown. Thorax and abdomen light brown throughout. Wing patches brown with contrasting tan veins. Wingtips black in most, unmarked in few.

329.1
Mayan Setwing
male—Presidio Co.,
TX, September 2004,
Martin Reid

329.2
Mayan Setwing
female—Nayarit,
Mexico, September
2001

Identification Only medium-small bright red-orange dragonfly with large basal wing patches. **Flame** and **Neon Skimmers** much larger and more robust, **Flame** more orange than red and with much larger wing patches, **Neon** redder than setwing and with wing patches more diffuse. Both KING SKIMMERS also with different perching behavior. **Red Saddlebags** larger, redder, and with quite different behavior. Other red to red-orange dragonflies in or near range (**Carmine Skimmer**, **Claret Pondhawk**, **Red-tailed Pennant**, **Spot-winged Meadowhawk**) with much smaller or no wing patches.

Natural History Males perch prominently on twigs and branches over shaded streams or fly rapidly along them. Also may hang vertically on hanging vines. Often perch in obelisk position with wings either depressed or greatly elevated. Females perch on twigs in clearings not far away. Copulation brief and in flight, then immediate oviposition with male in attendance. Oviposition erratic, moving rapidly over water and dipping down at random intervals. Often with Neon Skimmers, much aggression between them.

Habitat Rocky streams with pools, often in canyons.

Flight Season AZ Jul–Sep, TX Jul–Oct.

Distribution Locally common in Big Bend area of Texas, much more rarely seen in southeast Arizona. Ranges south in uplands to El Salvador.

Rock Skimmers *Paltothemis*

This is a small Mexican and Central American genus of robust stream-dwelling skimmers. It is related to setwings, clubskimmers, and sylphs but more tied to rocks for perching. Males vary from red to pruinose blue. Broad hindwings like gliders are adapted for gliding flight. World 3, NA 1, West 1.

330 Red Rock Skimmer *Paltothemis lineatipes* TL 47–54, HW 43–46

Description Large fast-flying skimmer of rocky streams in Southwest. *Male*: Eyes red above, some blue-gray below; face red. Thorax and abdomen dull reddish with black irregular markings. Wings with orange suffusion at bases. *Female*: Eyes reddish-brown above, gray below; face pale tan. Thorax light brown with irregular dark markings, somewhat like those of male but more heavily marked. Complexity of markings suggests Filigree Skimmer.

Identification Male much duller red than other large red skimmers that might occur with it. **Roseate Skimmer** brighter and more

330.1
Red Rock Skimmer
male—Cochise Co.,
AZ, September 2005

330.2
Red Rock Skimmer
female—Santa Cruz
Co., AZ, September
2005 (posed)

red-purple, **Carmine** and **Neon Skimmers** much more brilliant red. None of these rests on rocks. MEADOWHAWKS also might be mistaken for it, and **Variegated Meadowhawk** in particular rather similarly colored, but **Variegated** distinctly smaller and without orange wing bases. **Cardinal** and **Spot-winged Meadowhawks** much brighter red. Dull, heavily marked female not much like any other species found with it, but again note female **Variegated Meadowhawk** with complex abdominal pattern but smaller and with yellow spots on thorax and white spots on sides of abdomen. Among all similar species, only **Variegated** likely to rest on rocks.

Natural History Males defend 10–20 feet of stream with moderate current flowing over patches of fine gravel, at high density give way to other males during day. They perch flat on rocks and fly up- and downstream at intervals, apparently locating gravel that will serve as oviposition sites during brief inspection flights after arriving each morning. After brief copulation in flight, male takes female in tandem and flies back and forth over oviposition site, then releases her to lay eggs by tapping water repeatedly over gravel, male guarding above her. Females oviposit in slow flight, striking water at random, may be very localized at times. Stay very low between taps and usually oviposit for less than 2 min. Feeds like flier, especially females, cruising around well above ground. Sometimes in swarms of gliders and saddlebags, using gliding ability of rather broad hindwings.

Habitat Open, flowing rocky streams in upland areas.

Flight Season OR Jun–Aug, CA Mar–Nov, AZ Mar–Nov, NM May–Oct, TX Apr–Nov.

Distribution Ranges south in uplands to Costa Rica. Midsummer "invasions" of large numbers of this species reported in southern and central California, source unknown.

Sylphs *Macrothemis*

These mostly small skimmers fly up and down streams or feed, sometimes in little swarms, over clearings. Some species (three of ours) look like small clubskimmers, with expanded abdomen and spots on S7, and indeed the two groups are each others' closest relatives. Others, including one of ours, have a long, slender, uniformly colored abdomen. The genus is characterized by having an especially large tooth at the end of the tarsal claws, but a few species, including Jade-striped, do not show this characteristic. World 41, NA 4, West 4.

Description Small racket-tailed skimmer of Texas streams with pale-green-striped thorax. *Male*: S7–9 forming wide, flattened club. Eyes blue; face dark brown, metallic blue-black above. Thorax brown on front with two pale greenish T-shaped stripes, pale green on sides with two wide brown stripes. Abdomen mostly black, much expanded toward tip, with pale bluish dorsolateral spots on S2–5 and S7. *Female*: Eyes blue-gray with brown highlights above; face cream-colored, metallic blue-black above. Colored otherwise as male, but abdomen not expanded. Female

331.1
Jade-striped Sylph male—Caldwell Co., TX, August 2007, Greg Lasley

331.2
Jade-striped Sylph female—Caldwell Co., TX, August 2007, Eric Isley

wings all clear, or one or both pairs with dark tips. Orange basal area may be present in any of the three tip types.

Identification Only one other small skimmer with racket-tipped abdomen flies up and down Texas streams. **Ivory-striped Sylph** distinguished from this species by more heavily spotted abdomen (no gap between spots on midabdomen and "racket") and dark brown thorax with white spots on front and single white stripe on sides. Females more similar, distinguished by thoracic pattern and lack of spots on middle abdominal segments of **Jade-striped**. Superficially similar CLUBSKIMMERS are much larger. Color pattern not unlike various TROPICAL DASHERS, but flying and perching habits should distinguish them. Also, mature sylphs have blue eyes, DASHERS green eyes.

Natural History Typically perch on tops of leaves but also hang vertically from twigs and stems. Males fly up and down streams, often in very short beats over pools at about knee height. Females not often seen, likely to be hanging up in shrubs and trees within woodland. Copulation at rest. Female seen to hover a foot or more over water and dip abdomen in oviposition movements, not seen if eggs deposited.

Habitat Small forested streams, rocky or not and typically with little current, or pools in larger streams.

Flight Season TX May–Nov.

Distribution Ranges south to Venezuela.

Description Small racket-tailed skimmer of Texas streams with white-spotted thorax. *Male*: Abdomen with S7–9 forming wide, flattened club. Eyes blue; face dark brown, metallic black above. Thorax dark brown with pair of ivory to bluish-white spots on front and long and short jagged white stripe on each side. Abdomen dark brown with paired rows of whitish streaks and spots down  length, culminating in large spots on S7. *Female*: Eyes blue; face brown. Thorax dark brown with small pale areas on front and broad white long and short stripes on either side. Abdo-

332.1
Ivory-striped Sylph male—Tamaulipas, Mexico, October 2006, Robert A. Behrstock; inset male—Bexar Co., TX, May 2005, Martin Reid

332.2
Ivory-striped Sylph
female—Frio Co., TX,
July 2005, Martin Reid

men mostly tan to whitish on S1–6 with dark carinas and sutures, S7 with pair of white spots, S8–10 black. Female wings usually clear.

Identification Blue eyes, incessant flight over water, and hanging up habits like CLUBSKIMMERS but much smaller with darker, white-marked thorax. Most like **Jade-striped Sylph**, which has T-shaped markings on front and three wide pale stripes on sides of thorax. **Jade-striped** has white spots on abdomen sparser, not present on S6 (continuous from base to S7 on **Ivory-striped**). Females distinguished by same characters as males. **Ivory-striped** has a bit of dark brown on base of all wings, not present in **Jade-striped**.

Natural History Males fly short beats up and down over shaded pools in rivers at about knee height. Females not often seen, likely to be hanging up in nearby shrubs and trees. Feeds in light and airy flight over land near breeding habit.

Habitat Clear rivers and large streams with some current, usually rocky. Those in Texas mostly in open but also lives in forested regions.

Flight Season TX Jun–Nov.

Distribution Ranges south to Argentina.

---

333 White-tailed Sylph *Macrothemis pseudimitans* TL 40–43, HW 29–32

Description Small racket-tailed skimmer of southeast Arizona with pale-spotted thorax. *Male*: Abdomen with S7–9 making wide, flattened club. Eyes blue; face dark brown, metallic black above. Thorax dark brown with pair of ivory to bluish-white triangular spots on front and four irregular spots of same color on each side. Abdomen black with paired rows of pale bluish streaks and spots becoming increasingly smaller, tiny on S5–6; large triangular to squarish pair on S7. In some but not all males (perhaps about half), entire top of abdomen from S6 to S9 becomes whitish pruinose. *Female*: Eyes blue; face brown. Patterned more or less like male, but spots on front of thorax less well-defined and narrower, extended forward in long points. Abdomen thicker overall with less well-developed club. Forewing tips brown.

Identification Blue eyes, incessant flight over water, and hanging up habits like co-occurring CLUBSKIMMERS, but much smaller with darker, pale-spotted thorax. Males with pruinose abdomen unmistakable. Not known to occur in United States with similar **Ivory-striped** and **Jade-striped Sylphs**. **Ivory-striped** has prominent pale spots on S6 as well as S7, **Jade-striped** has stripes rather than spots on front and sides of thorax. Very different from co-occurring **Straw-colored Sylph**.

Natural History Males fly short beats up and down at about knee height over pools, even isolated ones, in streams; perch on leaves or rocks. No differences detected in spotted and white-tailed individuals. Females not often seen, likely to be hanging up in nearby shrubs and trees. Both sexes feed in slow back-and-forth cruising flight, typically from waist to head height, over clearings near breeding habit.

Habitat Clear rivers and large streams with some current, usually rocky. In open or forested country.

333.1
White-tailed Sylph
pruinose male—
Sonora, Mexico,
October 2006, Doug
Danforth

333.2
White-tailed Sylph
spotted male—
Sonora, Mexico,
August 2006

333.3
White-tailed Sylph immature female—San Luis Potosí, Mexico,
November 2006, Marion Dobbs

Flight Season AZ Oct.

Distribution Probably vagrant to United States, known only from individual collected on San Bernardino National Wildlife Refuge, Arizona, in 2007. Ranges south to Ecuador and Brazil.

334 Straw-colored Sylph *Macrothemis inacuta*   TL 42–48, HW 32–36

Description Slender-bodied brown skimmer of southwestern streams and rivers. *Male*: Eyes blue, face tan. Thorax patterned light and dark brown with dull whitish stripes, one anterior stripe somewhat T-shaped and two lateral stripes broken in middle. Abdomen very slender, light brown with darker rings at sutures. *Female*: Colored as male, abdomen slightly broader. Uncommon variants have hindwing base orange-brown and/or forewing tip brown.

Identification Nothing else looks like this species, whether flying up- and downstream or hanging up vertically or perched on rock. Other sylphs and CLUBSKIMMERS have club-shaped abdomens. DARNERS are larger, and none has straw-colored abdomen of this species. Perhaps most similar is **Hyacinth Glider**, but it has conspicuous dark basal hindwing markings and much shorter dark-tipped abdomen. Also perching by hanging, **Evening Skimmer** is even plainer brown than **Straw-colored Sylph**, with brown eyes and no markings on thorax; also not as slender-bodied. Another possibility for confusion is female **Band-winged Dragonlet**, also pale and slender-bodied, with unstriped thorax, much more distinct pattern of black on abdomen, and different habits.

Natural History Males fly incessantly up and down streams at around knee height, often difficult to see against dark water, but blue eyes conspicuous at close range. Cruising beat up to 60 feet or more. In most areas rarely seen perched, but at some localities often come to rest on large pale rocks in stream or sandy shore or hang from streamside twigs. Females

334.1
Straw-colored Sylph
male—Sonora,
Mexico, September
2006

334.2
Straw-colored Sylph
male—Pinal Co., AZ,
August 2005, Robert
Bowker

334.3
Straw-colored Sylph
female—Sonora,
Mexico, November
2002, Doug Danforth

and immature males usually seen hanging up away from water. Copulation brief and in flight, then female oviposits by tapping irregularly with male usually in attendance.

Habitat Small streams to good-sized rivers, mostly open but may have wooded banks; rocks usually but not always present. Also at large upland lakes with wave-washed shores south of border.

Flight Season AZ Jul–Nov, TX May–Nov.

Distribution Ranges south to Brazil and Argentina.

This group includes rather large skimmers with expanded abdomens reminiscent of some of the clubtails with narrower clubs. Similar to several groups of clubtails, especially leaftails, they have big pale spots on S7 that show conspicuously as they fly past. However, they are fliers rather than perchers and hang vertically or diagonally when they perch. The two species differ in the shape and size of the spots on S7. Two smaller species occur north to the northernmost Mexican states. World 14, NA 2, West 2.

335 Masked Clubskimmer *Brechmorhoga pertinax*   TL 50–52, HW 38–43

Description Large milky-winged flying skimmer of Southwest with long, slightly expanded abdomen. *Male*: Eyes dull blue; face light brown, top of frons metallic purple. Thorax dark brown and white striped. Abdomen black with fine white markings on S1–4 and small oval spots on S7. Wings with milky or glistening appearance, as if tips covered with pale film. *Female*: Colored much as male but
eyes duller, abdominal club less evident because abdomen base somewhat thicker. Milky appearance of wings less obvious. Wing color polymorphic, clear or with forewing tips brown; wings may become entirely brownish with age.

Identification Distinction in flight often possible because of milky appearance of wings beyond nodus, not evident in **Pale-faced**. Relatively narrow pale spots on S7 diverge from midline toward rear in males, whereas those on **Pale-faced** remain parallel and are larger. Thus, in flight, spot may look like single large spot in **Pale-faced** but paired spots in **Masked**. **Masked** looks overall darker than **Pale-faced**, with darker thorax and less pale color on abdomen; eyes also bluer in **Masked**. Females more similar, both with spots parallel to midline, but spots narrower in **Masked**.

Natural History While feeding, cruise well above ground in open areas. Behavior at water studied in Costa Rica. Much like that of Pale-faced Clubskimmer, males flying up and down riffles close to water, activity highest at midday. Any given male present only for short periods, less than 15 min, but may return on same day. Cruising beats may be quite long, although commonly 6–25 feet, and high level of aggression to other males, sometimes in

335.1
Masked Clubskimmer male—Cochise Co., AZ, July 2007, Doug Danforth

335.2
Masked Clubskimmer
male—Cochise Co.,
AZ, July 2007, Doug
Danforth

tight circling flights averaging 23 sec. Females visit water all day long, oviposit in shallow water with sand or gravel substrate defended by resident male. Copulation in flight for about 10 sec over oviposition site, egg-laying bouts last about a minute. Male remains near female during oviposition, chases away intruding males.

Habitat Rocky streams or rivers with moderate current. Banks wooded or open but must have sun.

Flight Season AZ Jun–Sep.

Distribution Known presently from two parts of Arizona, streams on both north and south rims of Grand Canyon and Cave Creek in Chiricahuas. Ranges south in uplands to Bolivia.

336 Pale-faced Clubskimmer   *Brechmorhoga mendax*   TL 53–62, HW 34–43

Description Large southwestern skimmer zipping up and down streams with long, slightly clubbed abdomen with conspicuous spots near tip. *Male*: Eyes light blue-gray; face light brown. Thorax striped dark brown and light gray-green. Abdomen black with whitish markings around S1–3 and whitish streaks on S4–5. Large whitish pair of spots almost touching and almost filling up S7. Variable, some males in Texas with dark metallic frons, belying common name (coined before variation noted). Often small pale spots at base of either side of S7. *Female*: Eyes light brown over blue-gray.  Colored as male, abdominal club less evident because abdomen base somewhat thicker. Wing color polymorphic, some individuals with both tips or only forewing tips amber or brown. Wings may become entirely brown with age.

Identification At rest, distinguished from **Masked Clubskimmer** by all pale face and relatively larger pale spots on S7 with inner margins almost reaching midline. Known to overlap only in Arizona, where **Masked** rare. In Texas, small percentage of **Pale-faced** have dark metallic frons, but abdominal spots still distinctive. In hand, male **Pale-faced** hamules evenly curved; male **Masked** hamules relatively straight, then suddenly curved near tip. Individuals in flight can be difficult to distinguish from CLUBTAILS with pale spots on S7 and similar habits of flying up and down streams. If it lands and perches horizontally, it is a CLUBTAIL; if it hangs vertically, a clubskimmer. Occurs with **Ivory-striped** and **Jade-striped Sylphs**, somewhat similar in color pattern and behavior but much smaller.

336.1
Pale-faced
Clubskimmer
male—Chihuahua,
Mexico, September
2005

336.2
Pale-faced
Clubskimmer
female—Tarrant Co.,
TX, August 2003,
Martin Reid

Natural History Males fly low and either slowly or rapidly up and down riffles, less often over pools between them. Beats often 15–30 feet long, probably dependent on length of riffle. Difficult to see under those conditions, sometimes only spots on abdomen apparent. Copulation brief, in flight over water while hovering or flying slowly. Females oviposit by flying along rapidly, striking water at fairly lengthy intervals. Both sexes perch by hanging vertically below chest height in shade in woods, flushed from perch one after another where common. Sometimes land on rocks. Feeds in flight in open areas from just above ground to well up in trees, sometimes with other swarm feeders such as gliders.

Habitat Shallow rocky streams with riffles and pools, some current. Banks wooded or open but must have sun.

Flight Season CA Apr–Nov, AZ Apr–Nov, NM May–Oct, TX Mar–Nov.

Distribution Ranges south in Mexico to Baja California Sur, Nayarit, and San Luis Potosí.

This small tropical genus of dusk-flying skimmers is thought to be related to gliders and saddlebags, although the wings are less broad and flight activity is mostly at dusk and dawn. The unusual anal loop is narrow, extending to the edge of the wing. Males of Old World species, *T. tillarga*, are red but display the same habits as our species. Egg-laying apparatus, unique to this genus, consists of not only standard subgenital plate but, behind it, an inverted trough with hairs that hold egg clusters within it. World 2, NA 1, West 1.

337 Evening Skimmer *Tholymis citrina* TL 48–53, HW 36–39

Description Slender brown tropical skimmer that flies actively at dusk and dawn. *Male*: Eyes dark gray; face dark gray, metallic blue on top. Entirely brown with front of thorax becoming dark gray. Golden-brown spot at middle of each hindwing, sometimes hint of same in forewing. *Female*: Similar but face, eyes, and thorax paler, light brown or yellowish (eyes may have bluish tinge). From above,

easily seen that female cerci diverge; males are parallel. Also, cerci of females long and curved downward. Wings become brownish in older individuals of both sexes.

Identification Tawny-winged Pennant most similar species in overall brown coloration and slender abdomen, but typical skimmer, perching horizontally in open during day and

337.1
Evening Skimmer
male—Veracruz,
Mexico, September
1999, Netta Smith

337.2
Evening Skimmer
female—Zapata Co.,
TX, November 2004,
Martin Reid

not performing evening flights. Almost all SHADOWDRAGONS, also brown and with similar evening flight, do not occur within **Evening Skimmer's** tropical range, but **Orange Shadowdragon** overlaps slightly. That species has blotchy wing markings mostly at base and anterior edge, not like central spot of **Evening Skimmer**. **Bar-sided**, **Pale-green**, and **Twilight Darners**, brown to greenish dusk-fliers, are much larger.

Natural History Roosts at forest edge, hanging like small darner; flies erratically when flushed and then suddenly lands again, sometimes in plain sight (but brown coloration renders it very cryptic). Feeding tends to be in short period before and at dusk, along with dusk-flying tropical darners and, if early enough, with gliders and saddlebags. Also probably flies at dawn as other species of genus. Feeding flight rapid and erratic. Breeding also takes place at those times, males flying back and forth over shallow water bodies with occasional hovering. Pairs of Old World *Tholymis tillarga* copulate briefly in flight, then male guards female as she oviposits. Females sometimes hit water to oviposit, then twist around 180° to repeat. Also lay eggs on leaves at water surface. Our species probably with similar reproductive behavior, but no observations.

Habitat Shallow ponds and marshes, often but not always associated with forest.

Flight Season AZ Jun, TX Sep–Nov.

Distribution Probably breeds in southern Texas; other records may be wanderers. Ranges south in lowlands to Argentina and Chile, also southern Florida and West Indies.

Pasture Gliders *Tauriphila*

Species of this small New World tropical genus exhibit the same incessant flight and hanging style of perching as rainpool gliders and saddlebags but are somewhat smaller and more slender. Closely tied to floating plants, they are probably most closely related to hyacinth gliders. Some species have brown spots at wing bases. Pasture gliders may be only vagrants to North America, without breeding populations. World 5, NA 2, West 1.

338 Aztec Glider *Tauriphila azteca* TL 42–50, HW 34–37

Description Small gliding skimmer with banded yellow abdomen, vagrant to Texas. *Male*: Eyes dark reddish brown above, blue-black below. Face metallic blue-black. Thorax dark metallic brown. Abdomen blackish at base, S4–9 dull yellow-orange with blackish basal median spot and irregular brown terminal rings, giving banded appearance. *Female*: Face brown with metallic blue  overtones. Thorax dark brown, somewhat metallic. Abdomen patterned as male, slightly less vivid.

338
Aztec Glider
male—Sinaloa,
Mexico, July 1976,
Sid Dunkle
(posed)

Identification Dark thorax and yellowish and black banded abdomen unique among gliders and good mark if visible at close range. If not capturable, watched gliders may eventually hang up in sight. Female **Black Pondhawk**, also scarcely present in North America, also has dark thorax, black and yellow banded abdomen, and dark spots at wing base, but it perches rather than hangs and engages only in short flights. **Aztec Glider** has abdomen slightly narrowed behind base, then spindle-shaped; **Black Pondhawk** has parallel-sided abdomen. Often flies with **Hyacinth Glider** in tropics, distinguished by larger size, dark rather than striped thorax.

Natural History Males territorial over floating vegetation, but this has not been seen in North America. Copulation in flight. Feeds in swarm flights with Hyacinth Gliders, rainpool gliders, and saddlebags. May roost inside forest.

Habitat Ponds of any size covered with floating vegetation, most usually water lettuce and water hyacinth.

Flight Season TX Jun.

Distribution Probably only a vagrant to North America, with a specimen and a sight record for south Texas. Common well south of border on both coasts and ranges south in lowlands to Costa Rica.

Hyacinth Gliders *Miathyria*

Hyacinth gliders are related to pasture gliders, saddlebags, and rainpool gliders, and with similarly broad hindwings, these small skimmers fly for long periods and then hang up more or less vertically. They are among the few dragonflies closely tied to particular plant species, primarily water hyacinth but also water lettuce. The other species in the genus, Dwarf Glider (*Miathyria simplex*), is smaller and red, with a round hindwing spot, and occurs throughout much of New World tropics. World 2, NA 1, West 1.

339 Hyacinth Glider *Miathyria marcella*   TL 37–40, HW 29–33

Description Small gliding skimmer of Southeast with dark hindwing saddles and orange abdomen, addicted to water hyacinths. *Male*: Eyes reddish-brown; face brown, blue-purple above. Thorax pruinose violet (plum-colored), including between wings, when fully mature. Abdomen yellow-orange with narrow black median line widening on S7–8; S9–10 black. Wing veins reddish, hindwing  with dark brown base. *Female*: Eyes reddish above, gray below. Thorax brown with whitish diagonal stripe running from front to below forewing, parallel whitish ventrolateral stripe, and broader blackish stripe between them. Abdomen orange-brown with black markings as male. Wing veins dark.

Identification No other small North American skimmer flies back and forth over aquatic vegetation beds or feeds with larger RAINPOOL GLIDERS and SADDLEBAGS in swarms in open country. Much smaller than its feeding associates, although superficially colored like narrow-saddled SADDLEBAGS. Could be mistaken for **Aztec Glider**, very rare in Texas. See **Straw-colored Sylph**.

Natural History Males fly back and forth, usually low, over beds of aquatic vegetation and search closely for females; then hang up in herbs, shrubs, or low trees, where females also perch. Perch like rainpool gliders, hanging on stems. Pairs in tandem and single females oviposit by dropping into tiny gaps in floating vegetation where water visible. Few skimmer species as closely tied to a few plant species as this one. Both sexes feed in groups at head height and above, often in tighter swarms than other fliers with which they associate such as pasture gliders, rainpool gliders, and saddlebags.

Habitat Ponds and lakes with abundant water hyacinths or water lettuce, less commonly over other floating vegetation. Has dramatically declined wherever water hyacinths largely

339.1
Hyacinth Glider male—
Highlands Co., FL, June 2003,
David McShaffrey (posed)

339.2
Hyacinth Glider female—
Sonora, Mexico, August 2006,
Netta Smith

eliminated in Southeast and could disappear from large parts of range with hyacinth-control programs.

Flight Season TX Apr–Nov.

Distribution Ranges south to Argentina, also Arkansas to Georgia, Florida, and Greater Antilles.

Saddlebags *Tramea*

These are gliding dragonflies with broad hindwings with conspicuous dark markings at their bases ("saddles"), more conspicuous than the spot on the rather similar Spot-winged Glider. They fly incessantly over open areas, even up to treetops, then perch at tips of twigs like pennants (often with front legs tucked behind eyes), or hang from branches like rainpool gliders. Even when perching horizontally, they may droop their abdomen almost straight down (the reverse of obelisking behavior) to avoid overheating at midday, and they often fly with abdomen pointing downward when it is hot. Pairs join in flight, then fly to nearby vegetation and perch for 10 min or so, then head for water in tandem flight, with unique oviposition behavior. The pair flies rapidly at knee height or lower and pauses at intervals for the female to be released, tap water, and be reclaimed by the male, almost too quickly to see details. A pair may fly all over a pond before stopping again. Distance between taps may be as little as a few feet, often longer. Eggs are held by the female with a large, divided subgenital plate. Saddlebags are easily divided into two groups from width of hindwing markings, "broadsaddle" with spot on each side wider than abdomen and "narrowsaddle" about as wide as abdomen. Any sighting should first determine which type is involved. Most individuals in the West are broadsaddle, but three species of narrowsaddle occur along the southern border. Distinctions can be difficult within each group. Worldwide distribution is primarily in tropics. World 21, NA 7, West 6.

Table 12   Saddlebags (*Tramea*) Identification

	A	B	C	D	E
Vermilion	1	2	1	2	2
Antillean	1	1	1	2	2
Striped	1	1	2	1	3
Red	2	2	1	2	1
Carolina	2	1	1	1	1
Black	2	1	1	1	1

A, saddle: 1, narrow; 2, broad.
B, frons color: 1, purple; 2, red.
C, thorax: 1, unmarked; 2, striped.
D, abdomen tip: 1, mostly black; 2, mostly red.
E, range: 1, widespread; 2, borderlands; 3, spreading north.

340 Vermilion Saddlebags   *Tramea abdominalis*   TL 44–50, HW 38–42

Description   Red-faced narrowsaddle saddlebags of extreme southern Texas. *Male*: Eyes above and face bright red. Thorax dull red, abdomen bright red with small black spots on top of S8–9. Appendages long and black, red at extreme base. Basal wing veins red. *Female*: Eyes red over gray, face reddish-tan. Thorax tan. Abdomen red with black spots as male. Appendages long, black. Wing veins black except red within dark hindwing saddles.

340.1
Vermilion Saddlebags male—Grenadines, June 2004, Mark DeSilva

340.2
Vermilion Saddlebags female—Curaçao, February 2005, Frank Jong

Identification Males of **Antillean** and **Striped Saddlebags** also entirely red with narrow saddles. **Vermilion** differs from **Antillean** in entirely red face (no trace of purple), but this can be very difficult to see in field. Shorter appendages of **Vermilion** might also be seen on perched male; not much longer than S9–10, considerably longer in **Antillean**. With even better look, longer hamules of **Vermilion** might be evident, longer than genital lobe (shorter in Antillean). Female **Antillean** and **Vermilion** impossible to distinguish in field unless really good look shows purple on top of frons of **Antillean**. In hand, subgenital plate extends to end of S9 in **Vermilion**, not that far in **Antillean**. Differs from **Striped** in untinted wings (wash of color in **Striped** not obvious against sky), unstriped thorax, and black spots at tip of red abdomen (**Striped** has most of abdomen tip black). See that species for additional differences.

Natural History Males fly at waist height or above along edge and over middle of breeding ponds. Both sexes feed, often in swarms and often mixed with other saddlebags species, in open areas away from water.

Habitat Vegetated ponds, including small and temporary ones. Also ditches and drainage canals, apparently able to coexist with fish.

Flight Season TX Jun.

Distribution Found at Santa Ana National Wildlife Refuge in 2005, presumably only vagrant to United States. Ranges south throughout lowlands to Argentina, also southern Florida and West Indies.

341 Antillean Saddlebags *Tramea insularis*   TL 41–49, HW 33–40

Description Purple-faced narrowsaddle saddlebags of southwestern border. *Male*: Eyes bright red above; face bright red, top of frons metallic purple. Thorax dull red, abdomen bright red with small black spots on top of S8–9. Appendages long and black, red at extreme base. Basal wing veins red. *Female*: Eyes red over gray, face reddish-tan. Thorax tan, abdomen red with black spots as male. Appendages long and black. Wing veins black except reddish within dark hindwing saddles.

341.1
Antillean Saddlebags
male—Pinal Co., AZ,
November 2006,
Doug Danforth

341.2
Antillean Saddlebags
female—Frio Co., TX,
July 2005, Martin Reid

Identification One of narrowsaddle group, very similar to **Vermilion Saddlebags** but distinguished by purple face. Less like **Striped Saddlebags**, which usually shows stripes on thorax and has much more black at abdomen tip. Male appendages in **Antillean** considerably longer (obviously longer than S9–10) than those of **Striped** and **Vermilion** (barely if at all longer than S9–10). Female **Antillean** and **Vermilion** extremely similar, distinguished in hand by entirely red frons of **Vermilion**, touch of purple on upper surface in **Antillean**. Subgenital plate of female **Antillean** shorter than S9, longer in **Vermilion**.

Natural History Males fly over breeding habitat at waist to chest height or perch near shore, often on dead tree branches. Solo females fly rapidly and tap water much more frequently than pairs, without hovering. Both sexes feed away from water, often in small swarms of their own and other species.

Habitat Males have been seen at lakes, cattle tanks, springs, and pools in rivers in Texas and Arizona, other habitats such as drainage canals elsewhere. Seems well adapted to life in small artificial water bodies.

Flight Season AZ Oct–Nov, TX May–Aug.

Distribution Texas population scattered but apparently widespread, oddly disjunct from main part of range throughout West Indies to southern Florida, and recently discovered southeast Arizona population even more disjunct. Also recorded from Campeche, Mexico.

342 Striped Saddlebags *Tramea calverti* TL 45–49, HW 39–42

Description Brown to reddish tropical narrowsaddle saddlebags with pale stripes on side of thorax and amber-tinted wings. *Male*: Eyes red over brown; face red with small metallic purple area at top of frons. Thorax brown, darker on sides with pair of dull yellowish stripes on each side. Abdomen bright red with most of S8, all of S9–10, and appendages black. Wing veins light brown, wings tinted

342.1
Striped Saddlebags
male—Hidalgo Co.,
TX, May 2000,
Robert A. Behrstock

342.2
Striped Saddlebags
female—Bexar Co.,
TX, June 2004, Martin
Reid

with amber. *Female*: Eyes red over gray; face tan with purple area smaller than in male. Thorax and abdomen light brown, thoracic stripes whitish. Wing veins darker than in male.

Identification Distinguished from other narrowsaddle saddlebags by pale stripes on sides of thorax. These stripes may be obscure in mature males, obliterated by overall reddish color of thorax. Such individuals distinguished from both **Antillean** and **Vermilion** by largely black S8–10 (other species have black spots on top of red S8–10). Rather pale wing veins and pronounced amber tinge to wings (not visible in some lights) also distinctive. Male appendages distinctly shorter than those of **Antillean**, same length as **Vermilion**. Note **Hyacinth Glider** has similar behavior and striped thorax, but differently colored otherwise and much smaller.

Natural History Males fly over water at waist height or lower, seem more likely to perch on branches at water than other saddlebags. Feeding flight mostly head height and above, often mixed with other saddlebags and gliders.

Habitat Shallow ponds with much open water and beds of vegetation.

Flight Season CA Oct–Nov, AZ Sep–Nov, TX Feb–Nov.

Distribution Sporadic and local within indicated range, normally resident only in southern Texas. Irregular farther north and west, present some years and not others but perhaps increasing; small numbers in Arizona each fall. Ranges south in lowlands to Argentina, also southern Florida and West Indies. Vagrant in East, up Atlantic Coast to Massachusetts.

343 Red Saddlebags *Tramea onusta* TL 41–49, HW 38–43

Description Red broadsaddle saddlebags common in southern plains and Southwest. *Male*: Top of eyes bright red; face bright red, extreme top of frons with slight purple tinge. Thorax reddish-brown, sometimes with scattered black markings. Abdomen bright red with black spots on top of S8–9; cerci long, reddish at base. Basal wing veins red. *Female*: Duller all over, with same black markings as male but slightly more reduced, in some scarcely evident. Cerci long and black. Immatures with all pale areas tan.

Identification One of two red saddlebags with broad saddles. Distinguished from **Carolina Saddlebags** by entirely red face (**Carolina** purple), slightly smaller size. Best field mark if it can be seen: black spots on top of S8–9 in **Red**, black markings on S8 extend to bottom of segment in **Carolina**. Occasional individuals may overlap, however, and more difficult in females, in which black less extensive. Check thorax for

343.1
Red Saddlebags
male—Hidalgo Co.,
TX, November 2005

purple highlights characteristic of **Carolina** and lacking in **Red**. Check wing pattern, similar except clear "window" (not always easy to see) on inside of wing patch in **Red Saddlebags** oval or circular, usually but not always larger than that in **Carolina**, larger than width of abdomen. Window in **Carolina** more typically a small acutely pointed triangle. See **Carolina Saddlebags** for discussion of male abdomen base. From below, different-sized subgenital plates might be evident in females.

Natural History Often seen far from water, feeding in swarms with Black Saddlebags and rainpool gliders. Both sexes roost on dead twigs high in treetops. Males fly incessantly along shore and over open water, defending territories as large as 100 by 30 feet. Pairs oviposit in tandem as typical of saddlebags, often on algal mats. Ovipositing pairs only harassed by lone males when female released to oviposit. Female may also oviposit solo in very different way, moving slowly and tapping constantly in vegetation bed, scarcely rising between taps; perhaps true of all saddlebags.

Habitat Lakes and ponds, also ditches and canals and large pools of slow rivers. Probably breeds most successfully in fish-free waters such as rainy-season ponds.

Flight Season CA Mar–Oct, AZ Jan–Nov, NM Mar–Oct, NE May–Oct, TX Feb–Dec.

Distribution Ranges south to Venezuela, also northeast to southern Ontario and New Jersey; sparse through most of Southeast except southern Florida.

344 Carolina Saddlebags *Tramea carolina* TL 48–53, HW 44–45

Description Purple-faced red broadsaddle saddlebags of the Southeast. *Male*: Eyes dark reddish-brown; face reddish with top of frons dark purple. Thorax dull reddish to brown, sometimes with scattered black markings. Abdomen bright red with much of S8–9 black; cerci long, reddish at base. Basal wing veins red. *Female*: Eyes red over brown, face tan with metallic purple just at base of top of frons. Duller all over, with same black markings as male except a bit more reduced, at least covering only upper halves of S8–9. Cerci long and black. Immatures with all pale areas tan.

Identification One of two large broadsaddle saddlebags with red coloration. See **Red Saddlebags** for differences. Easily distinguished from all-black **Black Saddlebags**, as red usually evident even when backlighted. Male broadsaddles can be distinguished with good look in side view. In **Red**, hamule long and narrow, extending past genital lobe; in **Carolina**, hamule and genital lobe same length, hard to distinguish; in **Black**, hamule shorter than genital lobe, forming two bumps.

344.1
Carolina Saddlebags
male—Laurens Co.,
GA, July 2007

344.2
Carolina Saddlebags
female—Clay Co.,
FL, November 2007,
Marion Dobbs

Natural History Males fly along shorelines at waist to head height, circling out over open water and back along shore. Also perch on dead twigs near water, often above flight height. Perching normally horizontal at twig tips but may hang up like rainpool glider. Copulation at rest in herbaceous or woody vegetation for 8–10 min. Most oviposition among emergent vegetation. Both sexes feed in open areas from low over ground to tree-tops, singly or in small to moderate swarms and often with other species. Feeds at all times of day, including until dusk, for example when feeding with Regal Darners.

Habitat Ponds, both marshy and open, and lakes with much submergent vegetation.

Flight Season TX Mar–Aug.

Distribution Occurs east to Atlantic Coast, north to Wisconsin and Maine. As in all saddle-bags, northerly records may be vagrants outside breeding range.

345 Black Saddlebags   *Tramea lacerata*   TL 51–55, HW 45–47

Description Widespread black broadsaddle saddlebags. *Male*: Eyes dark brown, face dark purple. Thorax dark brown to blackish, sometimes with scattered metallic black markings. Abdomen black, often showing yellow squares on S7 indicative of immaturity. Cerci very long (more than twice as long as epiproct) and black. *Female*: Duller all over, usually showing pale markings on S7; cerci long, black. Immatures dark brown, with yellow markings along most abdominal segments to S7.

Identification Other broadsaddle saddlebags are red; only other black saddlebags is **Sooty Saddlebags** (*Tramea binotata*), a narrowsaddle species that might occur in Texas as a vagrant. Windows on inside of saddlebags of **Black Saddlebags** variable, overlap with both **Carolina** and **Red**. Immature **Widow Skimmer**, lacking white bands on wings, might be confusing, but perches and flies like king skimmer, not saddlebags, and has patterned abdomen.

345.1
Black Saddlebags
male—Columbia Co.,
WI, May 1998,
David Westover

345.2
Black Saddlebags
female—Decatur Co.,
GA, October 2005,
Giff Beaton

345.3
Black Saddlebags
pair—Hamilton Co.,
OH, July 2007,
William Hull

Natural History Males cruise just out from shorelines and over open water, typically at waist height and often rather erratically; length of territory may exceed 100 feet. Copulation brief and in flight or somewhat lengthier on perch, followed by rapid tandem flight low over water, hovering briefly and then moving again, until some stimulus for oviposition causes them to drop to the water, male releasing female for one tap, then rejoining and moving on to do it again. Very tentative at this time, not surprising as fish such as bass follow them underwater and strike when they drop to surface. Both sexes roost on dead twigs high in treetops. Highly migratory, migrants appearing in northern part of range in summer and breeding, their offspring apparently migrating back to South. Large southbound flights reported from southern Great Plains and Texas coast.

Habitat Shallow open lakes and ponds with much aquatic vegetation; wanders far and wide away from water.

Flight Season BC Jun–Aug, WA Jun–Sep, OR Jun–Sep, CA Mar–Nov, AZ Mar–Nov, NM Mar–Oct, NE May–Oct, TX Apr–Nov.

Distribution Ranges south in lowlands of Mexico to Baja California Sur and Veracruz, also Yucatan Peninsula. In East north to Wisconsin and southern Quebec.

Rainpool Gliders *Pantala*

These broad-winged skimmers are the champion gliders of the odonate world, on the wing for hours and even days at a time as they wander even across oceans. Because of this, one species is the only dragonfly with worldwide tropical distribution. Perching is always by hanging up, unlike most skimmers. The very broad hindwings represent an important adaptation for gliding, as does the ability to deposit fat and then use it for energy during a long flight just as a migratory bird does. Abdomen is drooped in flight at midday to reduce heat load from the sun. Saddlebags are similar in many ways and are presumably close relatives. World 2, NA 2, West 2.

346 Wandering Glider *Pantala flavescens* TL 47–50, HW 36–42

Description Gliding skimmer that looks yellow to yellow-orange in flight. *Male*: Eyes reddish; face orange. Thorax and abdomen yellow, upper part of abdomen orange. Darker orange median line on abdomen, expanded on each segment and forming black spots toward rear, on S8–10. Cerci black, obviously pale at base. *Female*: Much as male, but lacks reddish-orange colors of that sex. Sexes shaped very similarly, but female cerci slightly longer and more slender than in male, less pale color at base.

Identification Nothing else gives quite the impression of this species in flight. **Spot-winged Glider** similar in size and habits but darker, looks more brownish in flight, and spot at base of hindwing visible with good look; may have to pass overhead to see spots clearly. Perching behavior distinguishes both gliders from most other skimmers and from all others that are yellowish (especially **Golden-winged** and **Needham's Skimmer** and immatures of other KING SKIMMERS). A bright orange-brown **Four-spotted Skimmer** flying overland could cause confusion, but the two species overlap little in distribution and none in habits and breeding habitat.

Natural History Feeding and territorial patrol in rapid and sweeping flight, at or away from water. Hangs up at a slant or almost vertically when not active. Typically perches low in grass or weeds, sometimes up in trees. Males patrol over or near water, rapidly back and forth over large water bodies. Patrol area 30–150 feet in length, at head height. Sexes meet there or nearby for brief copulation (30 sec to 5 min) and then go into tandem and fly rapidly in a fairly straight course, tapping water at intervals of a few feet. Female can lay up to 800 eggs during one mating. Oviposition may be early in morning before sun hits water. Lone females often tap surface of shiny automobiles, eggs apparently harming paint job! Perhaps much

346.1
Wandering Glider
male—Hidalgo Co.,
TX, November 2005

346.2
Wandering Glider
male—Maricopa Co.,
AZ, April 2006,
Robert Bowker

346.3
Wandering Glider
female—Chiriquí,
Panama, November
2005, Dan and
Kay Wade

egg laying in inappropriate waters. Pairs as well as singles seen far from water. Feed singly or in small to large swarms, sometimes clearly aggregated with no individuals for some distance away from swarm. Individuals have been seen to chase large darners flying near a feeding group. Alternate short feeding flights with hanging in shade on hot days. Stop feeding well before dusk, unlike some other swarm-feeders. Prey presumably small swarming insects. Highly migratory and with great dispersal powers, including far out over ocean, where they fly at night as well as during day. Not seen in numbers as great as some other migratory species in western North America, but some concentrations have been reported in Southwest. Presumably breeding individuals at high latitudes have migrated up there from lower latitudes, and recently emerged immatures may be seen in late summer before they head south. In tropics apparently fly toward areas of lower pressure, where rain likely. Flight season tends to be better correlated with rainy season than with "summer."

Habitat Seen anywhere in feeding flight over open country, attracted for breeding to temporary wetlands in newly filled basins, including drainage ditches. Often breeds in artificial ponds, even swimming pools or garden ponds, and may appear at new wetlands immediately. Fishlessness probably prerequisite for breeding habitat, as larvae conspicuous in waters without vegetation. Also oviposits in canals and large shallow pools of rivers in rainy season.

Flight Season WA Jun–Aug, OR Jun–Aug, CA all year, MT Aug, AZ Apr–Nov, NM May–Oct, NE Jun–Sep, TX all year.

Distribution Sparse at north end of summer range, only scattered individuals north of that. Ranges east to Atlantic Coast, north just to far southern Canada, south throughout tropics of both hemispheres. Very widespread, even on most oceanic islands in tropical latitudes, but absent from Europe.

Description Robust gliding skimmer, overall brown to brownish-red with prominent dark spot at base of each hindwing. *Male*: Eyes mostly gray with reddish cap; face dull red. Thorax brown with two pale gray stripes on each side. Abdomen brown with complex but not conspicuous pattern of fine lines on sutures, also black median line slightly expanded on each segment, expanded into spots on S8–10. Dark brown basal spot on hindwing blends with abdomen in flight. *Female*: Colored and shaped as male but face lighter reddish. Slightly longer and more slender cerci, if visible, good criterion for sexing. Color pattern on abdomen in immature may be more contrasty, actually quite striking.

Identification Distinguished from rather similar **Wandering Glider** by darker color, hindwing spots when they can be seen, but not always easy distinction, as spots not very visible in many views. **Wandering** typically looks yellow, **Spot-winged** more reddish or brown, but bright male **Wandering** also quite orange in front. Could be mistaken for **Striped Saddlebags**, both with restricted markings at hindwing base and stripes on side of thorax. Close scrutiny or capture may be necessary.

347.1
Spot-winged Glider male—Hidalgo
Co., TX, June 2005

347.2
Spot-winged Glider
female—Travis Co.,
TX, July 2004

One difference is that saddlebags often perch horizontally at twig tips, rainpool gliders always hang down. Even more dissimilar, nevertheless, **Four-spotted Skimmer** in rapid flight over open area could be mistaken for this species. **Red Rock Skimmer** of similar size and shape could be confused, especially when in feeding flight, but lacks dark hind-wing spot.

Natural History Feeding and territorial patrol in flight, hangs up at a slant or almost verti-cally when not active. Typically perches in trees, higher than Wandering Glider but some-times hanging from same branch. Often flushed from edge of woodland trails. Flight rapid and erratic, more so than Wandering, difficult to follow visually and a real challenge to catch or photograph. Swarms of dozens of individuals often seen, flying back and forth from knee to head height but often higher. Individuals may be well spaced while moving all over landscape. Discrete swarms may move uphill during day. Oviposition usually by fe-male alone, unlike Wandering Glider, but pairs seen flying cross-country in tandem and also ovipositing that way. Pairs make rapid approach to water, hit it once, then off again to swoop around and come back for another try; solo females more leisurely, often guarded by male. Also oviposits in large shallow pools of rivers in rainy season. Migratory, some-times spectacular southbound flights encountered from midsummer to fall in Texas, more rarely elsewhere; may dominate such flights or be thoroughly mixed with other species. Smaller northbound flights in spring also seen, mostly in East.

Habitat Seen anywhere in feeding flight over open country, attracted for breeding to tem-porary wetlands in newly filled, usually shallow and open basins. Often breeds in con-structed wetlands, even garden ponds or swimming pools, as long as they lack fish.

Flight Season BC Jul, WA Jun–Sep, OR May–Aug, CA Mar–Oct, AZ Mar–Oct, NM May–Oct, NE Jun–Sep, TX all year.

Distribution Moves up Pacific coast in late summer in some numbers, small groups to coastal Washington. Sparse in Northwest, boundary of known records shown except iso-lated records at Juneau, Alaska, and in southeast Manitoba. Ranges south to Argentina, east across United States north to far southern Canada.

Coastal Pennants *Macrodiplax*

This small genus is at times placed in a family separate from skimmers, along with several other Old World genera that share its very open venation. Most workers consider it a skimmer, and in habits it is much like other pennants, all of which are bona fide skimmers. A red species, *Macrodiplax cora*, replaces this one in the Old World tropics. It too is common in coastal habi-tats and disperses widely among islands. World 2, NA 1, West 1.

Description Medium-sized black or brown pennant, mostly of coastal areas, with prominent hindwing spot. *Male*: Eyes dark brown; face shiny black. Thorax and abdomen black. Prominent basal black spot in hindwing. *Female*: Eyes red-brown over blue-gray; face cream. Thorax brown in front, whitish on sides with prominent W-shaped black marking. Abdomen light brown, black-ish on sutures and low on sides; tip of S8 and S9–10 black. Hindwing spot dark brown.

Identification There should be no confusion with other species, especially in the typical open, often coastal, habitat of **Marl Pennant**. Males of other black species in its range, **Double-ringed Pennant**, **Seaside Dragonlet**, and **Swift Setwing**, have much smaller hindwing spots. No female SMALL PENNANT or other similar-sized skimmer duplicates the wing pattern of **Marl**. Thoracic pattern of brown front and pale sides with black W also distinctive. **Checkered Setwing** somewhat like female but has black and white banded abdomen and large color patch at base of forewings as well as hindwings. Female could be mistaken for **Hyacinth Glider**, **Spot-winged Glider**, or SADDLEBAGS at first glance when in feeding flight.

348.1
Marl Pennant
male—Wakulla Co.,
FL, June 2004

348.2
Marl Pennant
female—Wakulla Co.,
FL, June 2004

Natural History Characteristically perches at tips of twigs, grass inflorescences, and tall leaves such as cattails and bulrushes with abdomen horizontal or up in obelisk position. Wings at times raised like other pennants. Often perched among larger numbers of Four-spotted Pennants at coastal lakes, and like them sometimes in dragonfly feeding swarms. Males may spend long periods flying over open water. Also common, especially females and immatures, in open areas near breeding habitats. Pairs oviposit in tandem in open water, dipping to surface in straight and rapid approach for one tap, then slowing and preparing for another run.

Habitat Lakes and large ponds in open areas, from fresh to brackish but often at saline ponds at interior locations. Stonewort often an indicator of potential breeding site. Can be quite common in restricted habitat. Individuals (wandering?) of both sexes often seen at more typical odonate habitats or away from water.

Flight Season CA May–Sep, AZ Jun–Oct, NM May–Oct, TX May–Nov.

Distribution Ranges south along Gulf and Caribbean to Belize and Venezuela; also in western Mexico south to Sinaloa, along Gulf Coast to Florida, also Greater Antilles.

Species Added to the Western Fauna in 2008

The year 2008 proved again that there is a steady northward trickle, even a flow, of tropical species from Mexico into the United States. Six species were newly recorded from the Lower Rio Grande Valley of Texas, four of them new for North America north of Mexico, and there was another new record for the United States in southern Arizona. This brief account may assist in their identification when they are found again.

Blue-striped Spreadwing   *Lestes tenuatus*

Males of this woodland-based species have the thorax mostly bright blue, with broad black median and humeral stripes. Only S9 becomes pruinose, a good mark. Some females have a blue-and-black striped thorax just like males, others have the same areas pale blue and metallic brown. A population was found in Santa Ana National Wildlife Refuge, Texas, in September 2008.

Caribbean Darner   *Triacanthagyna caribbea*

This species, about the size of Pale-green and Blue-faced Darners, can be distinguished by its green thorax striped with black. The tip of the male abdomen is pale greenish, unlike larger species such as Regal and Swamp Darners with green-and-black-striped thorax. Neither of those species occurs regularly in the Lower Rio Grande Valley. Caribbean Darners were found at Santa Ana National Wildlife Refuge in November 2007 and September 2008.

Straight-tipped Ringtail   *Erpetogomphus elaps*

This is a small ringtail with a bright green thorax, its only conspicuous marking a half-stripe at T1. The stripe along the top of the abdomen from its base to S7 is light green, the only Arizona ringtail so colored. A male of this species was found at Parker Canyon Schoolhouse, Arizona, in September 2008.

Metallic Pennant   *Idiataphe cubensis*

This small, slender skimmer is dark brown with metallic overtones. About the size of the larger small pennants (*Celithemis*), it has a tiny brown spot at the base of the hindwings but no hint of the bluish pruinosity of a male Double-ringed Pennant. The sexes look alike, females with a slightly thicker abdomen. It could be mistaken for a setwing except for its perching habits, typically on the tips of vertical twigs and stems. Several individuals were seen at Bentsen-Rio Grande Valley State Park, Texas, in June 2008, after prolonged southerly winds.

Mexican Scarlet-tail   *Planiplax sanguiniventris*

Males have a pruinose blue-black thorax and scarlet-red abdomen, much like male Fiery-tailed Pondhawks, but the base of the abdomen is also red (black in the pondhawk), and the legs are exceptionally long. Females are probably brown but are not described! Male scarlet-tails typically perch and fly actively well out from shore over lakes and ponds. A small population was detected at a flooded resaca in Bentsen-Rio Grande Valley State Park in June 2008, a few persisting at least until early September.

Slender Clubskimmer   *Brechmorhoga praecox*

This species is a typical clubskimmer, most likely to be seen in flight or hanging from a shaded branch, but it is slightly smaller than both Pale-faced and Masked Clubskimmers. There are only two rows of cells basal to the anal loop (three in the others), and in males the club is distinctly less prominent than in the larger species. A single individual was found at the NABA Butterfly Garden near Mission, Texas, in February 2008.

Arch-tipped Glider   *Tauriphila argo*

Like other gliders, this species flies incessantly and then hangs up to perch. Between saddlebags and Hyacinth Glider in size, it could be mistaken for any of them, but it differs from the others in being entirely red, females duller than males, with a narrow but prominent hindwing spot. Along with the scarlet-tails, numbers of Arch-tipped Gliders appeared at Bentsen-Rio Grande Valley State Park in June 2008. An individual of the same species was seen at Santa Ana National Wildlife Refuge in September of the same year.

Appendix: Dragonfly Publications and Resources

Books on Dragonfly Biology and Natural History

Brooks, S. *Dragonflies*. 2003. Washington, DC: Smithsonian Books.

Corbet, P. *Dragonflies: Behavior and Ecology of Odonata*. 1999. Ithaca: Cornell University Press.

Miller, P. *Dragonflies*. 1995. Slough: The Richmond Publishing Co.

Silsby, J. *Dragonflies of the World*. 2001. Washington, DC: Smithsonian Institution Press.

North American Guides (some technical)

Dunkle, S. *Dragonflies through Binoculars*. 2000. New York: Oxford University Press.

Garrison, R., N. von Ellenrieder, and J. Louton. *Dragonfly Genera of the New World*. 2006. Baltimore: Johns Hopkins University Press.

May, M., and S. Dunkle. *Damselflies of North America, Color Supplement*. 2007. Gainesville: Scientific Publishers.

Needham, J., M. Westfall, and M. May. *Dragonflies of North America*. 2000. Gainesville: Scientific Publishers.

Nikula, B., J. Sones, D. Stokes, and L. Stokes. 2002. *Stokes Beginner's Guide to Dragonflies*. Boston: Little, Brown, and Company.

Walker, E, and P. Corbet. *The Odonata of Canada and Alaska*, 3 volumes. 1953–1975. Toronto: University of Toronto Press.

Westfall, M., and M. May. *Damselflies of North America*. 2006. Gainesville: Scientific Publishers.

Western Regional Guides

Abbott, J. *Dragonflies and Damselflies of Texas and the South-Central United States*. 2005. Princeton: Princeton University Press.

Acorn, J. *Damselflies of Alberta, Flying Neon Toothpicks in the Grass*. 2004. Edmonton: University of Alberta Press.

Biggs, K. *Dragonflies of California and Common Dragonflies of the Southwest, A Beginner's Pocket Guide*. 2006. Sebastopol: Azalea Creek Publishing.

Cannings, R. *Introducing the Dragonflies of British Columbia and the Yukon*. 2002. Victoria: Royal British Columbia Museum.

Gordon, S., and C. Kerst. *Dragonflies & Damselflies of the Willamette Valley, Oregon*. 2005. Eugene: CraneDance Publications.

Hudson, J., and R. Armstrong. *Dragonflies of Alaska*. 2005. Anchorage: Todd Communications.

Manolis, T. *Dragonflies and Damselflies of California*. 2003. Berkeley: University of California Press.

Paulson, D. *Dragonflies of Washington*. 1999. Seattle: Seattle Audubon Society.

Dragonfly Societies

Dragonfly Society of the Americas (http://www.odonatacentral.org/index.php/PageAction.get/name/DSAHomePage)

Foundation Societas Internationalis Odonatologica (http://bellsouthpwp.net/b/i/billmauffray/siointro.html)

Worldwide Dragonfly Association (http://ecoevo.uvigo.es/WDA/)

Regional Dragonfly Listserves

CalOdes (http://pets.groups.yahoo.com/group/CalOdes/)

Odonata-l (https://mailweb.ups.edu/mailman/listinfo/odonata-l)

SoWestOdes (http://pets.groups.yahoo.com/group/SoWestOdes/)

TexOdes (http://tech.groups.yahoo.com/group/TexOdes/)

Dragonfly Websites

These sites are as varied as their names, and they should be browsed to see what they have to offer. Many of them will link to the abundant sites that contain photos and still more information on western odonates. Or just enter "dragonflies" in a search engine and stand back.

Arizona Odonates (http://www.azodes.com/main/default.asp)

California Dragonflies and Damselflies (http://www.sonic.net/dragonfly/)

Digital Atlas of Idaho, Dragonflies & Damselflies (http://imnh.isu.edu/digitalatlas/bio/insects/drgnfly/dragfrm.htm)

Digital Dragonflies (http://www.dragonflies.org/)

Dragonfly Biodiversity, Slater Museum of Natural History (http://www.ups.edu/x5666.xml)

International Odonata Research Institute (http://www.iodonata.net/)

Nebraska Dragonflies and Damselflies (http://www.geocities.com/ne_odes/index.html)

Ode News (http://www.odenews.org/)

Odonata Central (http://www.odonatacentral.org/)

Odonata—Dragonflies & Damselflies (http://www.windsofkansas.com/Bodonata/odonata.html)

Odonata of the Lower Rio Grande Valley (http://www.fermatainc.com/nat_odonates.html)

The Oregon Dragonfly and Damselfly Survey (http://www.ent.orst.edu/ore_dfly/)

Dragonfly Collecting Gear

BioQuip Products (http://www.bioquip.com/default.asp)

Rose Entomology (http://www.roseentomology.com/)

Glossary

abdomen: last segment of body, the long and slender one, with 10 segments; often thought of as the "tail"

anal loop: foot-shaped cluster of cells in hind wing of skimmers and emeralds

anastomosing: branching and coming back together, like a braided river

angulate: edge of structure forming angle

Anisoptera: suborder to which "true" dragonflies belong

anteclypeus: narrow part of face between labrum and clypeus

antenodal: crossveins proximal to the nodus

anterior: toward front

apical: at the tip of a structure

appendages: structures at end of abdomen: 2 in females, 3 (2 superior, 1 inferior) in male dragonflies, 4 (2 superior, 2 inferior) in male damselflies; distinctive of species

auricle: projection from either side of abdominal segment 2 in males of most dragonflies (not skimmers), may be used to guide female abdomen during copulation

bar: a marking crosswise to the axis of a structure (body, wing, leg)

basal: at or near the base of a structure

carina: narrow keel running along or across a structure; prominent on front of thorax and along top and sides of abdominal segments in skimmers, across segments in all families

caudal lamellae: three leafy appendages at rear of abdomen in larval damselflies, for respiration and locomotion; also called caudal gills

cercus (pl. cerci): paired appendage at end of abdomen, upper pair in males, only appendages in females

clypeus: middle segment of "face"

congener, congeneric: species in the same genus

conspecific: in the same species

convergent: coming together

costa: anteriormost wing vein, forming front edge of wing

costal stripe: narrow colored area running along front margin of wing

crepuscular: active at dusk and, often, at dawn

cryptic: camouflaged, not easily detected

cuticle: outermost covering of insect

dimorphic: of two different shapes or colors, as in sexual dimorphism

distal: more toward tip of structure

divergent: going apart

dorsal: above, on top

dorsolateral: where top and sides meet

elongate: lengthened in one direction

emerge: to leave water and undergo metamorphosis into an adult; emergence is thus both from water and from exuvia

emergent vegetation: plants growing on bottom in shallow water that extend above water surface

endemic: occurring nowhere else

endophytic oviposition: laying eggs into plant tissue

ephemeral: not permanent; refers to ponds that fill up in wet season, go dry in dry season

epiproct: unpaired inferior abdominal appendage in male Anisoptera

exophytic oviposition: laying eggs onto water or land

exoskeleton: outer hard part of insect, including legs and wings

exuvia (pl. exuviae): cast skin from any larval molt (including transformation into adult)

femur (pl. femora): first long segment of leg, starting from base

flight season: period during which adults occur

frons: uppermost part of "face"

frontoclypeal suture: prominent line between frons and clypeus on face

genital lobe: projection from abdomen at posterior end of genital pocket on segment 2

genital valve: valves on either side of blade of ovipositor

Greater Antilles: large islands of West Indies—Cuba, Jamaica, Hispaniola, and Puerto Rico

hamules: paired structures that project from genital pocket under second segment and hold female abdomen in place during copulation

herbaceous: small plants with no woody tissue

immature: adult past teneral stage but still not with mature coloration; usually not at water

instar: a larval stage; most larvae go through 10–13 of them

labium: jointed mouthpart below mandibles, visible from bottom of head, sort of a lower lip; also "lip" of larva that is extended during prey capture

labrum: lowermost part of "face," just above mandibles, sort of an upper lip

larva (pl. larvae): immature stage of Odonata

lateral: on the side(s)

lateral carina: keel that runs across middle of abdominal segments

Lesser Antilles: extended chain of small islands bordering eastern Caribbean

linear: like a line; usually lengthwise, from front to back or along a structure

maiden flight: first flight of teneral away from water

mandibles: what a dragonfly bites with

mature: of reproductive age, fully colored

mesostigmal laminae (sing. lamina): paired plates, one on either side of front end of pterothorax that engage cerci of male during copulation and tandem

mesostigmal plate: another term for mesostigmal lamina

mesothorax: middle segment of thorax, bears second pair of legs and first pair of wings

metamorphosis: process of changing from larva to adult; happens within larval exoskeleton

metathorax: rear segment of thorax, bears third pair of legs and second pair of wings

middorsal carina: keel that runs down middle of top of abdominal segments

molt: each time exuvia is shed; larval growth can take place only when larva is briefly soft at this time

monotypic: only one member in group

naiad: another term for larva, commonly used for aquatic insects with incomplete metamorphosis

neotropical: occurring in the New World tropics

nymph: another term for larva, commonly used for insects with incomplete metamorphosis (no pupa)

obelisk: to hold abdomen pointing straight up, or the position of that abdomen

ocellus (pl. ocelli): one of three simple eyes between the large compound eyes; on the vertex

odonate: another term for dragonfly/damselfly

ommatidium (pl. ommatidia): one division of compound eye

oviposition: act of laying eggs

ovipositor: complex structure at posterior end of female damselflies, darners, and petaltails that functions in endophytic oviposition; also "spike" of spiketail; loosely applied to enlarged subgenital plate in emeralds and skimmers that hold clump of eggs

paraproct: paired inferior abdominal appendage in male Zygoptera

polymorphic: occurring in more than one form or color within one sex and age class

posterior: toward the rear

postnodal: crossveins between nodus and stigma

prothorax: first segment of thorax, the "neck"; bears first pair of legs

proximal: more toward the base of a structure

pruinescent: becoming pruinose

pruinose: exhibiting pruinosity

pruinosity: powdery bloom on odonates that exudes from cuticle and turns it light blue, gray, or white, deposited on mature individuals (more commonly males) of many species of odonates

pterostigma: technically correct word for stigma

pterothorax: obvious part of thorax, fused meso- and metathorax

radiation: can be used to describe the evolution of numerous similar species from a single common ancestor

relict: formerly more widespread and common, now much restricted

rendezvous: where the two sexes normally meet to mate

satellite: male allowed in other male's territory because of nonaggressive behavior and usual failure to mate

sclerite: small segment of exoskeleton, quite evident around wing bases

seminal receptacle: sperm-storage organ in female reproductive tract

seminal vesicle: sperm-storage organ in second abdominal segment of male

sexual patrol flight: characteristic flight of male odonates at water; either actively searching for females or defending a territory that females may enter

sinuous: a curved line, like a snake's body

sperm transfer: when male transfers sperm from tip of abdomen to accessory genitalia on second abdominal segment

stigma: thickened structure at front of each wingtip in most Odonata

stripe: a marking running along the axis of a structure (body, wing, leg), also used for markings on thorax that run between wings and legs, thus up and down

subapical: not quite at the tip

subgenital plate: plate below S8 that holds bunches of eggs when enlarged; variable enough in shape to be of value in identification

submergent vegetation: plants growing below the water surface

sympatric: occurring in the same geographic area

synthorax: same as pterothorax

tandem, tandem linkage: position taken when male has grasped female by head or thorax

tarsus (pl. tarsi): third part of leg, with many segments and paired tarsal claws

tibia (pl. tibiae): second long segment of leg

transverse: crosswise, from side to side

truncate: cut off squarely

tubercle: a bump

ventral: below, underneath

ventrolateral: where sides and underside meet

ventromedial carina: keel that runs along lower edge of abdominal segments

vertex: conical tubercle on top of head just in front of eye seam; bears the ocelli

vulvar lamina: subgenital plate

vulvar plate: another term for vulvar lamina

vulvar spine: spine projecting from rear edge of underside of S8 in some female pond damsels

wheel: the copulatory position in odonates

Index